SOTHIS PRESS

Sothis Press aims to completely reimagine the public domain book with an emphasis on universal access to captivatingly designed print editions. We believe that technology will enable every single book ever published to be perpetually accessible in beautiful print form. Sothis is beauty in ubiquity.

www.sothis.press
1 Waterhouse Square, London, ECIN 2ST

Builders' Hardware

A Manual for Architects, Builders
and House Furnishers

Clarence Howard Blackali

SOTHIS PRESS

ISBN 978-1-0368-0128-1

Builders' Hardware
is in the public domain.

This edition is published by
Sothis Press © 2023. The company
is registered in England and Wales.
The Company Number is 14900389.

BUILDERS' HARDWARE

A MANUAL FOR

ARCHITECTS, BUILDERS AND HOUSE FURNISHERS

BY

CLARENCE H. BLACKALL

ARCHITECT

ILLUSTRATED

BOSTON
TICKNOR AND COMPANY
211 Tremont Street
1890

S. J. Parkhill & Co., Printers, Bos

TABLE OF CONTENTS.

INTRODUCTION.

THE subject of Builders' Hardware is one which has a direct bearing upon the work of the architect, and is, at the same time, of vital interest to the builder. This book is, however, written primarily for the architect. Most builders, we fancy, are already pretty well posted on the subject of the hardware which they are called upon to use in the construction and finish of a house; but our experience has led us to believe that architects, as a rule, seldom go any deeper into the subject than is necessary to decide whether knobs shall be of one material or another, or whether some particular pattern of sash-fast will be satisfactory to the owner. It is needless to say that a more extended acquaintance with the subject would do no one any harm, and might even be conducive to much good, if only in the way of providing more fittingly for the needs of the client. This work is not intended, however, to be over-critical in its nature, nor necessarily so exhaustive as to embrace all the inventions and arrangements comprised in the general term of builders' hardware, though an attempt has been made to discriminate between what is merely novel and what is really suitable, and so far as possible the best of everything is noticed under the various heads, and an effort made to represent as nearly as possible the conditions and limitations of the builders' hardware market, as well as to show what is valuable for the uses of the architect.

It must be remembered that the statements are made from an architect's standpoint rather than from that of the builder or the manufacturer. The object in view is to show what can be obtained for special cases, and how it is necessary to be specified in order that there shall be no doubt in the minds of both builder and client as to what is called for. It is not intended to present abstracts of trade-catalogues, and if the book fails to represent all the wares in their proper light, or at times seems to ignore some valuable inventions or place undue stress upon articles that do not receive the heartiest commendation from those who use them, we can only plead in extenuation the difficulty of finding out everything that is in the market. Many of our best appliances have only a local and limited fame; and as they are advertised sparingly, it is often quite difficult to say exactly that such and such an appliance is absolutely the best.

For example, some Boston dealers consider that the best locks in the country are the hand-made goods turned out by such firms as Enoch Robinson's Sons. Without discussing the peculiar merits of the Robinson locks at this point, the assertion may safely be ventured that any one studying the subject in San Francisco would have as much difficulty in an investigation of Boston hand-made goods as we have had in our endeavors to deal with hundreds of valuable inventions which are hardly known at all in this part of the country. The hope is, therefore, expressed that this publication may call out facts concerning many articles of builders' hardware which at present are unknown to the ordinary trade.

It may also be hoped that the volume will suggest to architects the possibility of more definite specification, and of some better way of selecting and defining the necessary hardware than to merely insert a clause in the specification to the effect that a certain sum per door and a certain price per dozen is to be allowed in the contract for fixtures. This we know is a very common method of disposing of the question, and is, no doubt, the easiest way out of the difficulty, relieving the architect entirely from any necessity for exhaustive explanations in regard to subjects upon which he generally is not over-well posted. With an honest builder, or with one who has the fear of the inspector before his eyes and knows that all will be scrutinized, this may be a method sufficiently exact to suit most cases; but where the market is so full and there are so many really excellent appliances to choose from, it would seem certainly as if more exact specifications ought to be prepared. How many architects, for instance, ever insist upon a particular style of lock; and how many architects, even after many years experience with the best work, are able to tell a client what is the best sash-lock or what is the best style of hinge, without referring to a trade-catalogue? We cannot hope to meet all emergencies, but it is believed that, at least, a desire for a better selection of hardware than is generally possible can be stimulated.

As to the limitations of the work, the term "builders' hardware" is assumed to include, generally speaking, metal-work of every description entering into the construction and finish of a modern building from the nails and bolts used in the rough work, to the door furniture and brass lock and plate work of the finish.

The illustrations, which include nearly every article referred to, have been prepared as simply as possible. A great deal of care has also been taken in collecting the prices. The ordinary trade-catalogues are very unsatisfactory in this latter respect: many of them give no prices at all, and when the cost is stated it is always subject to discounts so uncertain and varying as to give absolutely no real information of value to the architect or the purchaser. The prices hereafter given will represent the retail cost of the articles in the Boston market, as nearly as we are able to ascertain it. These prices are, of course, somewhat fluctuating in their nature, but will, at least, serve as a means of comparison, even if not exact for all localities and seasons. Most of the prices were collected during the summer of 1888.

In the preparation of the subject matter which is to follow, the writer has had the heartiest coöperation from the leading dealers and manufacturers, and desires to especially acknowledge the courtesies of Messrs. Burditt & Williams, and Nichols & Bellamy, of Boston, the Hopkins & Dickinson Manufacturing Co., and the Yale & Towne Manufacturing Co.; without whose technical assistance any such work would lack the value which it is hoped this will possess.

BUILDERS' HARDWARE.

CHAPTER I.

Metals and Varieties of Finish.

BEFORE beginning a study of any portion of the subject, an attempt will be made to consider the metals used for builders' hardware and the various styles of finish possible in connection with them.

Wrought-iron.

Iron is naturally the first material thought of. In its purest form it is known as wrought-iron, a term which is derived from the fact that in the earliest processes the iron was beaten while hot and the impurities forced out by the blows. Also the name indicates, to an extent, the nature of the material, which is yielding and pliable rather than brittle or hard. Wrought-iron is the most available material for constructive metal-work. Nails, bolts, ties, anchors, etc., are more naturally made of wrought-iron than of any other material. Nearly all strap-hinges and the best makes of common butts are made of it; also, for some forms of common bolts and any article of hardware which permits the metal being punched or pressed into shape, wrought-iron is more suitable than cast, and is much used on account of its superior strength. For finished work, such as knobs, plates, etc., there is very little wrought-iron hardware in the market. A good deal of it is made to order in the shape of ornamental hinges, straps and braces, but the amount of work involved in

Chapter 1.

producing any given pattern is so largely in excess of what the same amount of work would produce in other metals that wrought-iron will always be a material for the artist or the amateur. Quite recently, however, there has been made an important discovery which bids fair to bring about a change in the use of this metal. Wrought-iron melts at a very high temperature and it has, therefore, long been practically impossible to cast in the fine moulds; but by the addition of a small percentage of aluminium to wrought-iron when in a highly-heated condition an alloy is formed which melts at a greatly reduced temperature and gives the metal the degree of fluidity necessary to successful casting. It is claimed that the wrought-iron castings obtained in this way have all the sharpness and clearness of cast brass and at the same time retain the structural qualities of wrought-iron. So far as we can ascertain, cast wrought-iron is not in the market.

Steel.

When wrought-iron has mixed with it a proportion of carbon exceeding 0.5% but not greater than 2% the structural nature of the metal is entirely changed and it becomes known as steel. So far as relates to finished hardware, steel is used only for springs and portions of detail, as the metal does not cast readily, and the surface will not keep bright. In fact, there are the same objections to it that there are to wrought-iron, and almost the only use to which it is now put in hardware is in the manufacture of nails and screws and some brands of butts, which will be described later on.

Cast Iron.

When iron contains in its composition over 2% of carbon it becomes known as cast-iron, a metal which melts at a comparatively low temperature and is capable of being cast with very sharp lines. It is used a great deal for finishing work, though from its brittle nature and lack of tensile strength it is not utilized to any great extent for constructive purpose.

Iron hardware is finished in a variety of ways. If left in its natural condition as it comes from the file or hammer, it will rust very speedily, and to prevent this, the most common method is to heat the iron and paint it while hot with linseed

Chapter I.

oil, colored with lamp-black and mixed with a quantity of dryer or turpentine. When this coating is dry, the iron is sometimes roasted in a kiln — fired, as it is termed. In this manner the paint seems to be dried into the pores of the metal, and it is left with a smooth, lustreless black surface, capable of resisting the action of the weather for quite a while, though by no means rust-proof.

Bronzed Iron.

If, instead of the foregoing, a bath is used composed of linseed oil, and gum-anime or copal, the iron, on being finally fired and polished, presents the appearance of bronze. The tones can be varied to a considerable extent by the addition of color in the shape of powdered alloys of copper and bronze, which are mixed with the oil. The "Tucker bronze," the "Berlin bronze," and the so-called "Boston finish" are all substantially of this nature. Only rarely is a thin film of bronze or composition spread on the iron by the aid of electrolysis. All of the bronze-faced iron hardware is treated with the hot-oil finish just described. The cheapest forms of iron hardware are japanned or even merely coated with ordinary black varnish, the quality of the japanning or varnish depending upon the grade of goods. Japanning is practically indestructible where the iron is not exposed to scratches or rubbing. Hinges and butts are finished in this way more than any other one form of iron hardware, though of late years the best machine lock makers have taken a great deal of care in japanning the outer casings of locks. Fine grades of pulleys are also sometimes sold with japanned frames.

Japan.

Galvanized Iron.

Iron is galvanized by simply immersing in a bath of a melted amalgam of zinc and mercury, containing a little sodium. The iron is first cleaned with sulphuric acid, and before immersion in the bath, it is usually dusted with sal-ammoniac powder. There is not a particle of real galvanic action about the process. It is sometimes thought that the zinc and mercury form a species of alloy with the iron which prevents it from rusting. Galvanizing is, indeed, the best preservative of iron applied in the form of a wash, but the color is not sufficiently pleasing to permit of its being used for nice work.

Chapter I.

None of the foregoing processes are especially suitable for iron, as they all effectually disguise the nature of the material. There has recently been invented a process for protecting iron from rust, which is in some respects the best thus far devised.

Bower-Barff Process.

It is known as the Bower-Barff process, a term used to indicate two processes by which the surface of the metal is converted into magnetic oxide of iron, in which condition it is absolutely rustless. In the Barff process the metal is simply subjected to the action of superheated steam. This process is peculiarly suitable to wrought-iron and highly finished work. In the Bower process the iron is successively subjected to the action of highly heated air and carbonic-oxide gas. The heat converts the surface of the metal first into red oxide of iron, which is finally reduced to the magnetic oxide by the action of the gas. Owing to the simplicity of the process, it is claimed that its cost is less than that of galvanizing. It may be applied to any kind or style of wrought or cast iron or steel. The surfaces so treated have a perfectly uniform blue-black color. The sharpness of the lines is not affected in the least, and when the work is polished the final color is a lustrous ebony black, such as can be obtained in no other way. This coating of magnetic oxide is so hard that it is removed with difficulty by an emery-wheel. A few of the leading dealers are beginning to keep in stock some fine grades of door hardware finished by the Bower-Barff process, but the only parties making use of the patents in the production of builder's hardware are the Yale & Towne Manufacturing Co. This concern is beginning to put in the market quite an extended line of Bower-Barff iron goods.

Copper and Nickel-plate.

Iron hardware is also found in the market finished with copper-plating, the raised surface of the pattern being buffed to show the natural copper color, while the background is left black or strongly oxydized. There is quite a variety of goods in this line. Similarly ironwork is nickel-plated, being left with either plain polished surfaces or with polished raised patterns on a black ground. Both nickel and copper plating, are laid on with the aid of a dynamo.

Brass.

Brass and bronze are terms which are often confounded when speaking of hardware, though the materials are quite different in composition and are usually dissimilar in appearance. Brass is an alloy of copper and zinc in varying proportions, the ratio for ordinary purposes being seven of brass to three of zinc. Brass has a light yellowish appearance, is susceptible of a high polish, and can be rendered more ductile by the admixture of a small quantity of lead, which at the same time will diminish the hardness of the alloy. Brass tarnishes very easily if handled or exposed to the weather, and is consequently generally protected by a coating of shellac, which, however, will not entirely prevent it from changing in tone. The color of the brass may be altered slightly by changing the proportions of the metals entering into the alloy, also by treating the finished castings with acids or hot chemicals in the same manner as will be described later for bronze.

Bronze.

Bronze is commonly an alloy of copper and tin in proportions varying from twelve parts copper to one part tin for metals to be used in the fabrication of mathematical instruments, to two parts copper and one part tin, for telescope or speculum metal. Bronze is sometimes more complex in its nature, lead being added in very slight quantity, and aluminium sometimes replacing the tin. It is less malleable but harder and more fusible than brass, and can be cast with greater ease and perfection. It is oxidized more easily than brass, and because of this it is possible to obtain a greater variety of colors for hardware goods.

Both brass and bronze can be cast with great ease, as compared with iron. The sharpness and delicacy of the casting depends, of course, entirely upon the care bestowed in preparing the pattern and the mould. There is a great difference in the quality of the work turned out by the various manufacturers. In the best of work the mould is formed with very fine sand and is double-faced, that is to say, after the pattern has been imprinted in the sand it is withdrawn, the matrix sprinkled with a still finer sand, and the mould repacked. After that, the moulds

Chapter I.

are suspended in the fumes of burning rosin, by which means they are coated with a fine layer of impalpable soot. When the molten metal is poured into the mould, the soot is pressed evenly into all the minute pores or interstices of the sand, and the resulting casting is so smooth and sharp as often to require but very little hand-chasing.

Brass goods are usually finished in the natural color of the metal with a thin coat of shellac. The same proportions of tin and copper are usually used for all the bronze work of a single firm, though the different manufacturers do not always adopt the same alloy. Bronze is, however, finished in a great variety of styles and colors, all produced essentially as follows:

Surface-finish.

The bronze on leaving the casting-room, is trimmed or chased as may be necessary, and is then immersed for a few moments in a strong acid bath which takes an almost infinitessimal film from the metal, leaving the surfaces entirely free from any oxidations or impurities. The piece of metal is then washed in weak alkali and clear water, to make it perfectly clean. Then the bronze is suspended in a bath of hot acids specially prepared with various chemicals to produce certain changes in the color of the metal. This operation is a very delicate one, requiring the care of trained workmen and a nice appreciation of the proper time to stop the action of the acids, as the appearance of the metal changes very rapidly, turning first from a bright copper red, to a pale gray and then into beautiful shades of browns and rich purples with ultimate dark tones depending upon the composition of the mixture or pickle. When the desired shade is reached the bronze is removed, dried in sawdust, and rubbed down to an even tone on a buffing-wheel. In some designs the raised portions are buffed down until the natural color of the metal appears, the pattern showing on a ground of the darker oxidized tone. Almost any color or shade can be had with bronze by a proper treatment. The Hopkins & Dickinson Manufacturing Co. has even produced a bronze as dark as iron, and in fact quite similar in appearance. Of course ordinary hardware is simply left the natural color of

Chapter I.

the metal, when made in plain bronze. It may be said here that plain bronze is as a rule more expensive than figured work. The plain surfaces require a very even grade of metal and smoothness of casting, whereas any imperfections are hidden by patterns, and a cheaper quality of workmanship does not expose itself. In the East there seems to be a decided preference for the plain goods, while in the West the figured styles are more commonly used. When exposed to rough usage a patterned piece of hardware will in general show wear less than a plain surface.

It must be remembered that the surface finishes produced in the manner just described are not to be considered as permanent. The color is in reality laid on, and is superior to paint or varnish only in that by the means of the hot pickle the color penetrates somewhat deeper than if put on with a brush. Bronze hardware is sure to change in time, no matter how it may be finished, and generally the stronger tones are the least satisfactory in the end, fading out to unpleasant musty hues. Shellac will hold the color, but as soon as it wears off, which it is sure to do on such pieces as knobs and door-handles, the exposed surfaces will turn while the shellacked spots do not. When time can be spared to keep the work polished brightly, a better plan seems to us to be to avoid everything but the natural bronze or brass color, omit the shellac, and polish by hand constantly. This, however, is seldom desirable. We have not yet reached the point where housekeepers will take as much care of the hardware as of the silver.

Oxydized Silver Finish.

The treatment of silver-plated bronze is of a similar nature. The knob or plate is immersed in a silver-plating bath attached to an electric dynamo. A very few moments suffice for the deposition of the silver. The article is then taken from the bath and treated with hot acids and solutions until any desired degree of oxidation is obtained, when it is dried in saw-dust and the raised patterns slightly brightened on a buffing-wheel. There is at present a great deal of oxidized silver hardware in the market. Most of it is oxidized but slightly, leaving the

Chapter I.

raised figures a clear frosty white, but it can be had quite black if desired. The Yale & Towne Manufacturing Co., has an oxidized silver finish which is as black as old oxidized gun-metal.

Copper-bronze.

Several very pleasing styles of finish are obtained by electro-plating the bronze with copper and then treating with acids. An old-copper color thus obtained is one of the most durable finishes in the market.

Gold-plate.

When the cost is not a consideration it is sometimes desirable to have gold-plated bronze hardware, especially for such things as knobs, which are exposed to constant handling. Gold-plate is no handsomer than some shades of natural bronze, but the gold will not tarnish or be influenced by the weather. Gold-plating increases the cost of bronze hardware nearly twenty fold, and being so expensive is executed only for special orders.

Nickel-plate.

Nickel-plating is applied more commonly to iron or steel than to bronze. This finish is but little used for nice work as the nickel will tarnish by exposure to the atmosphere, after which no amount of rubbing will restore its first appearance. Nickel-plating is best adapted for keys, springs, etc.

The following list will give an idea of the great variety of possible treatments of bronze hardware.

FINISHES FOR BRONZE HARDWARE.

I. *The Hopkins & Dickinson Manufacturing Co., finish:* —

A. Statuary. Dark background, light surface.
B. Statuary. Medium dark background and surface.
E. Ebony. Suitable only for Plain or Lined Design.
F. Statuary. Matted with Green.
G. Gilt or Gold-washed.
H. Gold-plated.
I. Gold-plated. Matted with Enamel, in all colors.
K. Gold-plated. Matted with Silver.
L. Silver-plated.
M. Silver-plated. Matted with Gold.
N. Nickel-plated.
O. Nickel-plated. Matted with Enamel.
P. Nickel-plated. Matted with Gold.
R. Nickel-plated. Matted with Copper.
S. Hand-plated. Silver, suitable only for perfectly plain patterns.

II. *P. & F. Corbin, finish:*—

No. 1. Natural Color, Light Bronze.
No. 2. Chemical Dark Brown or Statuary Bronze.
No. 3. Natural Color on Surface and Black Background.
No. 3½. Same as No. 3 Finish, with different arrangement of colors.
No. 4. Natural Color on Surface and Dark Brown Background.
No. 5. Nickel-plated.
No. 7. Nickel-plated Surface with Gold-plated Background.
No. 8. Gold-plated.
No. 10. High Polish, Natural Color.
No. 15. Sage Green Background and Natural Color on Surface.
No. 16. Terra-cotta Background and Natural Color on Surface.
No. 17. Steel Gray Background and Natural Color on Surface.
No. 18. Japanese Finish.
No. 19. Oxidized-silver Finish.
No. 20. Old Brass Finish.
No. 21. Oxidized-Iron Finish.
No. 22. Antique-Copper Finish.

III. *The Ireland Manufacturing Co., finish:*—

No. 1. Light Bronze.
No. 2. Light Brown.
No. 3. Black Background, Polished Surface.
No. 4. Brown Background, Polished Surface.
No. 5. Nickel-plated Surface and Background.
No. 15. Sage Green Background, Polished Surface.
No. 16. Terra-cotta Background, Polished Surface.
No. 17. Steel Gray Background, Polished Surface.
No. 19. Oxidized-silver.
No. 21. Iron Finish all over.

IV. *Nimick & Brittan, finish:*—

No. 1. Bright Surface, Natural Color.
No. 2. Dark Brown Surface and Background.
No. 3. Bright Surface and Black Background.
No. 4. Bright Surface and Brown Background.
No. 5. Bright Surface and Terra-Cotta Background.
No. 6. Bright Surface and Green Background.
No. 7. Nickel-plated Surface and Background.
No. 8. Nickel-plated Surface and Black Background.
No. 12. Gold-plated.
No. 13. Steel Gray Bronze.
No. 14. Old Gold Bronze
No. 15. Oxidized-silver Surface and Background.

Chapter I.

V. *Nashua Lock Co., finish:* —

No. 1. Natural Color, Light Bronze, Polished Surface.
No. 2. Dark Bronze.
No. 3. Light Bronze, Polished Surface, with Black Background.
No. 4. Light Bronze, Polished Surface, with Brown Background.
No. 5. Nickel-plated, Plain Surface.
No. 6. Nickel-plated Surface, with Black Background.
No. 15. Hand Finish.

Besides the finished work, both brass and bronze are used for screws, springs, lock-fittings, etc. A few locks are made with cases of cast-brass and some manufacturers are preparing to use copper-plated iron for the same purpose.

Brass and bronze together are used more than any other one metal for builders' hardware, and are the basis of nearly all door and window fittings.

Phosphor-Bronze.

An important addition to the list of metals available for hardware purposes has recently been made in the shape of a composition designated as Phosphor-Bronze, an alloy of which the constituent parts are not as yet made public, but which the patentees describe as being a phosphorized alloy of copper and tin. The chemical action of phosphorous on the metals composing the alloys is claimed to be two-fold ; on the one hand it reduced any oxides dissolved therein, and on the other it forms with the purified metals a most homogenous and regular combination, the hardness, strength, and toughness of which are completely under control. No other metal combines, in so high a degree as phosphor-bronze, the conditions of toughness. rigidity, hardness and great elastic resistance. Thus far phosphor-bronze has been used in the hardware trade only for screws and for lock-springs. In cities wherein soft coal is used, it has been found that brass springs soon lose their elasticity, owing to the action of free sulphuric acid in the air. Careful tests have shown that phosphor-bronze offers twice as much resistance to corrosion by acid as copper. Further experiments have been made, extending over a period of ten months, to determine its durability, when exposed to the weather, as compared with the best brass wire. After lying on the damp ground for that length of time the brass was found to be quite brittle and worthless, while the phosphor-bronze, under exactly similar conditions was practically unchanged. It will be readily seen then how valuable a metal this is. Unfortunately its degree of elasticity is less than that of brass, and the springs required for a lock when of phosphor-bronze are sometimes so large as to be impracticable. It is, however, by all odds the best material for springs and is used by the Hopkins & Dickinson Manufacturing Co. in their best grades of locks. Chapter I.

Aluminium. Aluminium has recently appeared as a possible substitute for bronze. It is a metal not unlike lead or platinum in appearance, but is very light, having a specific gravity of 2.56, equal to about one-third that of steel and one-fourth that of silver.

It is very strong, will not tarnish, and is almost indestructible. It melts at a comparatively low temperature and can be cast with sharp lines. Copper, brass, and bronze are improved in strength, color, and durability by the addition of ten per cent of aluminium. It is, however, a metal of the future and is not yet in the hardware market.

CHAPTER II.

NAILS.

Plate-nails.

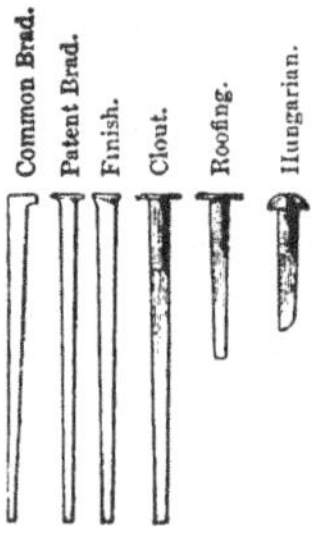

Fig. 1. Common Nails.
(Half Size.)

THE nails commonly used in connection with building operations are too well-known to require any description. They are specifically designated as plate-nails. Up to within a comparatively short time, nails were made by hand almost entirely by women and children, and it is one of the best arguments in favor of the introduction of machinery that the process by which nail-making has been perfected has released a vast multitude from the laborious and wearing occupation, besides giving a great deal better results. The application of machinery to the manufacture of nails is purely an American idea and so recent have been the innovations in connection with this industry that we imagine many people would be surprised to know the changes which have been made both in the form and in the character of the ordinary nails, during the past generation. It is only about fifteen years since iron nails were annealed, or capable of clinching without rupture.

In the manufacture of nails the iron is first rolled into plates having a thickness equal to the desired thickness of the nail and a width a little greater than the length of a finished nail. The plates are cut so that the length is at right angles to the grain, the idea being that when the nails are cut out the fibre will run lengthwise the nail and thus make up much stronger. Special machinery cuts the nails out in alternate wedge-shaped slices so that the metal is used without any loss. The wedges are picked up, held in a vise, the heads stamped on them, and the finished nails finally dropped out into the casks. The machinery is too complicated to allow of any description here, but the process is very simple and easily understood. Nearly all the common nails are made from plates. Hand nails are still made for special uses such as for horse-shoeing, but the cost is too great and hand-made nails are really no better than those which are made by machinery. Chapter II.

Nails are designated according to their length by pennies. The origin of the designation is generally assumed to be in the old system of weights, the nails being made with as many pennyweights of metal as the number indicates. This designation, of course, no longer holds good, as nails have changed materially both in size and weight, but it is still retained for common convenience and we believe the gradation is uniform with the various manufacturers. The weights run from two to seventy penny and the nails vary in length from one inch to six inches. Six-inch nails and larger are more properly designated as spikes, though the trade recognizes a special form of nail of somewhat stouter proportions, made in several of the larger lengths and technically sold under the name of spikes. Sizes.

Trautwine's "*Pocket-Book*," page 425, gives the sizes, etc., of nails as follows.

Name.	Length, inches.	No. to pound.
3d	1¼	557
4d	1½	336
5d	1¾	210
6d	2	163
7d	2¼	123
8d	2½	93
10d	3	66
12d	3¼	50
20d	4	32
30d	4½	19
40d	5	16
50d	5½	13

Chapter II.

Common nails are now very largely made of a low grade of steel, the cost being not over ten cents per hundred pounds more than in wrought-iron.

Strength of nails.

In regard to the strength of nails, Trautwine also states that boards of oak or pine nailed together by four to sixteen ten-penny common cut-nails and then pulled apart in a direction lengthwise of the boards and across the nails, tending to break the latter in two by a shearing action, averaged from three hundred to four hundred pounds per nail to separate them as the result of many trials. "*Johnson's Encyclopædia*" states that the rough surface of a cut-nail adds about twenty per cent to its holding power. The absolute resistance nails will offer to withdrawal varies so widely with circumstances, that no satisfactory results are available.

The sizes of nails used for specific purposes is largely a matter of judgment on the part of the builder, but the common custom is to use four-penny nails for shingling and slating, six-penny for clapboarding, sixes and eights for finish, eights and nines for flooring, nines and tens for boarding and forty-penny and upwards for framing. Architects sometimes consider it well to specify the sizes of nails to be used for bridging the floor-beams, and for slating, but on general principles we would suppose that a builder who would need any such restrictions would not be a man to employ under any circumstances.

Nails are commonly sold by the cask of one hundred pounds.

Common nails can be had galvanized in all the ordinary sizes. Galvanized nails cost 2½ cents per pound extra.

Canada wrought nails are sold for $16.00 per cask. Clinch nails (annealed) cost from $3.10 per cask for ten-penny, to $4.50 for two-penny. Swedes-iron nails are made from an extra quality of wrought-iron, and are especially used for slating, as they are supposed to stand the weather better than ordinary plate-nails. When made from genuine Swedish iron, four-penny nails are sold at $5.50 per cask. American-iron Swedes are $3.85 per cask. Architects usually find it advisable to specify tinned Swedes-iron nails for roofing-work.

Tinning adds from twenty to fifty per cent to the cost, depending on the number of nails to the pound. Chapter II.

The following are the net prices in the Boston Market.

Prices for Cut-steel Nails, June 20, 1888.

Common, Fence and Sheathing.	per keg.
50d — 60d — 70d	$2 50
12d — 20d — 30d — 40d	2 25
10d	2 35
8d and 9d	2 50
6d and 7d	2 65
4d and 5d	2 85
4d lt. and 3d	3 25
3d fine	3 75

Box.	per keg.
12d and larger	$2 75
10d	2 85
8d and 9d	3 00
6d and 7d	3 15
5d	3 35
4d	3 35
3d	3 75

Flooring.	per keg.
12d and larger	$2 75
10d	2 85
8d and 9d	3 00
6d and 7d	3 15

Fine Finishing.	per keg.
12d and larger	$3 15
10d	3 25
8d	3 40
6d	3 55
5d	3 75
4d	3 75
2d	3 75
2d fine	4 20

Slating.	per keg.
6d and 7d	$2 90
4d and 5d	3 10
3d	3 50

Spikes, of all sizes	$2 50

Finishing-nails are lighter and thinner than common plate-nails, and besides being made quite smooth, they have very slight heads, to permit of being easily countersunk in the wooden finish. They are made in a number of sizes, from $\frac{5}{8}$ to $2\frac{1}{2}$ inches in length. Next to finishing-nails are the common brads, made with a head in the form of a shank on one side. The sizes are from $\frac{1}{4}$ inch to 2 inches in length. Brads are used for small finish, tacking on panel mouldings, etc., the metal being quite thin and the brad driven edgewise of the grain so as not to split the wood. Swedes-iron patent brads are manufactured by the Stanley Works, and sold at from 18 cents to $1.25 per pound, with a discount. Common brads are listed at the same prices. Clout nails are made with broad, flat heads, and are sold in sizes varying from $\frac{5}{8}$ inch to $2\frac{1}{2}$ inches in length and

Finishing-nails.

Chapter II.

costing from 48 to 13 cents per pound, with a discount. They are used about a building chiefly for tacking gutters, etc. Hungarian nails are a species of large, rounded-headed tacks. They are made from $\frac{3}{8}$ inch to $1\frac{1}{8}$ inch long, and average 60 cents per pound, with a discount. Figure 1 illustrates the various special forms of common nails.

Wire Nails.

All of the foregoing may be classed as common or plate nails. Nails of a very different kind, manufactured from steel wire, have been in use for a number of years in America and for a longer period in Europe, and in both places they have been very favorably received and are fast superseding the common cut-nails for many purposes. The advantages of these over the common nails are many. For the same amount of metal they are much stronger; they can be driven into very thin boards without splitting them, and can be removed without leaving so unsightly a hole as is usually made by the common nails. Besides this, on account of their superior stiffness they can be driven into very hard wood, where much caution is necessary if common nails are to be used. They are also more easily produced and are handled with less labor. They are manufactured by a simple machine which is automatic in its action, a coil of the wire adjusted to it being cut off in even lengths, headed, pointed and, if necessary, ribbed according to the kind of nail which is desired. The same sizes prevail for these as for the ordinary plate-nails. The following table gives the lengths and number to the pound as listed by the Salem Nail Company, also the net retail prices per hundred pounds in the Boston market.

Other manufacturers occasionally classify the nails in a more natural way by lengths and numbers. The American Screw Company of Providence, R. I., manufactures a very extended line of these nails and sells them in lengths varying from three-sixteenths inch to twelve inches in length, with a thickness of wire varying from No. 22 to No. 0 wire-guage. The prices are by the pound. Everything above two inches, No. 9 wire, is sold at five cents per pound. For smaller sizes the prices increase up to fifty cents per pound for 3-16 of an inch, No. 22. For nails with special heads or special points add one cent per pound. For nails combining all these specialties add one cent per pound for each specialty. For tinning add fifty per cent. Chapter II.

TABLE OF BEST QUALITY OF STANDARD STEEL-WIRE NAILS.

Size.	Length.	Number of Nails to One Pound.	Price per Keg.
2d Common.	1 in.	1200	$5.00
3d "	1¼ "	720	5.00
4d "	1½ "	432	4.00
5d "	1¾ "	300	4.00
6d "	2 "	252	3.70
7d "	2¼ "	186	3.70
8d "	2½ "	132	3.35
9d "	2¾ "	105	3.35
10d "	3 "	87	3.15
12d "	3¼ "	66	3.00
16d "	3½ "	51	3.00
20d "	4 "	35	3.00
30d "	4½ "	27	3.00
40d "	5 "	21	3.00
50d "	5½ "	15	3.35
60d "	6 "	12	3.35

Besides the common wire nails the Salem Company makes a variety of patterns such as fence, clinch, smooth, box, casing, finishing, common brads, flooring, slating, shingle, fine nails, and wire spikes. The wire spikes are made in sizes from three to nine inches long averaging from fifty to four and a half nails per pound. Figure 2 shows the shapes of the different nails. Besides these there are several other special makes not relating to builders' hardware. The variety of nails manufactured from wire is very extensive, and the nails are deservedly popular. They do not hold quite as strongly against pulling out as the common cut-nails but in every other respect they are, on the whole, rather superior.

Picture-nails.

There are several kinds of ornamental nails manufactured for special purposes. Figure 3 shows the common porcelain-headed

Chapter II.

picture-nail. P. &. F. Corbin manufacture nails of this description from two and one-half to four inches in length at prices from $3.88 to $6.50 per gross. Brass-headed nails, Figure

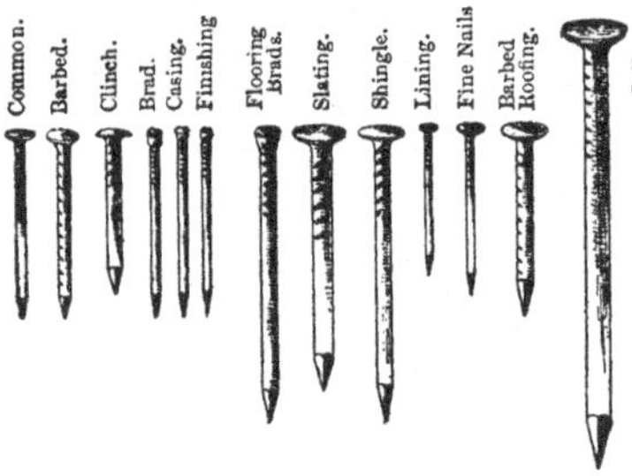

Fig. 2. Wire Nails. (Half Size.) Salem Nail Co.

Fig. 3.

Fig. 4.

4, are manufactured by the same firm in nine sizes, from one inch to four inches in length and from $1.05 to $4.00 per gross. These prices are with a discount.

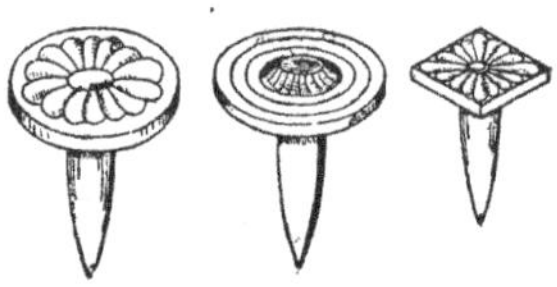

Fig. 5. Brass Door-Nails. (Half Size.) J. B. Shannon & Sons.

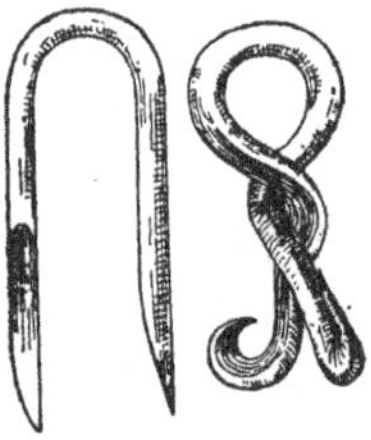

Fig. 6. Clinch Staples.

There are also a few styles of ornamental door nails manufactured. J. B. Shannon & Sons show three varieties in their catalogue, Figure 5. These are made in iron and bronze, at 50 to 72 cents per dozen, or in brass at 90 cents to $1.25 per dozen, list price.

Staples.

In connection with nails may be mentioned the common staples such as are used for blind slats, etc. These are sold in various sizes, from three-eighths of an inch to one and one-fourth inch in length, costing from forty-four to twenty-eight cents per pound. The Florence Tack Company, as well as several other manufacturers, carries a line of steel and iron clinch staples such as are shown by Figure 6. These are made in a number of sizes from five-eighths of an inch up, and cost from thirty cents a hundred and down. Chapter II.

TACKS.

Tacks.

Tacks are of comparatively little value to the builder, being used more especially for carpets, furniture, saddlery and the like. The various lengths of common tacks are designated by ounces, the size of the tack indicating the number of ounces put in a paper when packed full weight.[1] The lengths are always essentially the same with the different manufacturers, for the same weights. Tacks are cut by much the same kind of machinery as is used in the manufacture of plate nails. Steel, American-iron, Swedes-iron and copper are used for tacks. Iron is sometimes galvanized and occasionally nickel-plated or tinned. Figure 7 illustrates the relative sizes of tacks manufactured by the Stanley Works, and the following table gives the list prices of the same company.

Tinned tacks can be had in iron or copper, in all the above sizes, at an advance of from twenty to thirty per cent.

Double-pointed tacks have, within a few years been made from flat steel wire. Five sizes are made by the Florence Tack Company, from seven-sixteenths to three-fourths inches long, varying from $1.20 to $1.80 per dozen boxes. They can be had plated, tinned or in copper. The advantage of these tacks is that they hold more tightly in the wood and at the same time are more easily removed without breaking off in the hole.

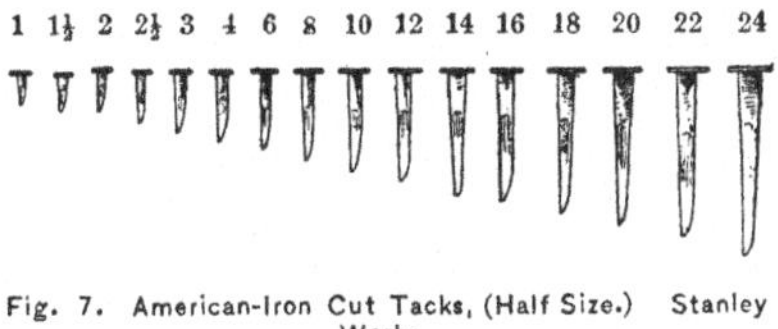

Fig. 7. American-Iron Cut Tacks, (Half Size.) Stanley Works.

[1] This statement is made in the catalogue of the Florence Tack Company.

Chapter II.

TABLE OF TACKS (STANLEY WORKS).

Size.		AMERICAN-IRON.			SWEDES-IRON.			COPPER.		
		Price per doz.		Price per lb.	Price per doz.		Price per lb.	Price per doz.		Price per lb.
Ounce.	Inch.	Full weight.	Half weight.	Pound boxes, or bulk.	Full weight.	Half weight.	Pound boxes, or bulk.	Full weight.	Half weight.	Pound and half-pound boxes, or bulk.
1/2	1/8	....	....	....	$0 60	$0 30	$1 60	$1 12	$0 56	$3 00
3/4	5/32	....	....	....	70	35	1 25	1 12	56	2 00
1	3/16	$0 80	$0 40	$1 00	80	40	1 00	1 12	56	1 50
1½	7/32	90	45	80	90	45	80	1 30	65	1 15
2	1/4	90	45	63	1 00	50	66	1 40	70	95
2½	5/16	1 00	50	55	1 10	55	58	1 60	80	85
3	3/8	1 10	55	50	1 20	60	52	1 80	90	80
4	7/16	1 20	60	44	1 40	70	46	2 00	1 00	66
6	1/2	1 30	65	34	1 60	80	36	2 52	1 26	56
8	9/16	1 50	75	30	1 90	95	32	3 36	1 68	56
10	5/8	1 60	80	28	2 20	1 10	30	4 20	2 10	56
12	11/16	1 80	90	26	2 50	1 25	28	5 04	2 52	56
14	3/4	2 00	1 00	24	2 80	1 40	26	5 88	2 94	56
16	13/16	2 20	1 10	23	3 00	1 50	25	6 72	3 36	56
18	7/8	2 40	1 20	22½	3 30	1 65	24	7 56	3 78	56
20	15/16	2 60	1 30	22	3 60	1 80	24	8 40	4 20	56
22	1	2 80	1 40	21	3 80	1 90	23	9 24	4 62	56
24	1⅛	3 00	1 50	20	4 00	2 00	22	10 08	5 04	56

CHAPTER III.

SCREWS.

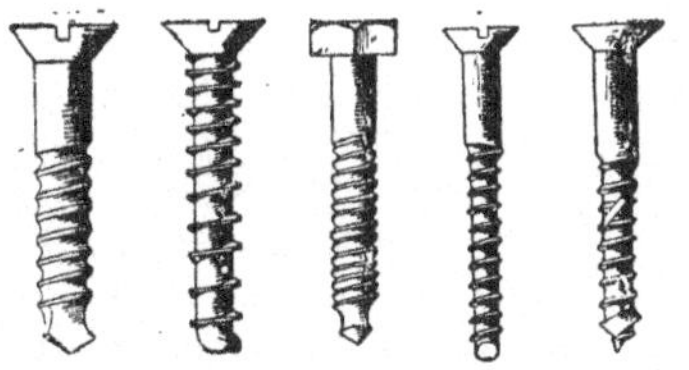

Fig. 8. Fig. 9. Fig. 10. Fig. 11. Fig. 12.

THE substitution of screws for nails in building operations is one of the most marked features of modern work, and is, in a way, indicative of the changes that have come about since Mediæval times. In those days men built for eternity; now, the object is to build so that it is possible to take the work apart; and nothing shows more clearly the extent to which this idea is carried than the variety of uses to which screws are put. There is, however, another way of looking at the the change, and a more practical one, too, for screws certainly have a great many advantages which nails never could possess. They are much more secure when in place; they are neater in appearance; they require but little more labor in driving than nails, and can, at any time, be removed without injuring the material into which they are screwed. Some who have had trouble in removing old screws which had rusted into hard-wood work may object to the last statement. An old carpenter however, once told us of a very simple way to remove even the most obdurate screw: if a red-hot poker is held against the head of the screw for a few moments, the heat will expand the metal, loosening it from its hold on the wood, after which it can be readily unscrewed. Screws.

Chapter III.

The improvements in the processes of manufacture of screws have been even more marked than those which have been noted in regard to nails. The blanks for screws were formerly cut out by hand. The first improvement was to cut them from rolled round iron, the heads being formed by pressing the blanks into a die while hot, and the threads cut with a file, a very laborious operation which resulted in a very poor quality of screw. The screw-working machinery, as at present in use, has very largely been designed by Americans. The blanks are cut and headed from coils of wire on one machine; another machine takes the pieces, gives the proper shape to the head and neck, turns the shank, and finishes the screw. It has been claimed that the entire operation necessary to turning out a perfected screw is so economical in its action, that the fluctuations in the first cost of the crude wire will often equal the total cost of manufacture.

Manufacture of Screws.

The most important improvement which has been made in their manufacture is the application of the gimlet point, by means of which a screw can be turned into the wood without the aid of a gimlet or auger. It seems so natural now to us that it is difficult to understand why the world was so long contented with the poor productions of half a century ago.

Patent Screws.

The form and style of the ordinary screw, as at present in use, seem so perfect in every respect that it would be difficult to suggest any improvements. There are, however, a few forms of patent screws in the market which may be of interest in this connection. Figure 8 illustrates a screw, patented in 1873, fitted with a drill point instead of the ordinary gimlet point. Figure 9 is a screw with a thread somewhat like that of a bit, the thread, however, diminishing in pitch from the bottom to the top. Figure 10 is a form of coach-screw, having threads of a curved cross-section and provided with a wedge-shaped point, which allows the screw to be partly driven into place with a hammer. Figure 11 is essentially the same as the ordinary screw, except that the point is flattened, and it has a circular cutting edge. Figure 12 is a wood-screw provided with a cutter and point in such a manner as to cut its way into the wood instead of pressing it to one side as is the case in the ordinary screw. None of these varieties, however, have any very extended sale, and we are unable to say by whom they are manufactured or controlled. Figure 13 illustrates about the best of the patent forms. It is a diamond-pointed steel screw, manufactured by Russell & Erwin. Screws of this form can be driven with a hammer their entire length into any hard wood, and then held by one or two turns as securely as the ordinary screw. The head is made convex to strengthen it and prevent its splitting when struck with a hammer. The thread is of the ratchet form, which permits its penetration without tearing the wood fibres.

Ordinary gimlet-pointed screws are made in four styles, depending upon the use for which they are intended. The shank and point are always practically the same. The head is either

Fig. 13.

Fig. 14.

Fig. 15.
Oval-head Screw.

Fig. 16.
Round-head Screw.

Fig. 17.
Piano-head Screw.

flat, Figure 14; oval, Figure 15; round, Figure 16; or of the form shown by Figure 17, which Russell & Erwin designate as having a "piano" head.

Chapter III.

TABLE OF SCREWS.

Abridged from Catalogue of the American Screw Co. Price per gross.

Gauge.	0	2	4	6	8	10	12	14	16	18	20	22	24	30
1/4	$.35 .52	$.35 .55	$.35 .60	....	....	....	....	....	....	......	......	......	......	......
3/8	.35 .52	.35 .56	.36 .62	$.37 .78	$.39	....	....	....	....	......	......	......	......	......
1/2	....	.35 .58	.36 .67	.39 .79	.40 .98	$.44 1.23	$.54	....	....	......	......	......	......	......
5/8	....	.35 .59	.36 .70	.39 .86	.42 1.09	.47 1.40	.56 1.75	$.66	....	......	......	......	......	......
3/4	....	.35 .70	.39 .74	.40 .93	.44 1.22	.51 1.57	.59 1.97	.71 2.43	$.84 2.96	......	......	......	......	......
7/8	....	.38 .90	.40 .95	.43 1.01	.48 1.35	.55 1.73	.65 2.18	.77 2.69	.92 3.29	......	......	......	......	......
1	....	....	.42 1.04	.47 1.10	.53 1.48	.59 1.91	.72 2.40	.86 2.97	1.03 3.62	$1.21 4.35	$1.43	......	......	......
1 1/4	....	....	.51 1.38	.55 1.44	.59 1.73	.68 2.13	.83 2.83	.99 3.51	1.17 4.29	1.39 5.16	1.65	$2.00	$2.45	......
1 1/2	....	....	.60 1.86	.64 1.92	.69 1.98	.78 2.58	.95 3.28	1.1 4.03	1.34 5.00	1.58 6.01	1.87 7.13	2.15	2.60	......
1 3/4	....	....	....	.73 2.45	.78 2.51	.88 2.91	1.05 3.72	1.27 4.64	1.52 5.68	1.82 6.84	2.09 8.13	2.30 9.53	2.77 11.07	......
2	....	....	....	.85 3.14	.89 3.20	.99 3.26	1.17 4.18	1.40 5.21	1.69 6.38	2.05 7.69	2.32 9.12	2.64 10.70	3.01 12.42	......
2 1/4	....	....	....	....	1.08 4.50	1.12 4 57	1.28 4.63	1.54 5.78	1.88 7.07	2.24 8.53	2.56 10.12	2.89 11.87	3.28	......
2 1/2	....	....	....	....	1.22	1.28 6.23	1.43 6.29	1.71 6.35	2.07 7.78	2.45 9.88	2.81 11.13	3.15 13.04	3.55 15.12	......
2 3/4	....	....	....	....	....	1.63	1.69	1.89	2.27	2.66	3.07	3.42	3.87	......
3	....	....	....	....	....	1.97 9.00	2.02 9.05	2.09 9.11	2.46 9.17	2.87 11.05	3.32 13.12	3.76 15.37	4.22 17.82	......
3 1/2	....	....	....	....	....	2.67	2.74	2.80	2.87	3.31	3.61	4.08	4.66	......
4	....	....	....	....	....	....	3.18	3.22	3.28	3.75	4.14	4.73	5.46	......
4 1/2	....	....	....	....	....	....	....	....	4.15	4.18	4.80	5.54	6.38	......
5	....	....	....	....	....	....	....	....	....	5.57	5.63	6.46	7.30	......
6	....	....	....	....	....	....	....	....	....	......	8.80	8.85	8.92	$12.67

The upper figures give the price of iron screws: the lower that of brass screws.
Discount: 60 to 75% on iron screws, 55 to 70% on brass screws.

Chapter III.

Wood-Screws.

The kinds of screws which are manufactured for various purposes are almost infinite in variety, but so far as relates to builders' hardware in general, it will be sufficient to consider only the wood-screws, that is to say, the screws made for working with wood. The materials of which screws are made for this purpose are iron, steel, brass, copper, bronze and phosphor-bronze. The screws commonly in use are of iron. Steel screws are comparatively little used on account of the cost. Brass, copper, and bronze screws are used in connection with finishing hardware. Phosphor-bronze screws are used only in special cases.

Screw Sizes.

Iron wood-screws are made in twenty different lengths, varying from one-quarter inch to six inches. Brass and nickel-plated screws are made only as high as three inches in length. Each length of screws has from six to eighteen varieties in thickness, there being in all thirty-one different gauges; so that altogether there are about 250 different sizes of ordinary

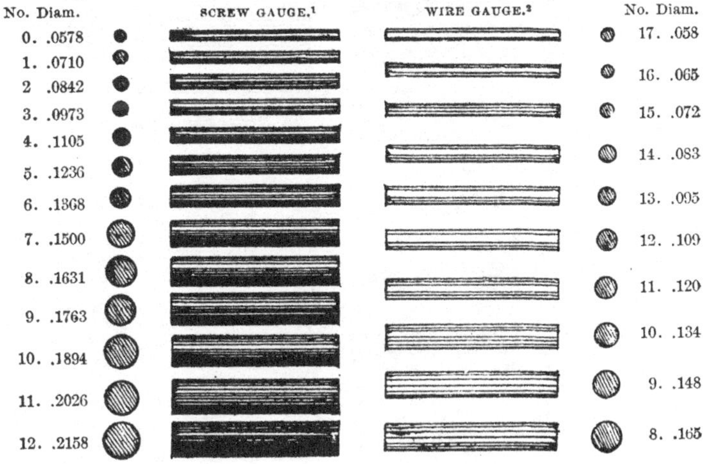

Fig. 18. Diameters of Wire. From American Screw Co.'s Catalogue.

[1] American Screw Gauge.

[2] Old Standard Birmingham Gauge.

Chapter III.

wood-screws in the market. Figure 18 gives the different gauges in use from zero to thirty. Iron screws are finished with either a bronze, japanned, lacquered on tinned surface. P. & F. Corbin also manufacture copper, bronzed, and silver-

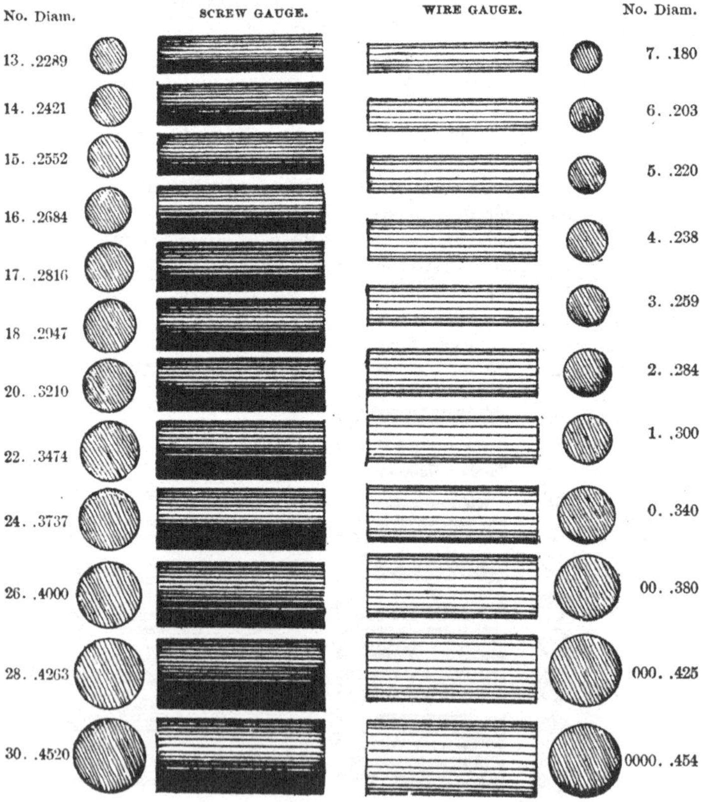

Fig. 18. Continued

plated screws. These and a few special varieties are kept in stock by most dealers. The preceding tables give the sizes, prices, etc., of iron, brass and nickel-plated screws, compiled from the catalogues of the American Screw Company, and P. & F. Corbin. Chapter III.

Nickel-plating increases the foregoing prices as follows:

Length.	Guage.	On Iron.	On Brass.
½	No. 4	$1.09	$0.98
1	" 6	1.03	1.10
2	" 9	1.49	1.72
3	" 14	2.91	4.34

Intermediate sizes approximately at the same ratio. *Discount:* 75 and 66 %.

Lag-screws.

Besides the ordinary wood-screws, the only other kinds used constructively to any extent by builders are lag-screws, and hand-rail screws. The former are more commonly known as coach-screws, and are manufactured in 128 different sizes

TABLE OF GIMLET-POINT COACH-SCREWS.

Price per hundred. *Discount:* 66⅔ %.

Diameter in inches.	Length under the Head.					
	1½	2½	4	8	10	12
¼	$2.70	$3.10	$3.70	..	..	..
5/16	2.70	3.10	3.70	..	..	..
⅜	3.10	3.50	4.10	..	..	..
7/16	4.00	4.50	5.25	$7.25	..	..
½	4.30	4.90	5.80	8.20	$9.40	$10.60
9/16	..	6.90	8.10	11.30	12.90	14.50
⅝	..	6.90	8.10	11.30	12.90	14.50
¾	..	10.00	11.50	15.50	17.50	19.50
⅞	..	..	16.50	22.50	25.50	28.50
1	..	..	22.50	30.25	34.25	38.25

Chapter III.

varying from $\frac{1}{4}'' \times 1\frac{1}{2}''$ to $1'' \times 12''$. In appearance the shank and the gimlet-point are the same as ordinary wood-screws, but the head is square, like a nut, and without any cross-cut, so that the screw can be turned up with a wrench. The preceding table gives the prices of a few of the sizes, as per the lists of the American Screw Co.

Joint-bolts.

Hand-rail screws or joint-bolts are usually made in two ways, either with one end cut with a wood-screw thread, and the other provided with a machine-screw thread and loose nut, or with a machine-screw thread and nut on each end. One nut is generally cogged so it can be turned up easily by a pocket wrench. Joint-bolts are of two diameters, either $\frac{5}{16}$ or $\frac{3}{8}$-inch, and the standard lengths are from four to six inches, though some manufacturers produce joint-bolts as long as fifteen inches. Joint-bolts with two nuts are sometimes made with $\frac{7}{16}$-inch diameter with a swelled centre. Figure 19 shows the various forms of joint-bolts, and the following table gives the prices per gross.

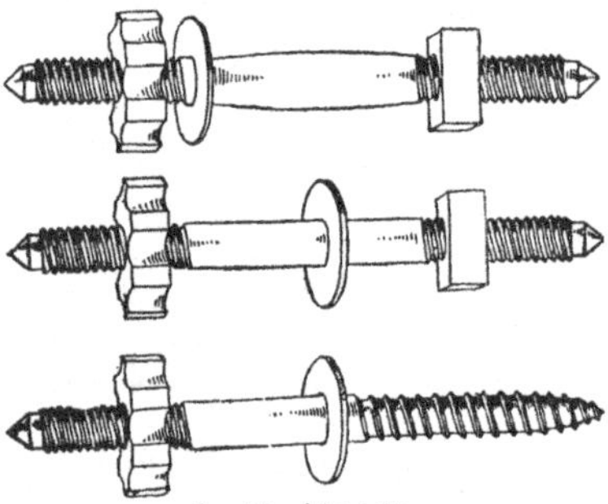

Fig. 19. Joint-bolts.

TABLE OF JOINT-BOLTS.

Discount : 75%.

Diam.	Length, 4 inches.			Length, 4½ inches.			Length, 5 inches.			Length, 5½ inches.			Length, 6 inches.		
	One nut.	Two nuts.	Swell'd centre	One nut.	Two nuts.	Swell'd centre.	One nut.	Two nuts.	Swell'd centre.	One nut.	Two nuts.	Swell'd centre.	One nut.	Two nuts.	Swell'd centre.
$\frac{5}{16}$	$9.00	$10.25	..	$10.00	$11.25	..	$11.25	$12.50	..	$12.25	$13.50	..	..	..	..
$\frac{3}{8}$	..	..	..	11.25	13.50	..	12.50	14.50	..	13.50	15.75	..	$14.50	$16.75	..
$\frac{7}{16}$	..	..	$16.75	..	..	$18.00	..	..	$19.00	..	..	$20.25	..	..	$21.50

Chapter III.

Screw-eyes.

Screw-eyes are too well-known to require illustration. They are made of steel, iron or brass wire, with a gimlet-pointed thread cut on the shank. Iron screw-eyes are made plain, bright, bronzed or nickel-plated. The diameters of wire used vary from wire guage 0, to 14, the lengths of the eyes being from $\frac{11}{16}$ to $2\frac{7}{8}$ inches. The list price in iron of the largest sizes is $9.00 per gross. The smallest size cost 95 cents per gross. Brass screw-eyes cost about twice as much as iron. These prices are with a heavy discount.

Screw-hooks.

Figure 20 shows the common forms of screw-hooks. The list prices for these are $6.00 per gross for No. 4 iron, and $22.00 per gross for No. 4 brass. About the same variety of sizes are listed for screw-hooks as for screw-eyes. The smaller sizes of hooks can be had in brass with washers or roses at a slight advance in price. The hooks are made as large as $4\frac{3}{4}$-inch, No. 0 wire.

Picture-hangers

Picture-knobs or hangers are intended to screw into the wall, through the plastering. They are made with a long screw

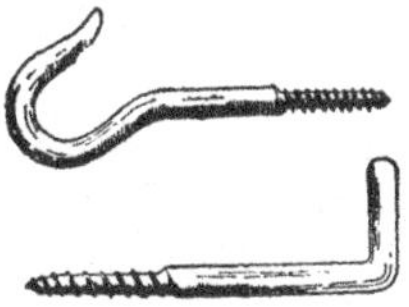

Fig. 20. Screw-hooks.

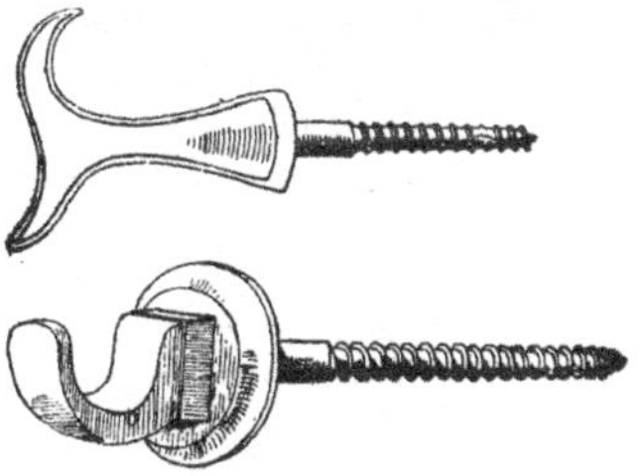

Fig. 21. Picture-hangers. (Half-size.)

shank provided with a knob of porcelain or metal, and are listed in four lengths, $\frac{5}{8}$-inch, $\frac{3}{4}$-inch, 1-inch and $1\frac{1}{4}$-inch; being sold at from $4.75 to $6.20 per gross, with a discount. Figure 21 illustrates one variety.

Picture-rod hooks are intended to support a rod on the wall, answering as a picture moulding. The list price (P. & F. Corbin) is from $2.00 to $4.50 per dozen, according to length and finish.

CHAPTER IV.

BOLTS.

Bolts.

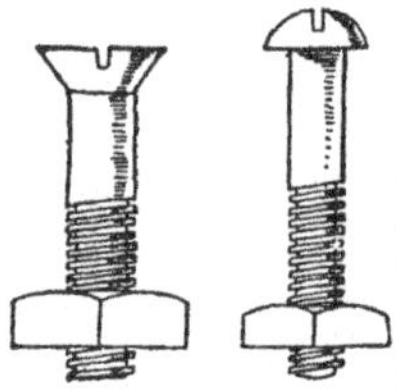

Fig. 23. Stove-bolts.

ASIDE from the coach or lag screws, and the stair-rail bolts already described, the only constructive bolts used by the builder are such as are necessary in joining header and trimmer beams. These are similar to the stair-rail bolts, but heavier and less finished. They are often made to order, but a few sizes are kept in stock by some dealers. Ordinarily $\frac{1}{2}$ to 1 inch bolts are used, 8 to 24 inches long, with a square head on one end and a thread and square nut at the other. In any other cases requiring the use of constructive bolts, lag-screws are generally found to answer every purpose, though stove-bolts, Figure 23, are sometimes useful. These are made with flat or round heads. They are manufactured in six diameters. from $\frac{5}{32}$ to $\frac{5}{8}$ inch, and thirty-two lengths, from $\frac{3}{8}$-inch to 7 inches. The manufacturers' list-prices are from $0.85 to $4.20 per hundred. Sink-bolts are similar to the stove-bolts except that the shank is threaded the whole length, and provided with two nuts. Tire-bolts are like flat-headed stove-bolts, but are without the cross-cut in the head. Many other forms of bolts are in the market, but even the foregoing are rarely used by builders.

DOOR–BOLTS.

Barrel-bolts.

Figure 24 shows the most common form of wrought-iron door-bolt, designated specifically as a "barrel-bolt." This is

made to screw onto the face of the door. The jamb-staple may Chapter
be plain, as in Figure 24; bent, Figure 25; or necked, Figure 26. The latter is for a door swinging out, which is to be bolted on the inside. All of these forms are likewise made in cast brass. The iron bolts may be japanned, tinned or bronzed, and the knobs are sometimes nickel-plated, tinned, or made of brass or porcelain. Neck-bolts, Figure 27, are used when the bolt-plate or staple cannot be put directly on the line of the face of the door. The style shown by the illustration is that manufactured by the Stanley Works, and is made additionally strong by a central rod running into the bolt and riveted to the edge of the bolt-plate as shown by the figure. A similar style of bolt with a flat bar and a raised end instead of a knob, Figure 28, has a flat spring between the bolt and the plate, serving to keep the former in position.

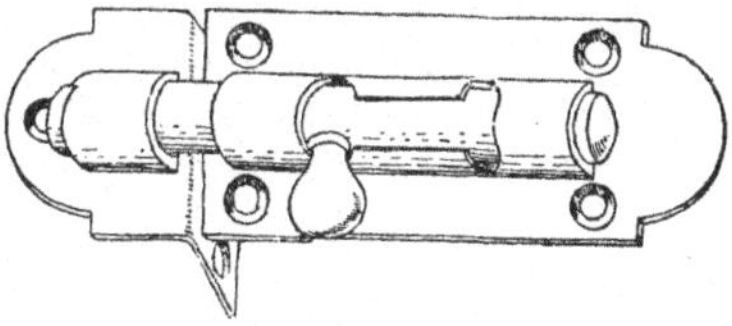

Fig. 24. Common Staple

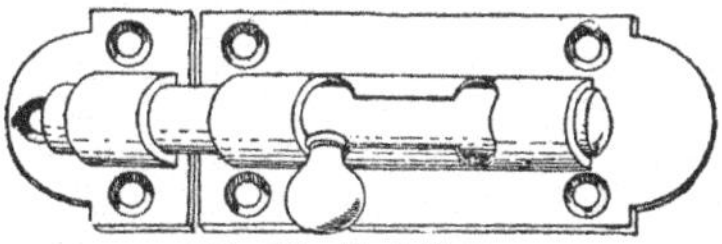

Fig. 25. Bent Staple.

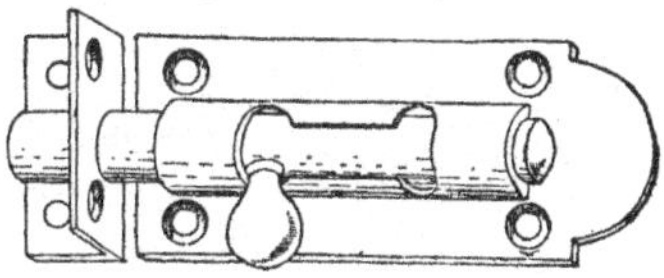

Fig. 26. Necked Bolt.
Barrel-Bolts.

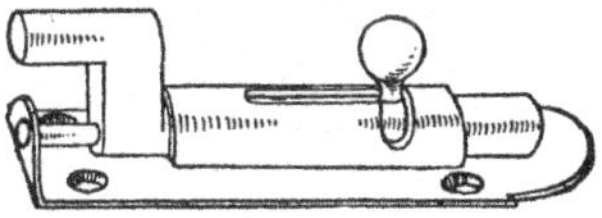

Fig. 27. Round Neck-bolts. Stanley Works.

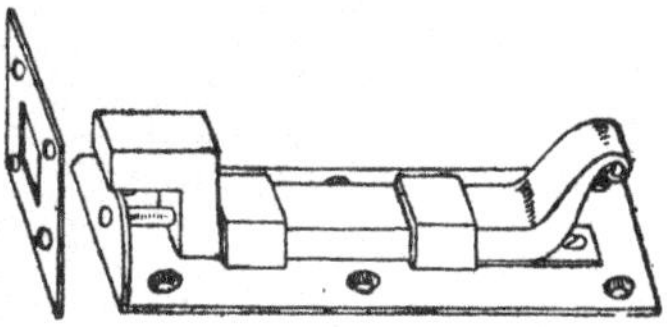

Fig. 28. Square Neck-Bolt. Stanley Works.

Chapter IV.

Figure 29 shows a form which is designated as a mosquito-bar bolt, and is used for a number of light purposes. It is made without any jamb-staple.

Spring-bolts.

Excepting Figure 28, the foregoing bolts are made without any springs. Much the same patterns are found in the market under the designation of spring-bolts, the bolt being held either open

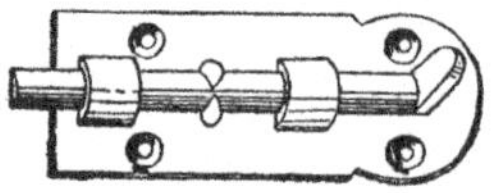

Fig. 29. Mosquito-bar Bolt.

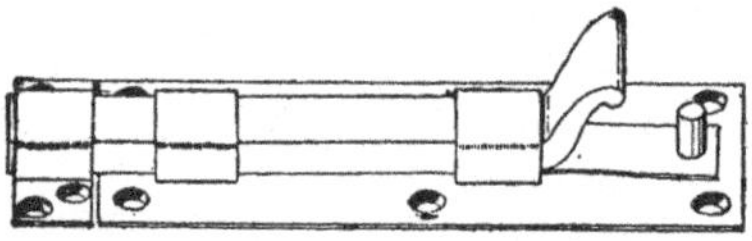

Fig. 30. Square Bolt, with Spring.

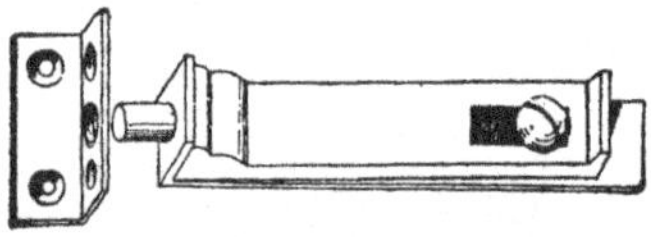

Fig. 32. Straight Cupboard-Bolt.

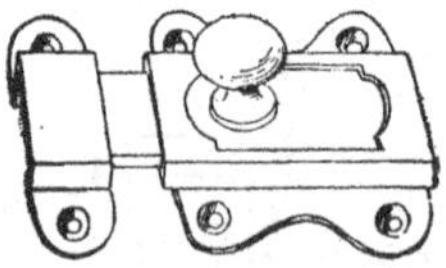

Fig. 33. Flat Cupboard-Bolt.

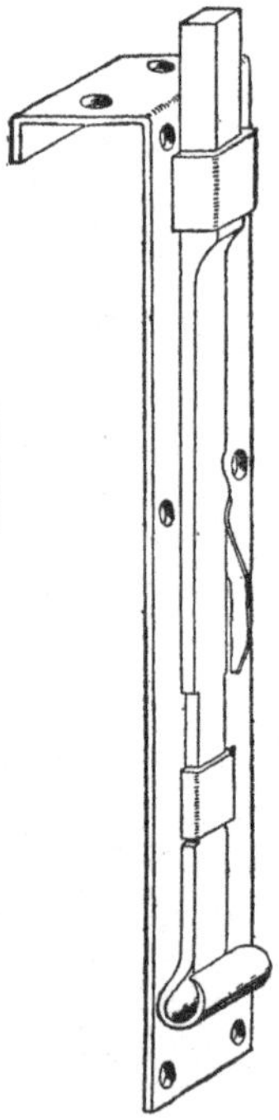

Fig. 31. Square Bolt, with Spring.

or shut by means of a spring inserted under the bolt against the bolt-plate. These are in a number of varieties, including neck-bolts, straight-bolts, square or round bolts, with porcelain knobs, brass knobs, etc. Figure 30 shows a form of square spring-bolt manufactured by the Stanley Works. There is also another form, Figure 31, in which the spring is on one side of the bolt, the notch in the shank holding the bolt either open or shut. Chapter IV.

Cupboard-bolts.

Straight cupboard-bolts, Figure 32, and flat cupboard-bolts, Figure 33, are manufactured in a variety of forms of which those shown are types. They are finished in the usual variety of styles. Figure 34 shows what is designated as a ship-bolt. Figure 35 is a variety of side flush-bolt adapted for chests, desk-tops, etc. Figure 36 and Figure 37 are two forms of bookcase-bolts. The former is screwed flush on the edge of the standing-door at the top, while a flat plate is attached to the edge of the swinging-door. On closing the latter, the brass plate strikes on the knob of the bolt and throws the bolt up into the door-

Fig. 34. Ship-Bolt.

Bookcase-bolts.

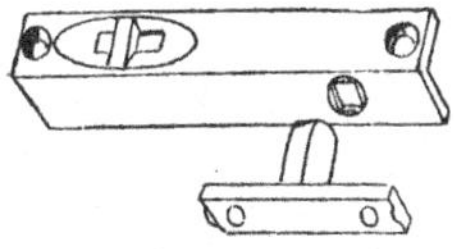

Fig. 35. Side Flush-Bolt.

Fig. 36. Bookcase-Bolt. P. & F. Corbin.

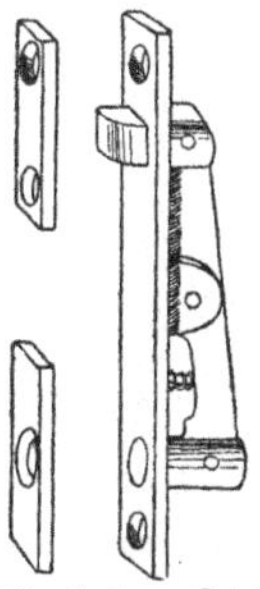

Fig. 37. Bookcase-Catch. P. & F. Corbin.

soffit, the knob shank following the oblique cut in the plate; a spring throws the bolt down when the door is opened. The

Chapter IV. action of Figure 37 is somewhat different. The bolt is mortised into the soffit or the bottom of the door-frame, and the two plates are screwed to the tops of the doors. For a bolt as

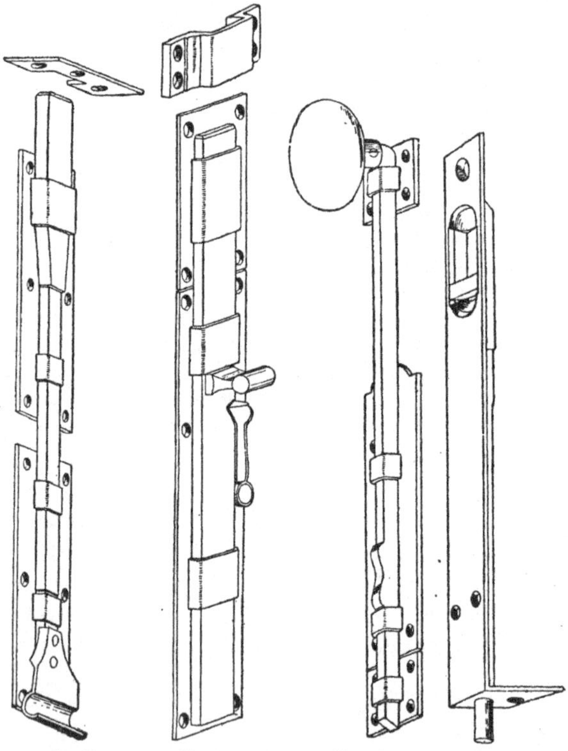

Fig. 38. Flat Tail-Bolt. Fig. 39. Lock Shutter-Bolt. Stanley Works. Fig. 40. Canada Bolt. Stanley Works. Fig. 41. Sunk Flush-Bolt.

shown by the figure, the right-hand door is closed first, when the other door is closed it strikes the bevelled connection of the bolt, forcing it up and consequently forcing the other arm of the bolt down into the plate on top of the right-hand door. The doors can then be locked together with a key or catch, though the friction on the striker will keep them closed. A spring forces the bolt up when the left door is opened. This form of bolt can be used for cupboards, wardrobes, etc., but we do not know of its having ever been applied successfully to large double-doors.

The following table gives the average retail prices of the bolts enumerated. Only the principal sizes are listed, but these will be sufficient to give an idea of the cost.

TABLE OF PLAIN BOLTS.

Prices per dozen.

Fig.	Length in inches.	2	2½	3	4	5	6	8	12	36	84
		$	$	$	$	$	$	$	$	$	$
24	Wrought-iron barrel-bolt common staples	....	....	1.10	....	....	....	3.25	....	....	....
25	Wrought-iron barrel-bolt bent staples	....	....	2.25	....	....	....	4.00	....	....	....
26	Wrought-iron barrel-bolt necked staples	....	....	1.15	....	....	....	3.40	....	....	....
27	Wrought-iron round, neck bolts	....	....	....	3.10	3.30	3.50	....	....	....	....
28	Wrought-iron square, neck bolts	....	....	....	2.45	....	3.65	4.70	....	....	....
29	Brass mosquito-bar bolts	1.05	....	....	....	....	....	....	....	....	....
30	Wrought-iron, square, spring bolts	....	....	1.20	1.50	1.60	1.70	2.40	....	....	....
31	Wrought-iron side spring bolts	....	....	....	....	....	....	5.95	....	....	....
32	Brass straight cupboard-bolts	1.50	....	2.15	....	....	....	....	....	....	....
33	Brass flat cupboard-bolts,	....	.88	....	....	....	....	....	....	....	....
34	Brass ship flush-bolts	2.40	....	....	....	....	....	....	....	....	....
35	Brass side flush-bolts	....	2.10	....	....	....	....	....	....	....	....
36	Brass bookcase-bolts, one size	2.10	....	....	....	....	....	....	....	....	....
37	Brass bookcase catch, one size	2.10	....	....	....	....	....	....	....	....	....
38	Wrought-iron flat-tail bolts	....	....	....	....	....	....	....	10.20	....	26.25
—	Wrought-iron shutter-bolts plain	....	....	....	....	....	1.40	1.83	2.68	....	....
39	Wrought-iron shutter-bolts with lock	....	....	....	....	....	1.49	1.92	2.77	....	....
40	Wrought-iron Canada bolts	....	....	....	....	....	2.80	....	....	4.32	....

Tail-bolts.

Flat-tail-bolts, Figure 38, are intended for high doors requiring to be bolted at the top, and are made in a number of different

Chapter IV.

lengths, from one to seven feet. When the bolt is shot it is kept from slipping down by a rebate in the shank which catches on the lowest staple, as seen by the figure. Figure 39 is a form of bolt used for shutters having a wide bearing on each side. It is provided with a locking lever at the upper

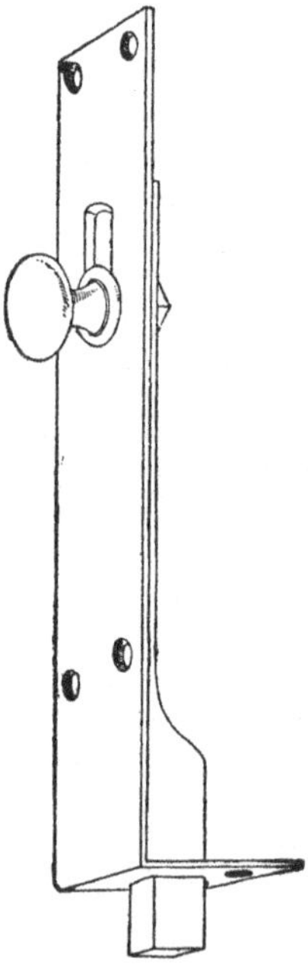

Fig. 42. Wrought-Iron Flush Bolt.

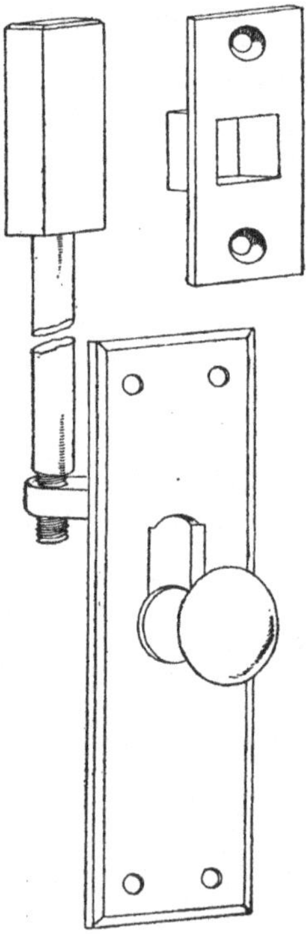

Fig. 43. Mortise Flush Bolt.
J. B. Shannon & Co.

side, catching in a notch on the bolt. The same form is made without the locking-lever. Canada-bolts, Figure 40, consist of a long, square shank or bolt, with mineral or porcelain knob. The bolt is kept from slipping by a short, flat spring underneath. These are sold with several varieties of staples. Chapter IV.

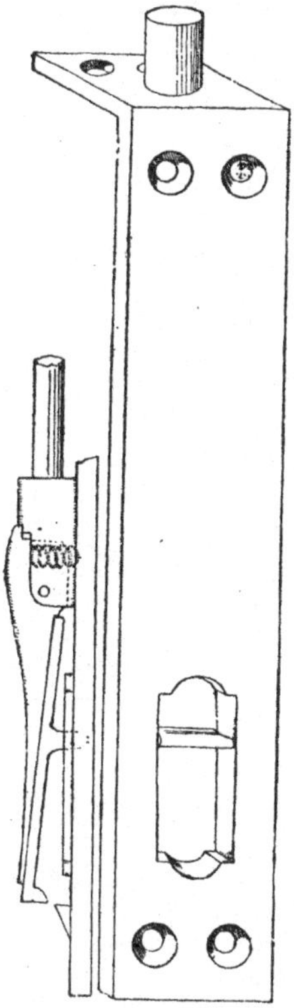

Fig. 44. Flush Bolt, with Patent Stop. P. & F. Corbin.

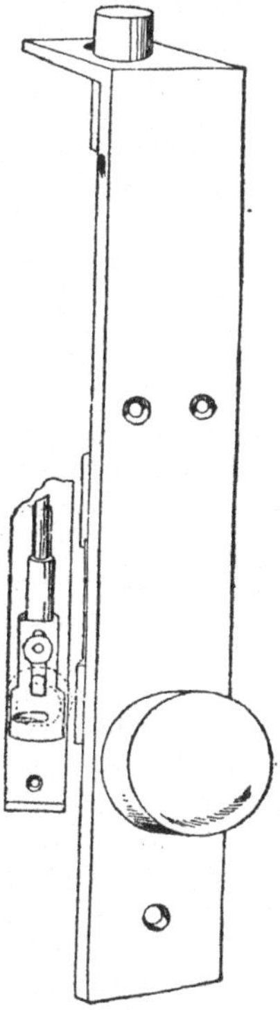

Fig. 45. Flush Bolt, with Patent Stop. Ireland Manufacturing Co.

For front and vestibule doors in two folds as well as for other double doors some form of mortise-bolt is required.

Chapter IV.

Sunk Flush-bolt.

These may be mortised into the edge or sunk flush with the face of the standing door. Figure 41 shows the ordinary sunk flush-bolt. This pattern is made with plates three-fourths inch wide and bolts from six to twenty-four inches long. The retail prices average from $5.60 to $9.52 per dozen in bronzed wrought-iron. Bolts of similar description but with a square rod, Figure 42, are made with plates one and one-fourth inches wide and fifteen to sixty inches long, costing $9.80 to $23.-60 per dozen in bronzed iron, with bronze knob. The same styles of bolts are also made in cast brass; a few dealers keep them in stock nickel-plated. The pattern represented by Figure 42, is, of course, used only on the face of a door.

Mortise Flush-bolt.

A mortise flush-bolt is one which is mortised into the thickness of the door and is operated by a knob or handle working in a face-plate. Figure 43 illustrates one variety. The bolts are made from nine to forty-eight inches long, and the retail prices are from $1.50 to $1.80 each, in bronze.

Figure 44, illustrates a form of self-locking flush-bolt. A lever on the bottom of the bolt catches over a shoulder on the face-plate when the bolt is shot. To release the lever it is simply pushed inward, a coiled spring at the top otherwise holding it in position. Figure 45 shows another device in which the bolt is thrown by turning the knob. A peg at the back of the knob works in a horizontal slot in a tail-piece attached to the bolt. Raising the bolt brings the peg in the line of the centre of rotation of the knob and so locks it.

Latch Spring-bolt.

There are several varieties of latch spring flush-bolts, in which the knob remains thrown out but can be drawn down by a pull within convenient reach of the hand. Figure 46 illustrates one such arrangement. P. & F. Corbin also manufacture a latch-spring bolt which is mortised into the edge of the door, and in which the latch bolt is released by pressing on the face-plate.

For store doors it is customary to use bolts applied to the face of the door instead of being mortised-in flush. In this case the upper bolt is attached to a chain which hangs about six feet from the floor. The lower bolt is held up by a spring, but can be pressed down into place with the foot, a spring catch on the face holding the bolt when down. Chapter IV.

Figures 44 to 46 inclusive, are types of a great variety of styles manufactured in several different metals with all kinds of finish and design. It is, therefore, impracticable to give for these any average prices which could serve as fair criterions.

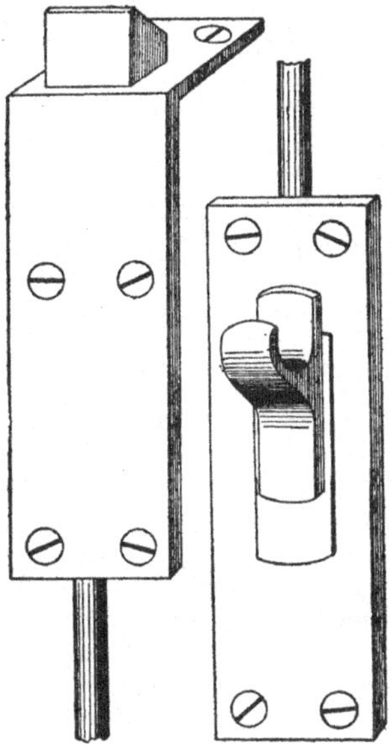

Fig. 46. Extension Latch-Spring Flush-Bolt. P. & F. Corbin.

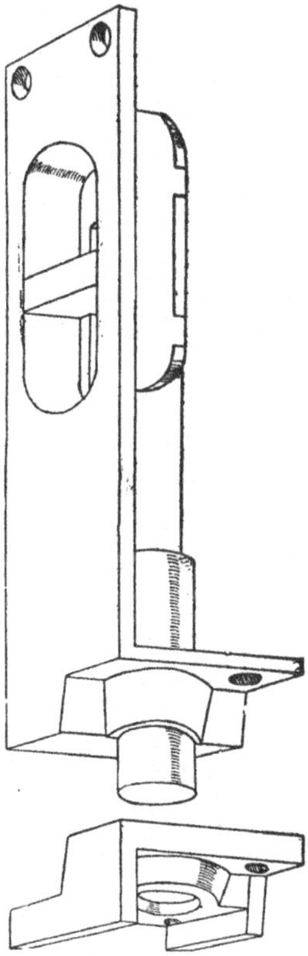

Fig. 47. Dutch-Door Bolt. Hopkins & Dickinson Manufg. Co.

The Hopkins & Dickinson Manufacturing Co., has recently

Chapter IV.

put on the market a form of flush-bolt intended specially for Dutch doors, that is to say, doors in two folds, horizontally. Figure 47 illustrates this. The bolt-plate is about seven inches long, and is rebated to match the rebates of the doors. The retail price of a single bronze bolt is $2.50. Figure 48 illustrates a chain and a foot-bolt.

Engine-house bolts are made in a variety of forms generally so as to permit of being opened easily, by a large catch or

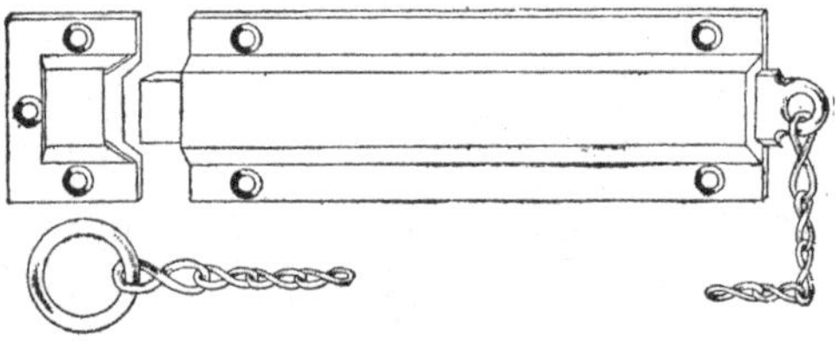

Fig. 48.

latch which throws the bolts up and down from bottom and top. These cost from $6 to $9 each, though it is impossible to give any fair general price as the bolts are made only to order.

FRENCH WINDOW-BOLTS.

French window-bolts are usually mortised into the centre style of one of the sashes and are so arranged that a crank or handle on the face of the sash will throw a bolt in each direction, so as to lock the window at the top and the bottom. Figure 49 is a form operated by a knob and spindle. In setting it, the gearing-box or mechanism of the bolts can be let into the door in the same manner as an ordinary mortise-lock, and the two rods dropped in through a hole bored the length of the sash, the rods screwing into the hubs on the gearing-box. The retail price of this appliance is $4.00 in bronze, including a bronze handle. Figure 50 is another form, similar in its action as regards the bolts, but intended to plant on the face of the sash. The crank handle drops into a catch on the opposite sash, and can be locked by turning the bar on top of the catch. The retail price in bronze is $4.00. Chapter IV

Espagnolette-bolts are arranged, like the foregoing, to Espagnolette-Bolts.

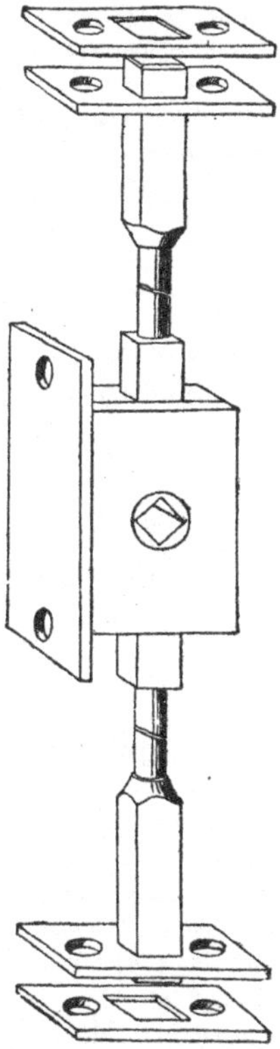

Fig. 49. Mortise French Window-bolt. J. B. Shannon & Sons.

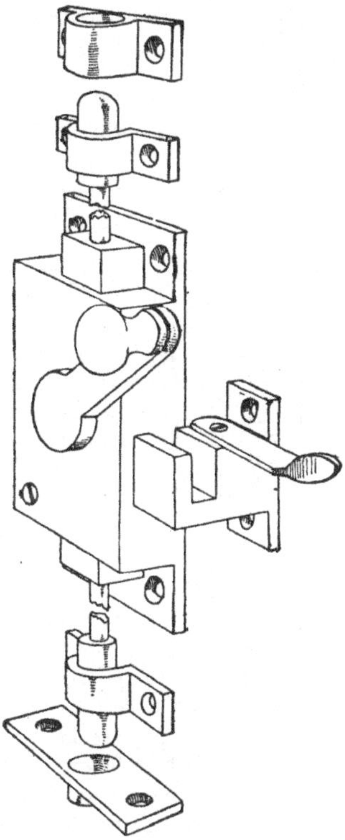

Fig. 50. French Window-bolt. Hopkins & Dickinson Manufacturing Co.

shoot up and down, but in addition the bolts are made to turn so as to hook onto a post or peg at the top and bottom. They

Chapter IV.

are much used in France for double windows, and have several advantages, as they not only lock the sashes, but also draw them up firmly against the window-frame, thus making them more secure against the weather. Figure 51 illustrates one form of espagnolette-bolt.

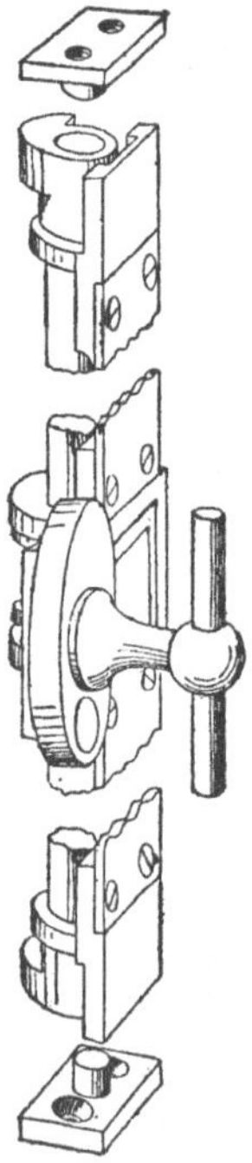

Fig. 51. Espagnolette-bolt. Hopkins & Dickinson Mfg. Co.

There are many other devices adopted for securing French windows. In the cheapest class of work an ordinary cup-board-catch is used; and in the best work they are sometimes secured with a regular key lock.

CHAIN AND CHECK BOLTS.

Chain-bolts.

At one time it was considered quite essential that a front door should be provided with a chain door-fastener, which would permit the door to be opened a few inches to inspect any doubtful character on the outside, while it could be opened no farther. Figure 52 illustrates a typical form of chain door-fastener, consisting of a slotted plate to go on the face of the door, and a chain secured to the door jamb, with a dog on the end of the chain which will slide freely in the slot of the plate. A holder is provided to which the chain can be attached when not in use. There are many varieties of chain fasteners. They would average about $1.00 per set in cast brass, and $2.50 silver plated. P. & F. Corbin manufacture a rim door-bolt which has a chain attachment, the dog of the chain working in a slot cut in the barrel or plate enclosing the bolt.

Instead of a chain, some form of hinged bar is often

employed, the fixture then being designated as a check-bolt or door-fast. Figure 53 illustrates one form of door-fast (Nichols, Chapter IV. Check-bolts.

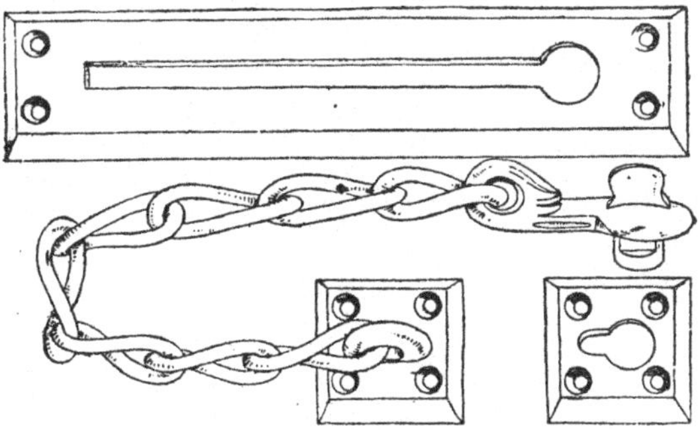

Fig. 52. Chain Door-fastener. P. & F. Corbin.

Bellamy & Co., agents). The staple-shaped bar or rod works in a standard which is screwed to the door-jamb, and fits over a knob secured to the door. The door, on being opened forces the knob along between the prongs of the bar until it can go no farther, permitting the door to open only about four inches. When it is not desired to secure the door, the bar is turned back against the wall. When the bar is turned at right angles to the wall, or midway between these positions, the shoulders are brought directly over the knob on the door catch, and the door is secured so that it cannot be opened at all. This fixture retails at $2.50, in bronze.

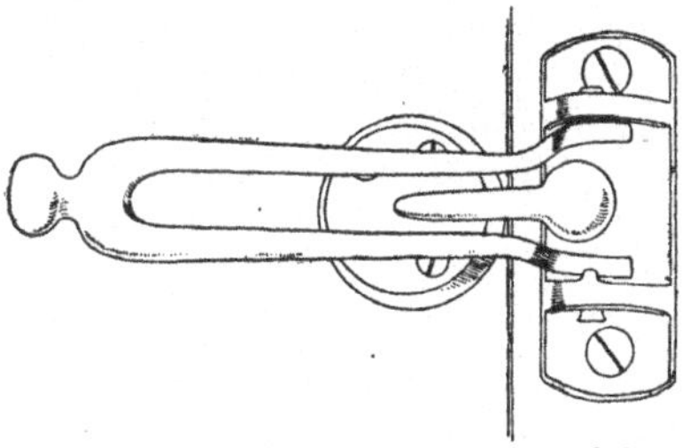

Fig. 53. French Door-fast. Nichols, Bellamy & Co.

Figure 54 is another form of door-fast consisting of a

straight bolt working through a hinged socket attached to the door. The bolt has shoulders at the end which fit into the

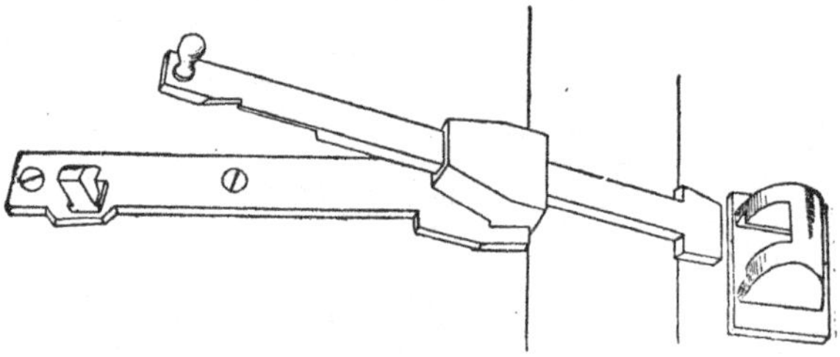

Fig. 54. Door-fast. Sargent & Co.

catch on the door-jamb in such a manner that when the door is opened and the bolt tilted the shoulders are held by the jamb-catch, the bolt slipping through the socket on the door. The retail price of this fixture, in bronze, is $2.00.

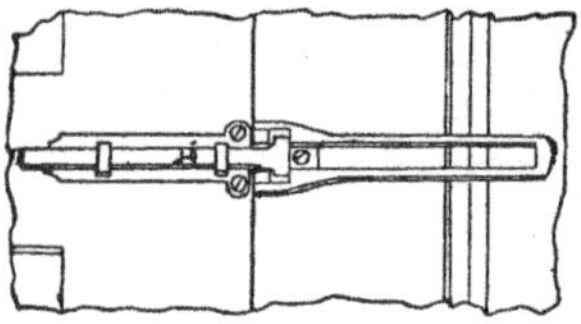

Figure 55 illustrates a form of door-check which combines some of the features of both of the foregoing fixtures, though taking up more space when applied. It retails at $1.00 in either nickle-plate or bronze.

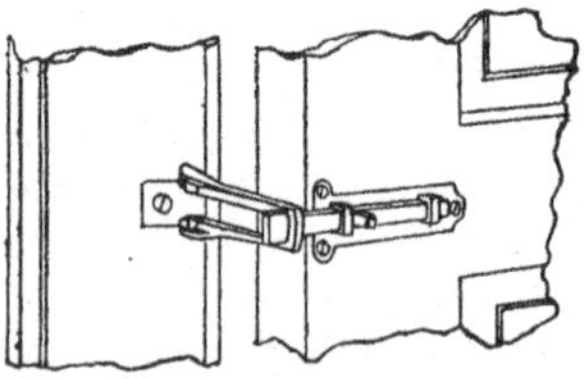

Fig. 55. Perkins's Door-check. Portsmouth Wrench Co.

MORTISED DOOR-BOLTS.

In addition to the ordinary lock on a door, it is sometimes found desirable to attach a plain bolt of some form, as an extra security. The form most commonly used is known as a mortise door-bolt, consisting simply of a barrel-bolt in a cylindrical case, which is mortised directly into the door-style and is operated by a spindle with turn-button or knob on the inner face of the door. In external appearance the various makes of mortised door-bolts are very much alike, though some are finished so as to require no other mortising than can be done with an auger, while others require more hand-work in the application.

Ireland Bolt.

Figure 56 illustrates a complete bolt, and also shows one form of internal arrangement. When the bolt is thrown, the shoulder on the follow, *B*, is turned so as to bring it very nearly on a line with the centre of rotation of the spindle, thus locking the bolt. The spring, *C*, tends to keep the shoulder from rising.

Gem Bolt.

The simplest and most ingenious mortise door-bolt which has come to our attention is the one manufactured by Sargent & Greenleaf, designated as the "Gem mortise-bolt," Figure 57. It consists simply of a solid bolt completely filling the diameter of the bolt-case. The cogged spindle works over the teeth cut across the bolt. The last tooth towards the back of the bolt is smaller and projects less than the others, and the last cut is wider than the intermediate ones; so that when the bolt is thrown the cogged spindle wedges into the wide cut and locks the bolt, making it practically impossible to throw it back by external pressure. The beauty of this bolt is, there is nothing about it that can possibly get

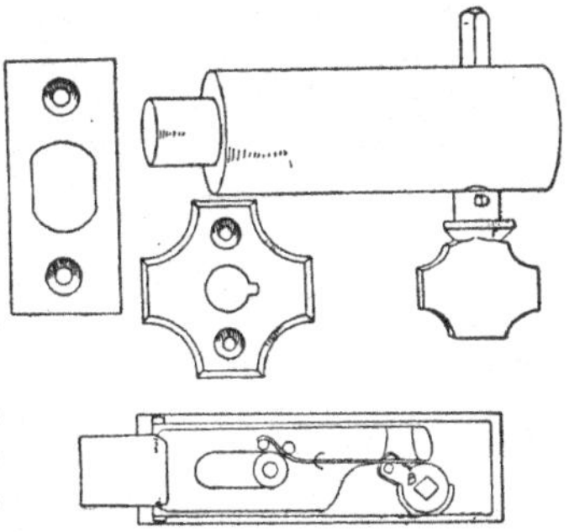

Fig. 56. Cylindrical Door-bolt. Ireland Mfg. Co.

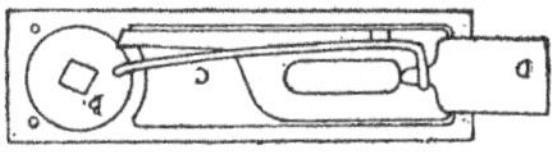

Fig. 56*a*. Ives Mortise Door-bolt. H. B. Ives & Co.

Ives Bolts.

Chapter IV.

out of order, and the only effect of wear would be to make it work more efficiently, if possible.

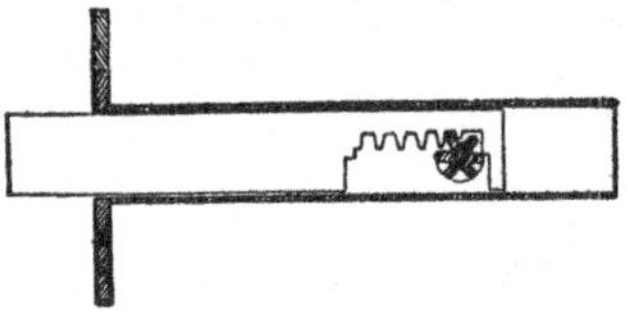

Fig. 57. Gem Mortise Door-bolt. Sargent & Greenleaf.

Figure 56*a* shows the internal construction of the Ives mortise door-bolt. The spindle, *A*, operates the circular follow or disc, *B*, to which is pivoted a heavy brass wire, *C*, which is fastened to the bolt. When the follow is rotated the wire forces out the bolt until the attachment of the wire is brought on a line with the centre of the bolt, when the latter is locked, and cannot be forced back by external pressure.

P. & F. Corbin, Russell & Erwin, the Stoddard Lock Manufacturing Co., and the Reading Hardware Co., manufacture door-bolts similar in the main to Figures 56 and 56*a*. The following table gives the average retail prices:

TABLE OF MORTISE DOOR-BOLTS.

Retail prices per dozen.

MANUFACTURER.	Plain Bronze.	Nickle-plated.
Ireland Manufacturing Company..	$2.25	$2.62
H. B. Ives & Company........................	3.00	3.50
Sargent & Greenleaf...............	5.50	3.50
P. & F. Corbin................................	2.70	3.00
Russell & Erwin........	2.55	2.55
Stoddard Lock and Manufacturing Company	*6.00	*6.00
Reading Hardware Company..................	....	2.00

* List Prices.

CHAPTER V.

Hinges.

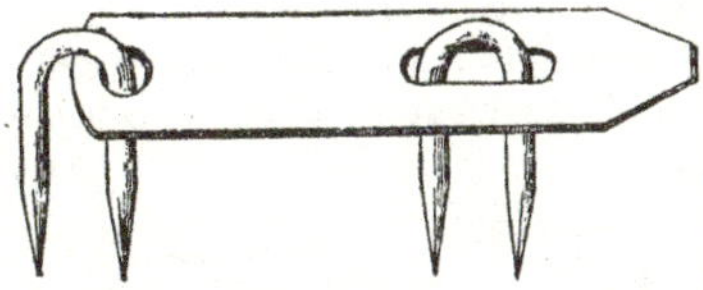

Fig. 59. Hasp and Staples.

Hasps and Staples.

Wrought-iron hasps and staples are not properly to be classed with hinges, but it seems convenient to introduce them at this point. Ordinary wrought-iron staples are made in thirteen sizes, from one inch to six inches long. They are made both in plain and galvanized iron and are used in building operations chiefly in connection with wrought-iron hasps. Figure 59 shows the commonest form, a plain hasp with two staples. Hasps are made in even inches from five to twelve inches in length. A variation from the common hasp has a latch on the hasp which catches into one of the staples in place of a padlock, as shown by Figure 60. A natural simplification of this devise is to do away with the hasp, connecting the staples by a wrought-iron hook, the staples being either driven independ-

Staples.

Hasps.

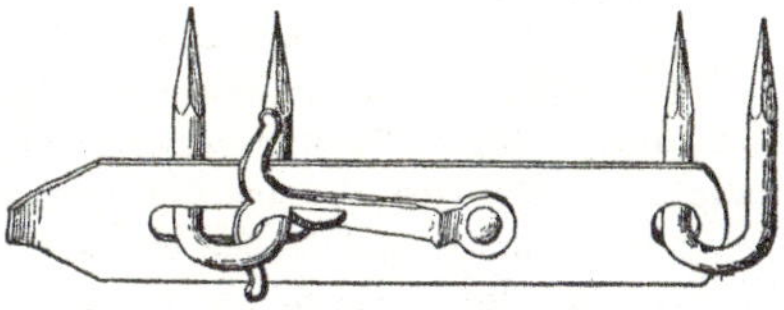

Fig. 60. Hasp and Staple with Double Hook.

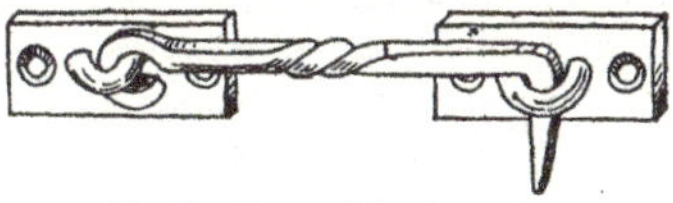

Fig. 61. Hasp and Staple on plates.

Chapter V.

ently, as in the previous examples, or riveted to plates, as shown by Figure 61. There is also a form of hasp and staple intended to be secured with a padlock, the locking-staple being swivelled on a back-plate which is screwed to the jamb.

Bent Hasps.

In place of the ordinary wrought-iron hasp and staple, Figure 59, hasps are made bent at right angles at the middle so as to lap around the edge of a box or a door, if necessary. Bent hasps

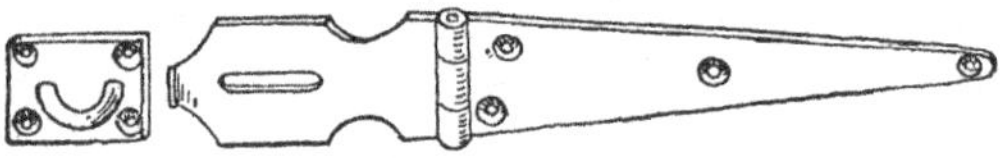

Fig. 62. Hinged Hasp.

can be had in the market from five to ten inches in length. Besides this, there are numerous special forms which are made by a few manufacturers, and as hasps are always of wrought iron or steel, they can be bent to any desired shape.

Hinged Hasp.

The connection between hasps and hinges can be readily illustrated by Figure 62, representing a hinged hasp. This is a natural outcome from the common hasp and staple, the hasp being hinged in the middle, one end screwed onto the door or box-top, while the other end has a slot through which is passed a staple for securing the padlock. The figure shows one of these, with an ordinary staple attached to a wall-plate. They are also manufactured with swivelled staples. These hinged hasps are made three, four and one-half, six, eight, ten and

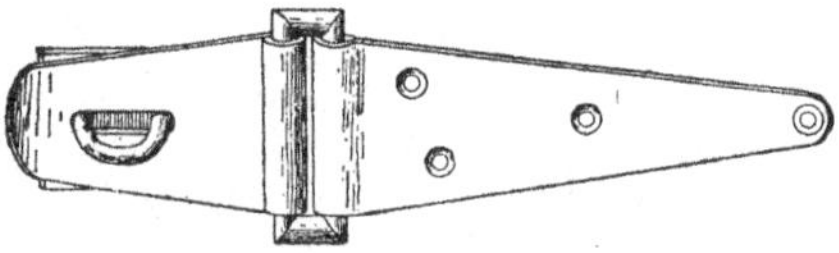

Fig. 63. Solid Link Hinged Hasp. Stanley Works.

twelve inches long. The price is the same for either the plain or the swivelled staples. Some manufacturers have in the market varieties of hinged hasps made in brass or bronze for fancy work. These are, however, not used very extensively, and the form is more strictly for rough work.

A form of hinge-hasp shown by Figure 63 is sometimes used for extra-heavy work, and for trap-doors. In this, the hasp works upon a solid link of wrought-iron, and considerable gain of strength is so acquired. Chapter V.

STRAP–HINGES.

Hinges, proper, may be divided into two general classes: First, those which are placed on the face of a door or shutter, and are known as strap-hinges; and second, those which are mortised into the butt edge of the door and against the frame, and are, consequently, designated as butts. Figure 64 shows the commonest form of a strap-hinge, such as is seen on barn-doors, etc. These hinges are made in even sizes from three to sixteen inches long, measured when opened flat. They are Strap-hinges.

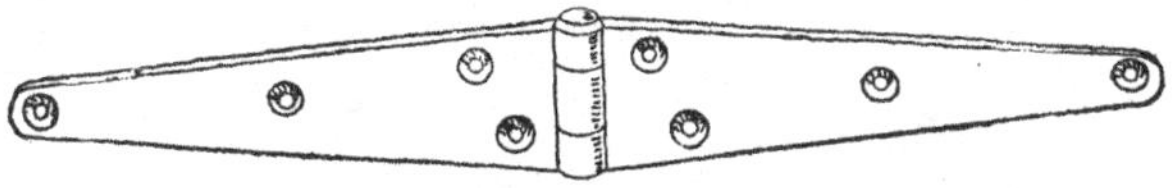

Fig. 64. Strap-Hinge.

made in various ways and widths to suit special necessities. A variety of strap-hinge is made by the Stanley Works, with the

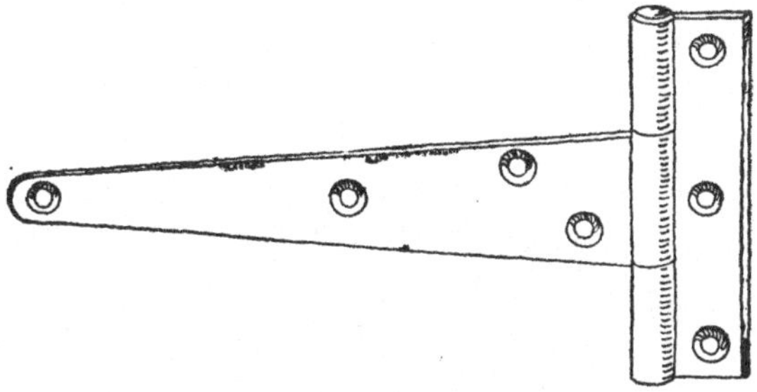

Fig. 65. T-Hinge.

same sort of solid link that has been described for Figure 63. This form of hinge can be used only where there is plenty of room both on the door and on the jamb for attaching the leaves

Chapter V.

of the hinges. When the width on the jamb is restricted, as is often the case, a T-hinge, Figure 65, is used. In some cases it is necessary to have the fixed plate bent, a form known as the chest-hinge being then used, Figure 66. The latter costs considerably more than the common form. The T-hinges are

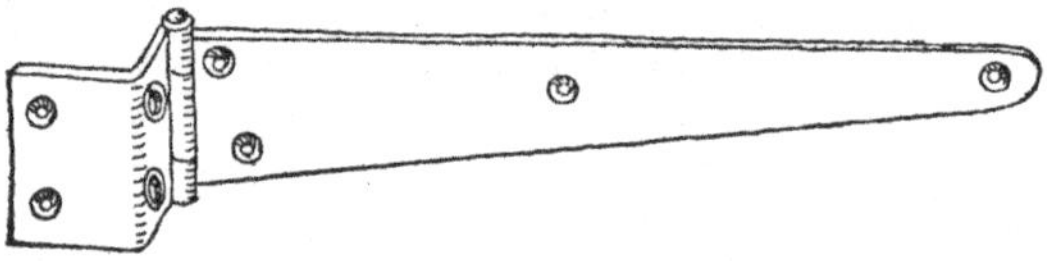

Fig. 66. Chest-Hinge.

about the same price as the ordinary strap-hinge. The Stanley Works manufactures a T-hinge with a braced leaf, which is very useful in some cases. This is shown by Figure 67.

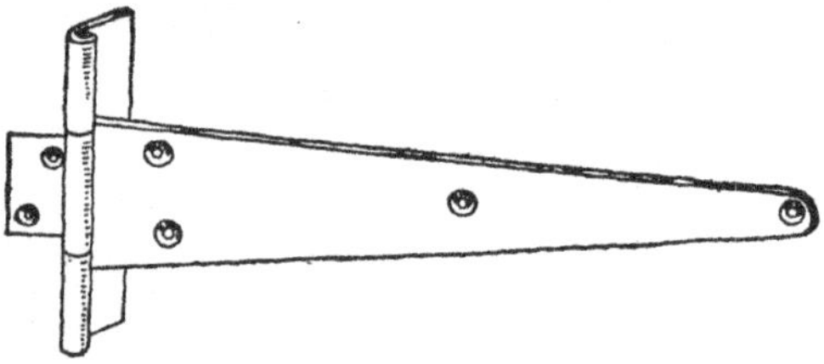

Fig. 67. T-Hinge with braced Leaf.

Reinforced Strap-hinges.

There are several special makes of strap and T hinges, which are reinforced so as to afford greater strength. One of these is the Wells patent hinge, Figure 68, in which the metal of each leaf of the hinge is carried completely around the bend and back onto itself, so that it would be almost impossible to tear it away. The Hart patent hinge is reinforced by a double thickness of metal about the pin, and

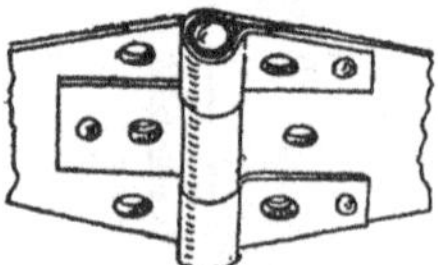

Fig. 68. Wells Patent Hinge. Stanley Works.

Chapter V.

the Record hinge is strengthened by two flange-plates, which are bolted to each leaf and attached to the pin, as shown by Figure 69. The prices of these reinforced hinges are the same for the different kinds.

Figure 70 shows a hinge which is used when it is desired that the pin should be well out from the door or shutter, so as to throw it open away from the jamb. This hinge is made in sizes from six to eighteen inches long.

The following table gives the average retail prices of the foregoing hinges, in a few of the leading sizes.

TABLE OF WROUGHT-STEEL STRAP-HINGES.

Prices per dozen pairs.

Fig.	NAME.	3 inch.	6 inch.	10 inch.	12 inch.	16 inch.
		$	$	$	$	$
59	Hasp and staple	....	.55	1.00	1.53	....
60	Hasp and staple with double hook	....	.56	1.00	1.54	....
61	Hook and staples on plate [1]	1.20	....	....	....	....
	Hasp and staple with swivel staple	....	.91	1.54	....	....
62	Hinged hasp	.96	1.24	2.61	3.85	....
63	Solid-link hinged hasp	....	8 inch.[3]	....	....	....
64	Strap-hinge [2]	.50	1.35	1.75	3.00	4.50
65	T-hinge	.50	.75	1.20	1.88	....
66	Chest-hinge	....	.88	1.25	1.50	....
67	T-hinge with braced leaf	....	....	1.79	2.75	....
68	Wells patent hinge	....	2.10	4.95	7.05	....
	Hart's patent hinge	....	2.10	4.95	7.05	....
69	Record's patent hinge	....	2.10	4.95	7.05	....
70	Raised strap-hinge	....	....	2.63	2.95	4.25

[1] These are made as small as ½ inch.
[2] Prices given are for light strap-hinges. Heavy strap-hinges are sold by the pound at 12 to 14 cents.
[3] $6.60.

Figure 71 shows a special form of hinge manufactured for trap-doors, permitting the door to be hung from the under side,

Trap-door Hinge.

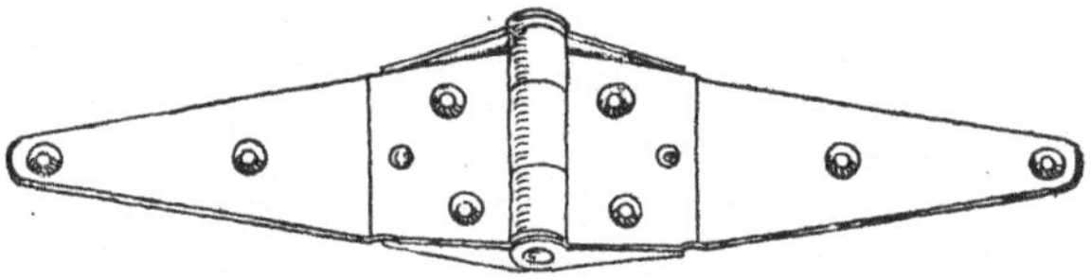

Fig. 69. Record's Patent Hinge. Stanley Works.

leaving the upper side free from obstruction, and flush with the

Chapter V.

floor, while at the same time the use of the full size of the trap can be had when the door is up; the hinges will hold the door in position. The working of this hinge will be seen by the figure. The retail price is $1 per pair. It is listed in the catalogues of several of the hardware manufacturers.

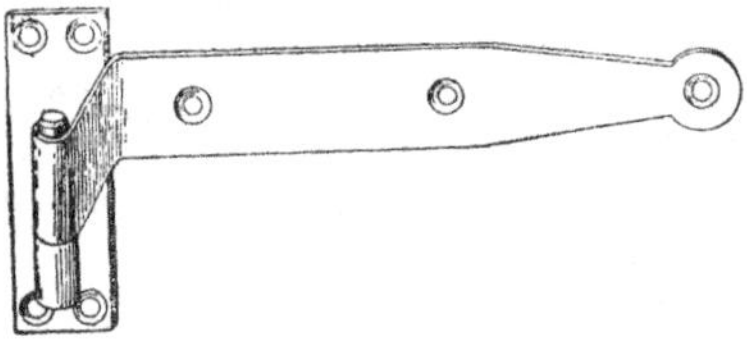

Fig. 70. Raised Strap-Hinge.

Figure 72 illustrates a species of rude hinge used quite

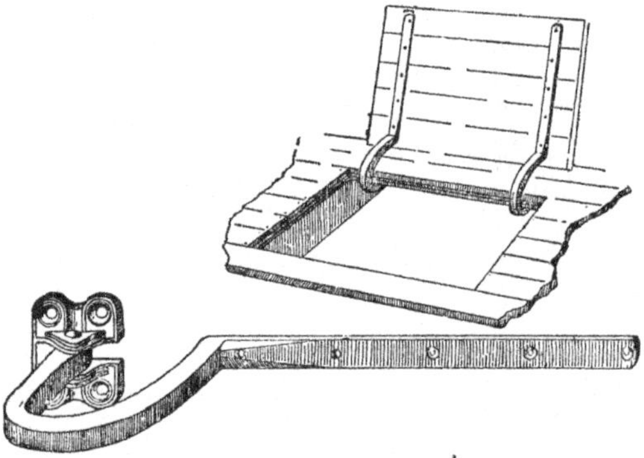

Fig. 71. Trap-door Hinge.

frequently for barn and warehouse doors, consisting of a hook to be driven into the door-post and a bar with an eye at the end to be bolted through the door. These are made of iron $\frac{1}{2}$, $\frac{5}{8}$, $\frac{3}{4}$, $\frac{7}{8}$ and one inch thick, and are sold by the pound, and at 25 cents for $\frac{1}{2}$ and $\frac{5}{8}$ inch, and 20 cents for the other sizes.

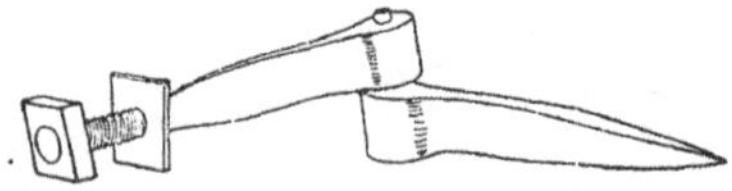

Fig. 72. Barn Door Hook and Eye Hinge.

BLIND–HINGES.

Blind-Hinges.

A variety of hinge which may not be called a strap-hinge, but which, nevertheless, partakes of its nature, is manufactured in a great many forms for outside blinds. The practice of hanging blinds differ in different parts of the country. In the vicinity of Boston the blinds are generally hung on the outside of the casing, and the hinges consist of a simple half hinge on the blind and a hook driven into the face of the casing. In New York the blinds are, almost invariably, set flush with the outside casing, requiring a different style of hinge. Throughout the West a number of forms of cast-iron hinges are used, which, in a measure, lock the blind when open or shut, as will be hereafter noticed. For blinds hung in the Boston fashion,

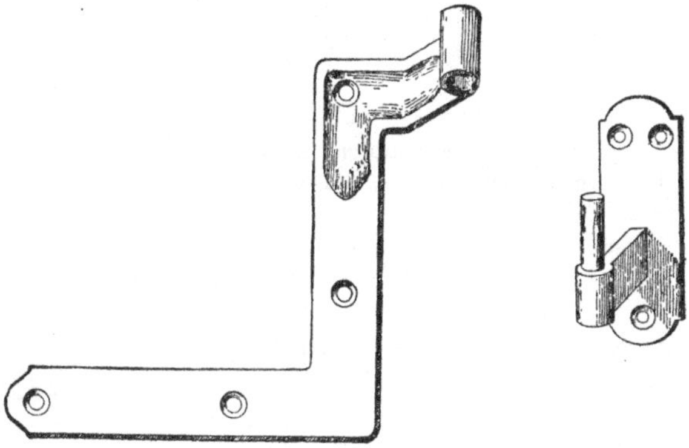

Fig. 73. Blind-Hinges, New York Style.

the commonest way is simply to attach a half hinge to the blind, as previously stated. These half hinges are made in two sizes, two and two-and-one-half inch, costing $5 to $7 per hundred sets. They are mortised into the edge of the blind. If additional strength is required, a longer strap-hinge is used, which is screwed onto the face of the blind. There are various forms of these strap-hinges used for this purpose, all of

Chapter V.

which are too commonly known to require description. Another variety of the same kind of hinge is made so as to throw the blind well out from the casing and away from the moulding. These are made with a two-inch throw, and others with a four-inch throw are also in the market for use in connection with brick buildings.

For the blinds attached in the New York manner, some of the foregoing forms can be used, if the butts are set out sufficiently to clear the face mouldings of the frame, but generally speaking strap-hinges are used, in most cases, of the form shown by Figure 73, the strap, which is bent so as to strengthen the frame of the blind as well as support it, being secured both to the rail and the style. Instead of the hook shown by the figure, some form of drive hook is often used, and the hinge, instead of being bent, sometimes consists of a straight face-plate or strap. Figure 74 is a form of malleable-iron hinge used for blinds hung in the New York style.

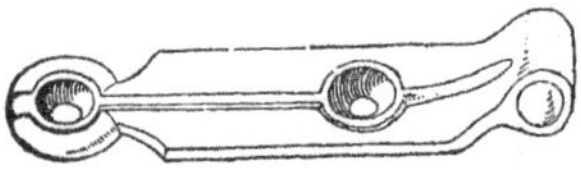

Fig. 74. Malleable Iron Blind-Hinge.

The hooks which are used for blinds are mostly made of malleable-iron. There are several varieties. A plain hook to drive into the wood, shown by Figure 75, is made three and a half inches long with a shorter length of the same style for light blinds hung flush with the casing. Figure 76 shows what is designated as a drive brace. Figure 77 is a form sometimes used, an iron screw-hook; and Figure 78 is the most common form of screw brace.

Fig. 75. Plain Drive Blind-Hook.

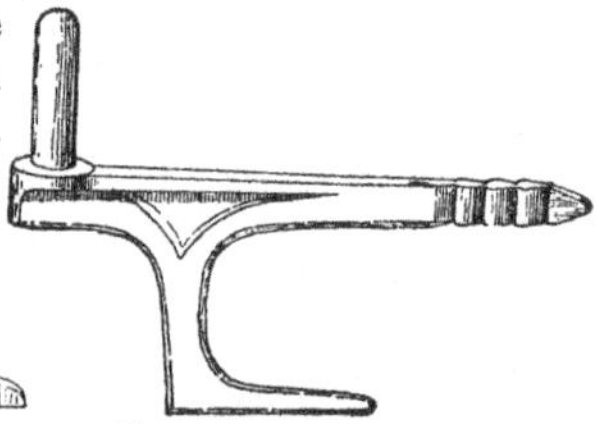

Fig. 76. Drive Brace Blind-Hook.

Chapter V.

The advantages of the styles of hinges previously described are that they are mostly made of wrought-iron and are not apt to break, while there is absolutely nothing to get out of order about them. The disadvantages are that they contain in themselves no principle which will hold the blind open or shut, and when it is secured in the ordinary way it takes considerable bending and twisting to close the blind after it is opened. To overcome these difficulties a number of forms have been devised, most of which are used more freely in the West than in the East. All of them are arranged to keep the blind from closing itself.

Patent Blind-Hinges.

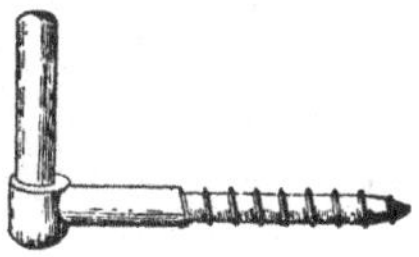

Fig. 77. Screw Blind-Hook.

They are generally made of malleable-iron, a feature which would be an objection in the eyes of Eastern builders. One of the simplest forms is the Seymour hinge, manufactured by P. & F. Corbin, Figure 79. The essential principle in this hinge includes a raised cone on the house-hinge working into a socket on the blind-hinge. The hinge is shown partially raised and in the position it takes when the blind is

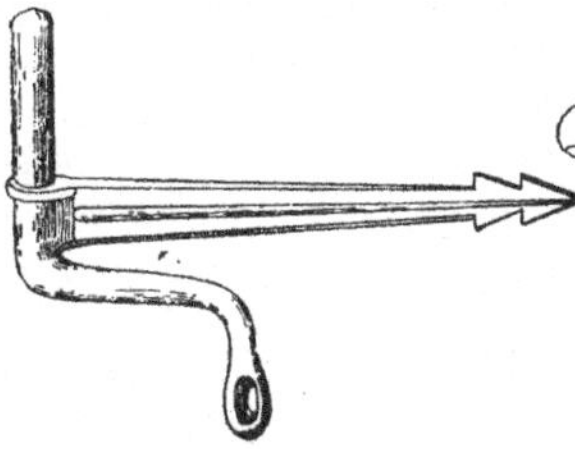

Fig. 78. Screw Brace Blind-Hook.

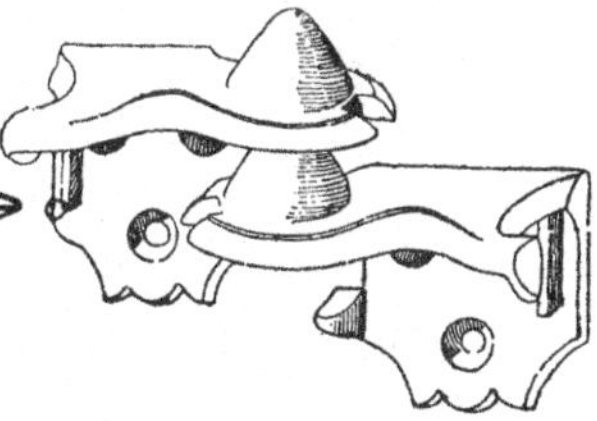

Fig. 79. Seymour's Blind-Hinge. P. & F. Corbin.

closed. When the blind is thrown back a lip on the upper hinge catches on a projection on the lower hinge and holds the blind firmly in place, so that it can be released only by raising the blind bodily.

The Shepard Hardware Co., of Buffalo, manufactures a num-

Chapter V.

ber of varieties of window-blind hinges which are used quite extensively. All of them are double locking and arranged so that the blind can be lifted off the hinge only in one position, thus obviating any upsetting of the blind when trying to close it. Most of the Shepard hinges close by gravity when once

Shepard Hinges.

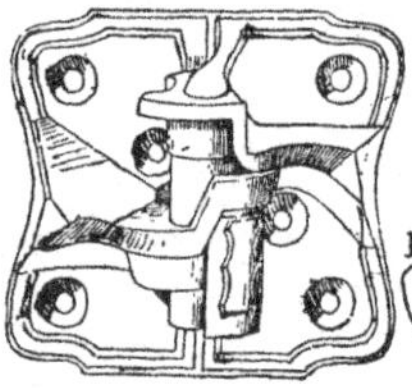

Fig. 80. Shepard's Noiseless Blind-Hinge. Shepard Hardware Co.

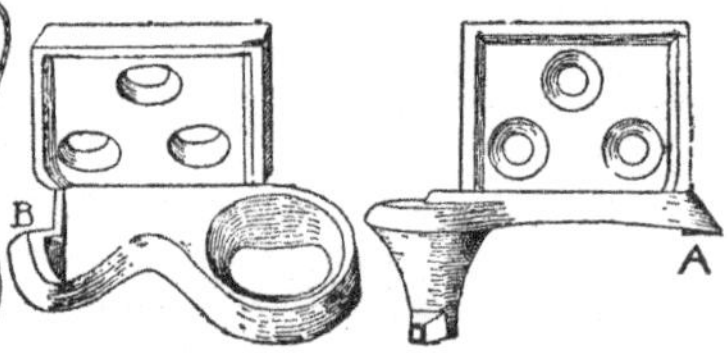

Fig. 81. Shepard's Standard Blind-Hinge. Shepard Hardware Co.

raised; that is to say, the surfaces of the upper and lower hinge are bevelled so that the blind will slide down of its own weight and so close. Figure 80 shows one of the best of these hinges and illustrates also the manner in which it closes by gravity. This hinge and nearly all of the Shepard make are planted on the face of the blind rather than mortised into the edges. A very simple form, and one quite good in its way is shown by Figure 81. The two parts of the hinges are shown separately so

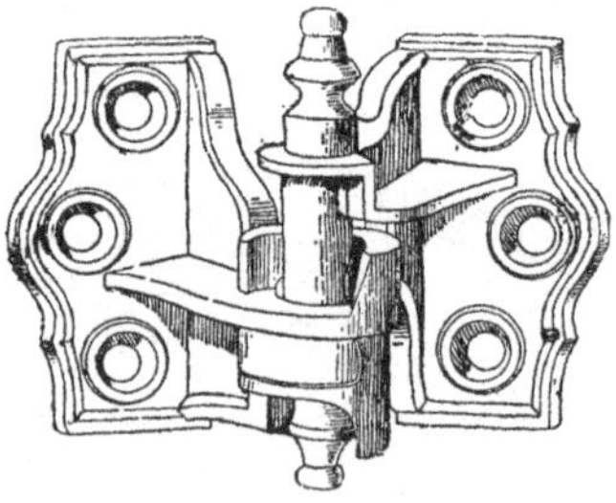

Fig. 82. Shepard's Gravity Blind-Hinge. Shepard Hardware Co.

a to represent it more clearly. The fold on the right is attached to the blind and the hook rests in the socket of the other fold of this hinge. The bottom of the socket is contracted to an ellipse and by reason of the lug on the blind hook, the blind can be lifted off the hinges only when standing at right angles to the house. When the blind is open the lug *A* catches into *B* and holds the blind securely. In order to close the blind it is lifted bodily until the lug clears the catch. Figure 82 is another variety of the Shepard hinge which can be used in case the blind is set on the face of the casing, the two arms of the hinge being unequal in length. All of the Shepard goods are very nicely finished and seem like very durable and serviceable articles. There are many varieties but the foregoing will answer for the purposes of general illustration. They retail at ten cents per set, or fifteen cents with screws.

Chapter V.

BUTTS.

Right and Left hand Butts.

As previously stated, a butt is properly a hinge which is screwed onto the butt edge of a door. The arrangement of the parts of a butt is governed somewhat by the direction in which the door is to swing, and in order to properly distinguish the doors, and consequently the corresponding butts, they are designated as being right or left hand. This distinction is not always clear even to those who are somewhat familiar with the subject of builders' hardware; but a very simple way to remember which is right and which left, is to bear in mind that when a door, in opening, turns on its hinges in the same direction as is followed by the hands of a clock, it is a right-hand door, and contrarywise a left-hand door. In other words, the distinction is the same as is made in physics between positive or right-hand, and negative or left-hand deflections.

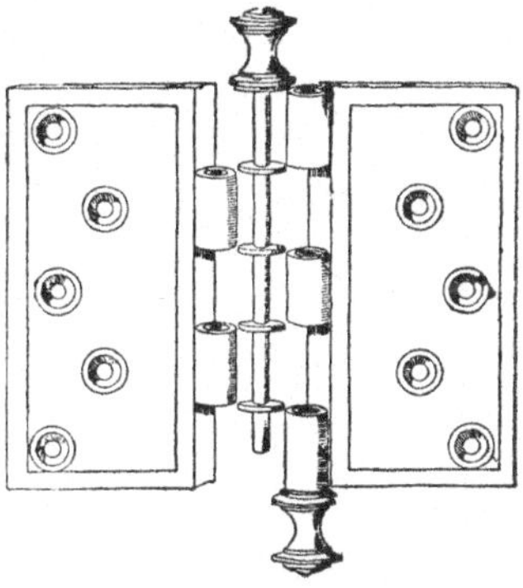

Fig. 85. Loose-pin Butt.

Chapter V.

Fast-pin Butt.

All of the commonest forms of butts are so made that the two leaves cannot be separated, the pin being riveted in place. This constitutes a fast-pin butt. Nearly all of the strap-hinges previously described are fast-pin. Such a form is not available for nice work, as it does not permit the door to be removed without unscrewing the hinge. The better class of butts are, accordingly, made with a loose pin, Figure 85, which can be readily withdrawn if the door is to be unhinged. A fast-pin butt would have exactly the same appearance as this when put together. Figure 86 is a third variety, known as a loose-joint butt. In this the pin is cast or attached solidly to the lower hub, working in the hollow hub of the other leaf, as shown by the dotted lines of the figure. This form permits the door to be removed with the greatest ease, it being simply swung clear of the casing and lifted bodily off of the hinge-pin. This

Loose-pin Butt.

Loose-joint Butt.

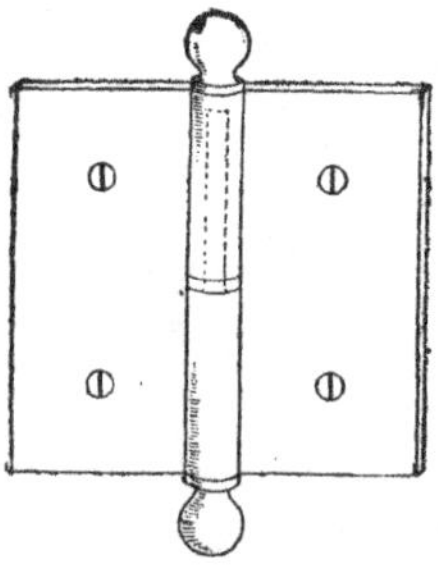

Fig. 86. Loose-joint Butt.

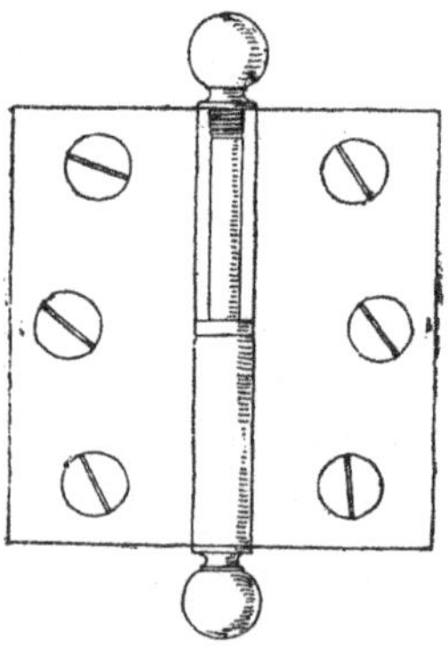

Fig. 87. Improved Loose-joint Butt. Yale & Towne Manufg. Co.

style of butt is usually preferred in New England, though elsewhere the loose-pin butt is more generally employed.

The obvious advantage of the loose-pin butt is, that the amount of bearing-surface is increased to a maximum, and as the pin is distinct from the leaves of the hinge, it can be made of a metal which will stand more wear than the ordinary bronze or iron of the butts. The loose-pin butt illustrated has eight bearing-surfaces. There are never less than four, but, even then the butt has a considerable advantage over the form shown by Figure 86. On the other hand, the loose-joint butt is more readily attached to the door and unhinged, and it costs less, though the difference in the latter respect is but slight. The difference in the amount of bearing is, in a measure, lessened in the best makes of loose-joint butts by accurately adjusting the length of the pin so that it will bear at its upper end against the top of the socket, which is usually closed by the false tip. Figure 87 shows an improved form of loose-pin butt made by the Yale & Towne Manufacturing Company, in which the pin and all the bearing-surfaces are of steel.

Washers.

It will be readily understood that, even with the most multiple form of loose-pin butt, the bearing-surfaces would soon give out if not reinforced with some more durable material than bronze. Accordingly, in all but the cheapest kinds of goods, the bearing-surfaces are fitted with some form of steel washer. In loose-joint butts the washers are exposed, as shown by the figures, and, besides taking up the wear, are useful in adjusting the butts to the doors, as two or three washers may be used on a butt if necessary, though, of course, a first rate mechanic would fit the butts properly without any washers. Loose-pin butts may have washers in the same manner, but the more general custom is for the joints to be bushed, or provided with washers which are countersunk in the hubs of the butt, so as not to appear externally. The Yale & Towne Manufacturing Company has a device by which the bushings are imbedded in plumbago, enabling the joints to lubricate themselves by their own motion. A pair of hinges so prepared has been attached to a motor, and turned back and forth a number of times equivalent to the use of over thirty years, without showing any signs of wear.

Butts are made of a variety of metals, the commonest grades being of malleable-iron. The next grade is of wrought-iron or wrought-steel. Iron and steel butts are left either with a plain

Chapter V.

bright finish, japanned, bronze-faced, Bower-Barffed, or nickel-plated. For nicer work butts are made of brass, bronze, or silver. All of these styles of finish are in the market, and the different manufacturers so closely agree in their goods that it would be impossible to make any comparison. There are great varieties in finish and design of the portions of the butts which show, and of the tips of the pins. Some of these will be considered in a subsequent chapter.

Steel Butts.

The best butts for common or cheap work are made of wrought-steel. The following table is compiled from the catalogue of the Stanley Works, which is about the largest manufactory of goods of this description. The figures given are the average retail prices in Boston.

TABLE OF WROUGHT-STEEL LOOSE-JOINT BUTTS.

Screw holes in each Butt.	Size of Screw.	Size open.	Steeple-tips, washers. Bronzed polished. Price per doz. pairs.	No tips, No washers,[1] Common finish. Price per doz. pairs.
4	No. 8	2 x 2	$3.94	$1.36
4	8	2 x 2½	4.07	1.44
6	9	2½ x 2	4.54	1.72
6	9	2½ x 2½	4.80	1.84
6	10	3 x 2½	5.18	2.00
6	10	3 x 3	5.55	2.16
6	10	3 x 3½	5.81	2.32
6	11	3½ x 3	6.41	2.50
6	11	3½ x 3½	7.12	2.68
6	11	3½ x 4	7.84	—
8	11	4 x 4	8.17	3.38
8	12	4 x 4½	8.55	3.44
8	12	4½ x 4½	9.49	4.08
8	12	4½ x 5	10.05	4.32
8	13	5 x 5	11.59	5.28
8	13	5 x 6	12.48	5.76
8	13	5½ x 5½	13.50	—
8	13	6 x 6	14.62	6.72

[1] For washers add from 35 to 50 cents per dozen pairs of butts.

Chapter V.

The finer grades of butts include such as are used for nice interior work. The pin is sometimes made of the same metal as the butt, but is better made of steel, rigidly inserted in the lower hub. Butts were formerly finished by hand almost entirely, but some most interesting machinery has been devised for centring, drilling, turning and finishing the work with almost mathematical precision.

Fine Butts.

The following table gives the average retail prices of the butts manufactured by some of the principal firms in the country. The goods referred to are perfectly plain, with simple ball-tips. Figured patterns are apt to be somewhat cheaper, though the designs vary too greatly to admit of any fair comparison. The prices are for two butts with the necessary screws. All of Robinson's goods are hand-made. The others are machine-made. All of the butts are supposed to be steel-bushed or have steel washers.

TABLE OF LOOSE-JOINT, BALL-TIP BUTTS.

Sizes open.	Bronzed Steel, Fine Polished. Stanley Works.	Bronze, Brass or Nickel. P. & F. Corbin.	Bronze or Bower-Barff Iron. Yale & Towne Mfg. Co.	Bronze. Chicago Hardware Co.	Bronze. A. G. Newman.	Bronze. Enoch Robinson.
3 x 3	$ .70	$ .95	$1.00	$1.00	——	——
3½ x 3½	.90	1.12	1.25	——	——	——
4 x 4	1.00	1.42	1.50	1.35	$2.00	$2.50
4½ x 4½	1.30	1.70	1.75	1.75	2.25	3.00
5 x 5	1.50	1.92	2.00	2.00	4.00	3.50
5½ x 5½	1.75	2.42	2.50	2.25	5.00	4.50
6 x 6	2.00	3.00	2.75	2.75	6.00	5.00

Butts are also made in irregular sizes, that is to say, of such dimensions that when opened out flat they will not be exactly

Chapter V.

square. The forms more commonly employed, however, are the square sizes, as given above. Larger sizes than 6 x 6 are seldom used, as it is found better to increase the number of butts, rather than the size. The Robinson butts listed are so made that the ball-tip can be unscrewed, to permit of greater ease in oiling the bearings, and the pin is made to bear on its point against the head of the socket.

The Yale & Towne butts here listed are of the ordinary loose-joint pattern. The special makes are sold as follows, the prices referring to a plain, ball-tip butt, in either bronze or Bower-Barffed iron:

TABLE OF YALE & TOWNE BUTTS — SPECIAL PATTERNS. PRICES PER PAIR.

	3 x 3	3½ x 3½	4 x 4	4½ x 4½	5 x 5	5½ x 5½	6 x 6	6 x 8
	$	$	$	$	$	$	$	$
Loose-joint, steel bearings, as per Figure 87.	1.75	2.00	2.50	2.75	3.25	3.50	4.00	6.50
Loose-pin, extra heavy, five steel bearings, self-lubricating washers	2.00	2.50	3.00	3.25	3.75	4.25	4.75	7.50

Too much care cannot be given to the selection of the butts or hinges which are to be used in the interior of a dwelling-house, for there is hardly any hardware about a building which is subjected to such constant and extreme wear as the door-hinge. Nor is there any department of house-trimmings in which so many poor grades of goods have been introduced, always excepting locks, however. Even the best manufacturers have been forced to meet the competition in cheap goods, and often two butts will be sold by the same house, of which one will be poor and the other first-class, though, to a superficial inspection, they present exactly the same appearance. The whole secret, aside from mere questions of design and external finish, lies in the nicety of adjustment of the bearing-joints; and as the only sure test is that of actual wear, a poor butt looking as well when new as a good one, the wisest plan for the architect is to select his butts entirely from samples. A reference to a catalogue is not sufficient to ensure the proper quality of goods, unless one possesses an acquaintance with the hieroglyphics of the trade — which few architects ever acquire. To be sure, many clients want cheap goods, and some would prefer periodical visits with an oil-can to all the squeaky hinges in a house, rather than to pay the extra price for such articles as the Yale & Towne self-lubricating butts. Still, the obligation is no less on the architect to acquaint himself with the best of everything, as well as to know how to get it.

SPECIAL HINGES.

Parliament Butts.

Parliament butts are shaped like Figure 88. They are intended to be used on very thin doors or shutters where considerable space is needed for the screws. They are made in several varieties of design and finish. Figure 89 is a hinge used almost exclusively for wash-trays. Figure 90 is a form of chest-hinge somewhat similar to that shown by Figure 67. There are also several forms manufactured for hanging inside-blinds. Figure 91 is the common form for ordinary blinds in two folds. The same is made with either fast or loose pin, and there are several varieties with ornamented surfaces. They are more specifically designated as "shutter flaps." Figure 92 is a form of shutter-hinge used for shutters which fold back over each other, as shown by the cut.

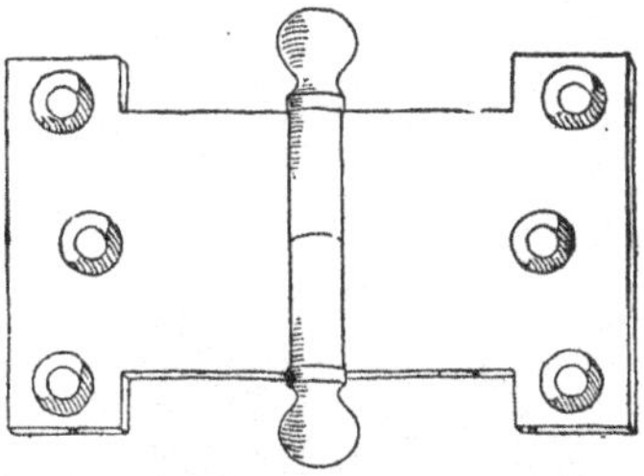

Fig. 88. Parliament Butt.

Shutter Flaps.

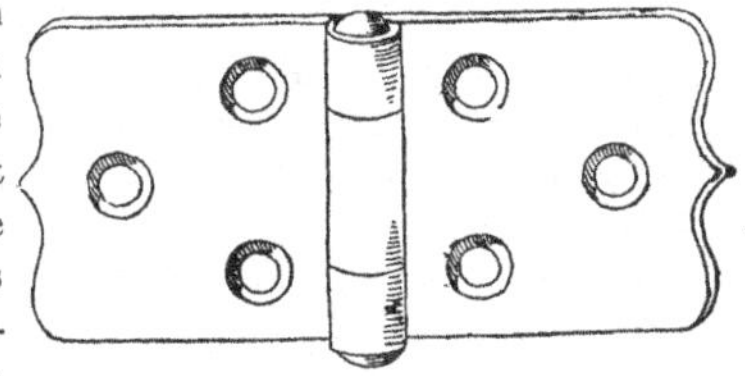

Fig. 89. Wash-Tray Hinge.

W. C. Hinges

Figure 93 is a hinge which is very convenient to use for water-closet seats, double-hinged lids, etc., the central flap being screwed to the seat, while one of the outer flaps serves for the seat cover and the other is attached to the frame at the back.

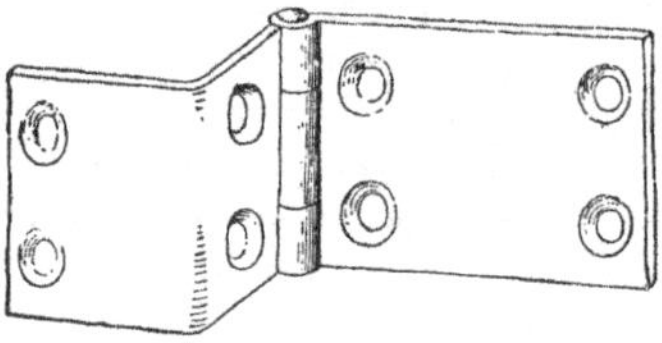

Fig. 90. Chest-Hinge.

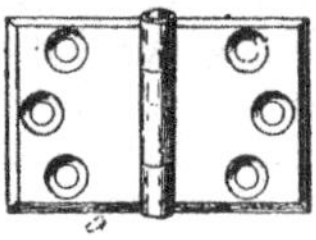

Fig. 91. Inside Blind-Hinge.

Special forms.

Figure 94 is a pivot or centre hinge to go underneath and above a door. Figure 95 is used for much the same purposes. Figure 96 is a form of wardrobe hinge. Besides these there are many special forms of hinges used in connection with furniture and a few for more strictly building purposes, none of

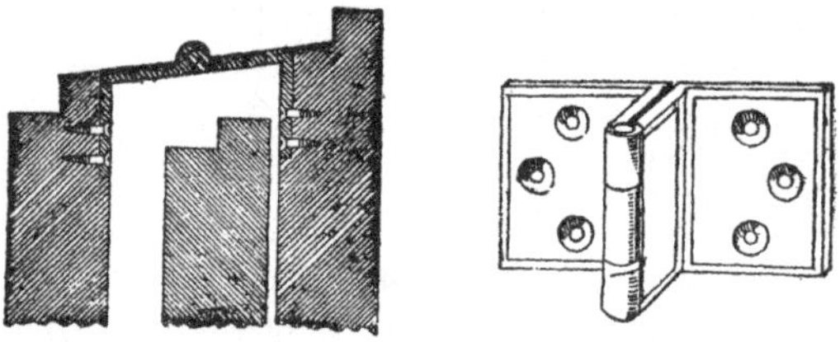

Fig. 92. Three-fold Shutter Flap.

Screen-butts.

which, however, are of any special value or interest, except the hinges used for double-acting screen-doors, that is to say, doors which swing both ways. Most of these are fitted with springs and will be considered later on, but the form represented by

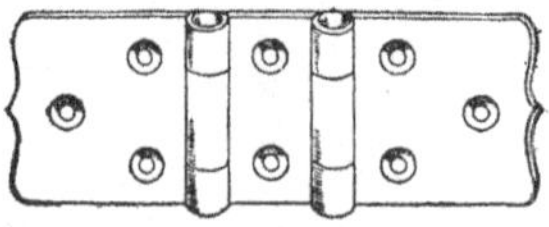

Fig. 93. Water-closet Seat Hinge.

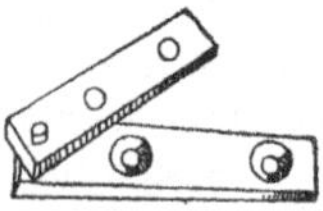

Fig. 94. Pin or Centre Hinge.

Figure 97 is peculiar to itself. It is rather hard to appreciate it clearly from the drawing, but the sectional plan will make the arrangement more easily understood. The figure shows the hinge as it would appear when opened out. It really consists of three separate double-acting hinges, each as shown by the

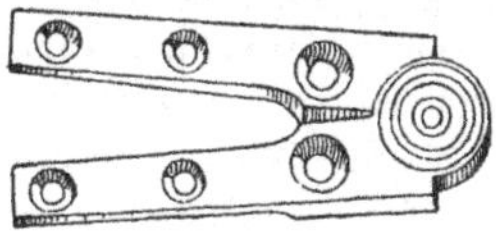

Fig. 95. Quadrant Hinge.

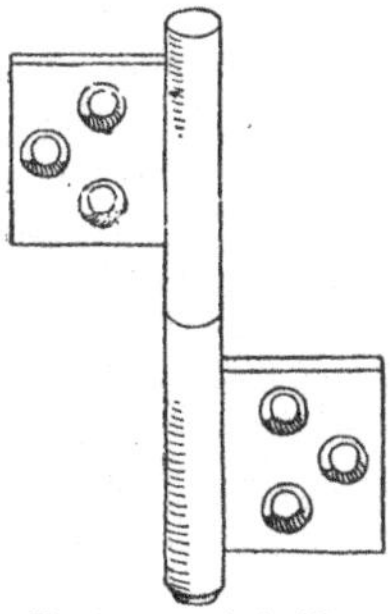

Fig. 96. Wardrobe Hinge.

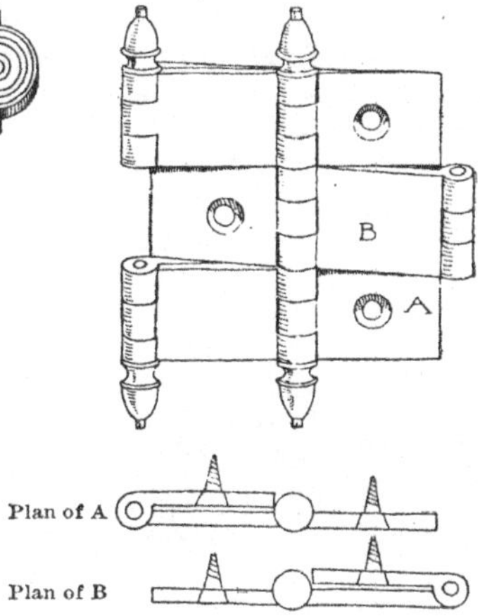

Fig. 97. Screen-Butt. A. G. Newman.

plan. A similar hinge is made with two sections instead of three.

Figure 98 is even more puzzling, at least, no drawing can

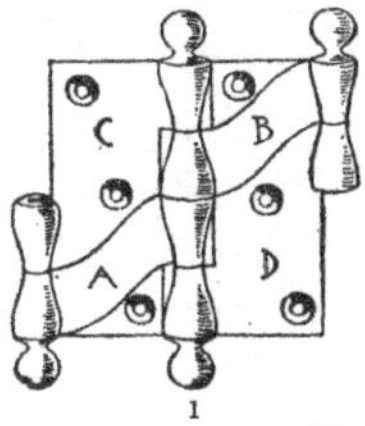

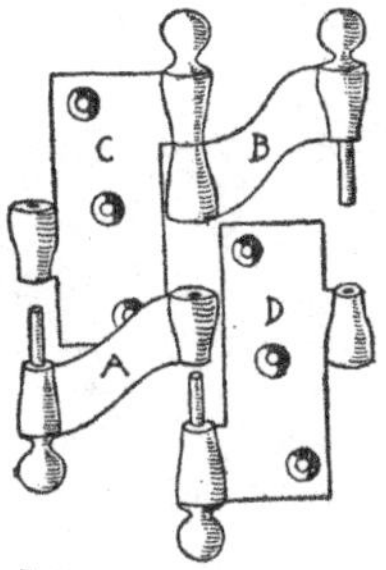

Fig. 98. Screen-Butt.

illustrate it clearly, though the thing itself is easily understood.

Chapter V.

Its action is on exactly the same principle as Figure 97. The curved connecting pieces are between the two main hinge plates when the door is shut. For clearness the hinge is shown both in sections and put together. If the plate *C* be fastened to the door frame, when the door is opened towards the left the plates *D* and *B* will revolve together about the axis 1, 1, taking the position shown by the figure. If, however, the door is opened in the contrary direction, the revolution is about the axis 2, 2, the plates *D* and *A* turning together.

SPRING–HINGES.

Garden City Spring-hinges.

One of the simplest forms of spring-hinge is that shown by Figure 99, consisting of a single spiral coil spring about a hinge pin, so arranged than when the door is opened the spring is twisted more tightly, and by its resistance tends to close the door. The lower flange is secured to the door frame, and in setting the hinge the spring is brought to a proper tension by turning up the hinge pin, holes being pierced in the bottom of

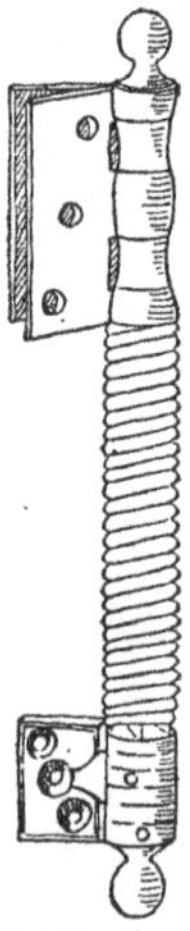

Fig. 99. Garden City Spring-Butt. Chicago Spring-Butt Co.

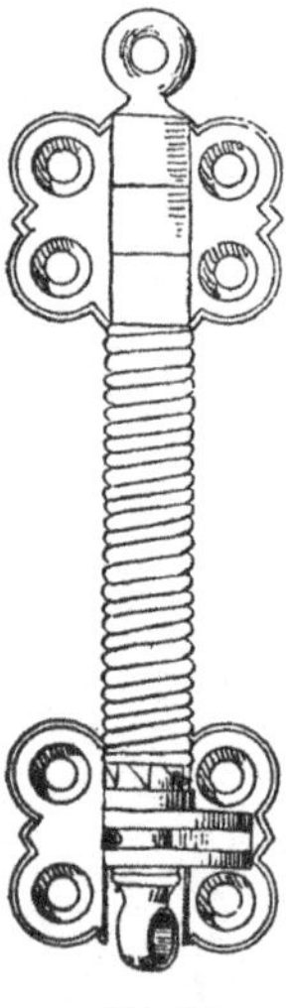

Fig. 100. Garden City Spring Butt. Chicago Spring-Butt Co.

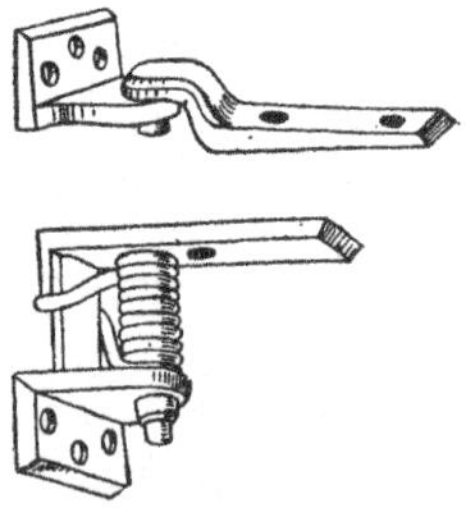

Fig. 101. Keene's Double-acting Saloon-Door Hinge. Chicago Spring-Butt Co.

the pin, as shown, to facilitate the use of a lever, while a shoulder, dropped into one of the holes, bears against the fixed flange and prevents the spring from uncoiling. Figure 100 is a spring hinge to be planted on the face of the door and the door frame, the spring being turned up by using a bar in the hole through the bottom of the pin, and kept from uncoiling by the ratchets shown in the lower portion. Both of these hinges are for single swinging doors. Chapter V.

Figure 101 is a light form of double-acting spring hinge Keene's Hinge.

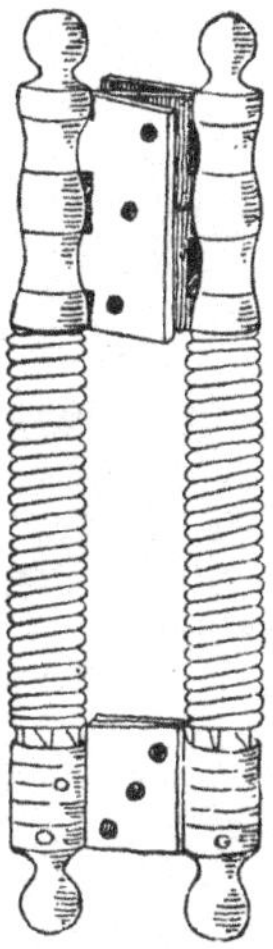

Fig. 102. Garden City Double-Acting Spring-Butt. Chicago Spring Butt Co.

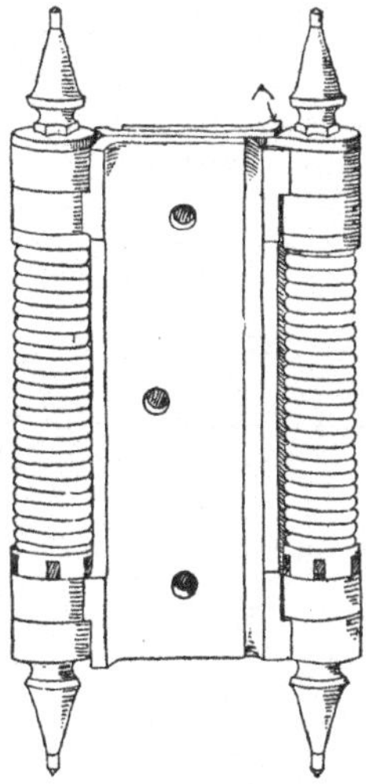

Fig. 103. Empire Spring-Hinge. Van Wagoner & Williams Co.

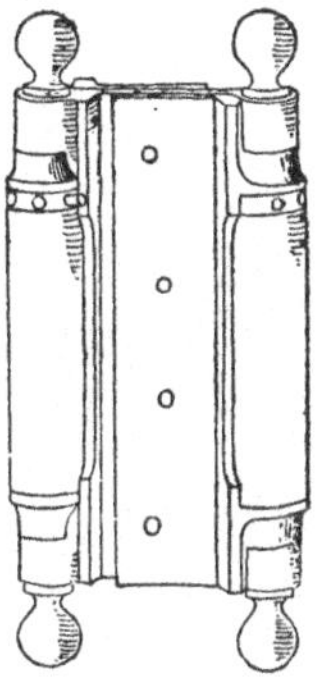

Fig. 104. Star Spring-Hinge. Van Wagoner & Williams Co.

suitable for fly doors which are set up from the floor, and do not extend to the top of the door frame. Its action will be readily understood by reference to the figure.

Figure 102 represents a double-acting spring butt of the same general form as the single-acting butt, Figure 99. The springs are turned up and secured in essentially the same man-

Chapter V.

ner and the appearance is the same. This form really embodies the principles of nearly all the varieties of spring-hinges, the differences being in appearance and in compactness of construction rather than in the workings. Two styles, the "Empire" spring hinge, Figure 103, and the "Star" spring-hinge, Figure 104, will fully serve as types of a great variety of double-acting spring-hinges listed in the catalogues of the various manufacturers.

Empire Hinge.

In the "Empire" spring-hinge the coiled springs are exposed to view, and are tightened by inserting a lever in the cogs at the bottom of the hinge and drawing it around until the catch *A* on the top of the hinge pin is held by a little projection on the hinge plate.

Star Hinge.

In the "Star" hinge the springs are encased, though they are set up in essentially the same manner. Both these hinges are excellent of their kind and are about as good as anything in the market.

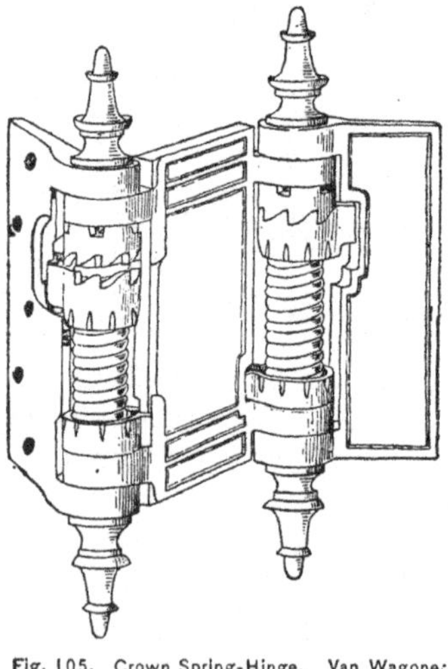

Fig. 105. Crown Spring-Hinge. Van Wagoner & Williams Co.

Crown Hinge.

Figure 105 shows the "Crown" hinge, a form which permits of a little nicer adjustment than the others in that the ratchets about the spiral spring are quite close together and the tension can be increased to any desired strength. This is a strong and durable hinge, although none of the foregoing hinges should be used for outside work, as they are liable to rust and clog.

Chapter V.

All of the preceding double-acting hinges necessitate two coil springs. Figure 106 shows a form of double-acting spring butt in which the force is derived entirely from a single strong coil, which is concealed when the door is closed, working in the thickness of the door. This hinge has generally proved very satisfactory in use and is much called for, being very neat and tidy in appearance, and it is especially adapted for light interior doors, where it is desirable that the hinges shall be as inconspicuous as possible.

Chicago Spring-butt.

The simplest double-acting hinge, however, is the "Champion" spring-hinge, shown by Figure 107. This is about as convenient and satisfactory a door-spring as is in the market, consisting of a single spring operating for either swing

Champion Hinge.

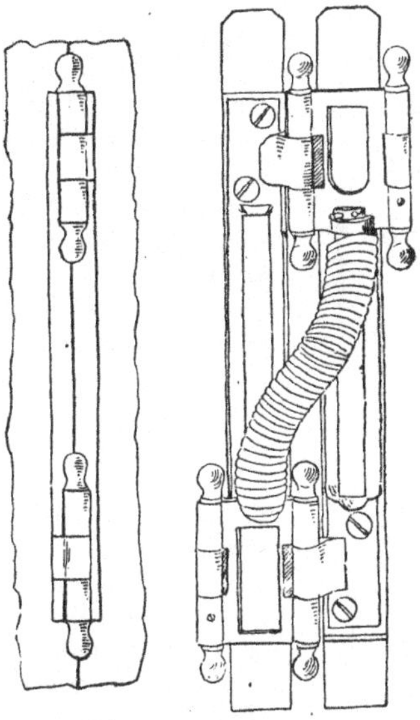

Fig. 106. Chicago Spring-Butt. Chicago Spring-Butt Co.

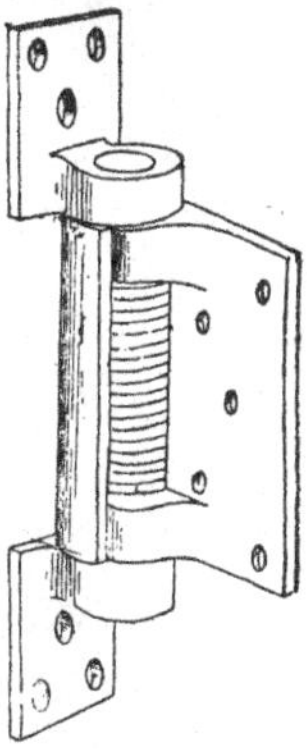

Fig. 107. Champion Spring-Hinge. Chicago Hardware Co.

of the door. A catch on the hub of the lower jamb plate resists the door in one direction, while a corresponding catch on the upper plate resists the tension in the opposite direction. These

Chapter V.

hinges look neat in place and are very easily applied, and, having no complicated machinery about them, are not likely to get out of order.

Jewett Hinge.

A hinge which has the appearance of considerable complication is the Jewett spring-butt, shown by Figure 108. This butt, however, has many excellent qualities. It consists of four separate springs, two on each side. It will be seen by the plan that when the door is opened the tension is brought on the springs by aid of the push-bar or pin, A, which fits on the shoulder between the two springs and on a ratchet attached to the central hinge-plate. The springs can easily be released from their tension without taking off the door, by simply removing the push-bar, and, owing to the manner in which the springs are attached, they exert their greatest power when the door is closed, the force gradually decreasing as the door is opened. Another good quality is that, as there are four springs, if their combined power is too great for the door, any one of them can be released singly, without affecting the action of the others, and the tension of the spring brought to any point. It is also possible to replace a spring without taking the butt off the door, something which can be done with very few other

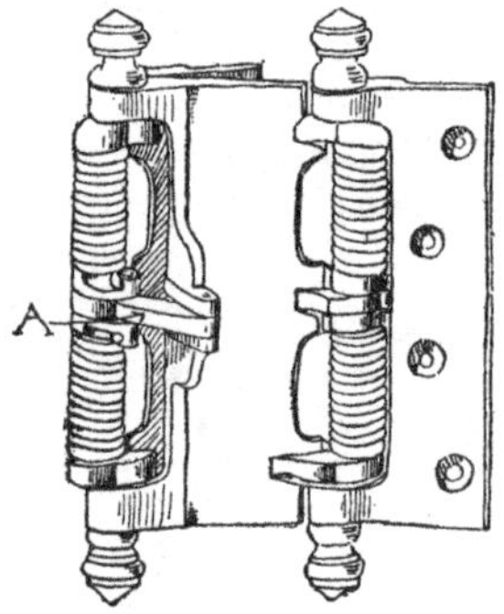

Fig. 108*a*.

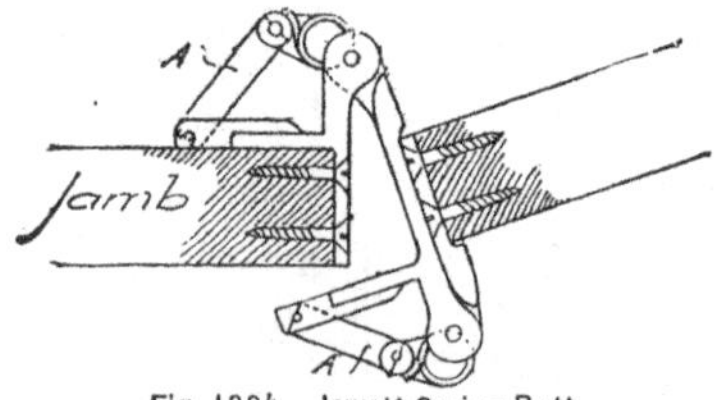

Fig. 108*b*. Jewett Spring-Butt.
Smith & Egge Mfg. Co.

makes. The only objection to the hinge in our mind is that it Chapter V.

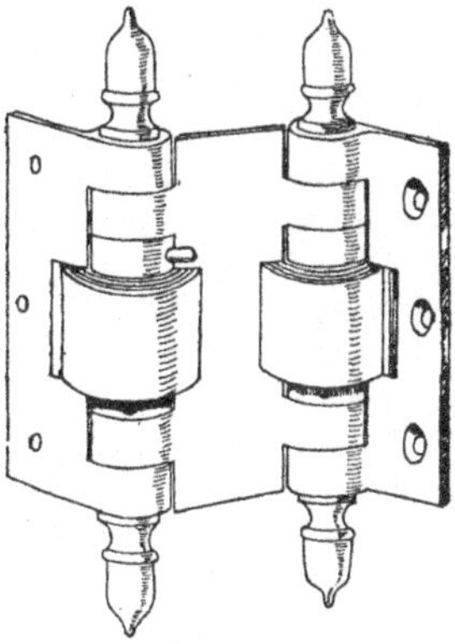

Fig. 109. Union Spring-Hinge. M. W. Robinson.

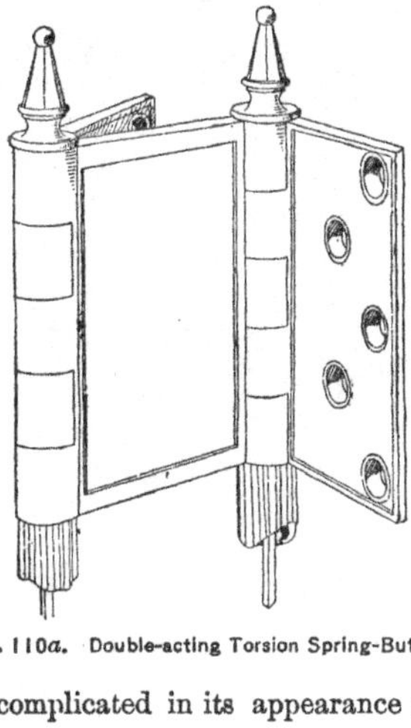

Fig. 110a. Double-acting Torsion Spring-Butt.

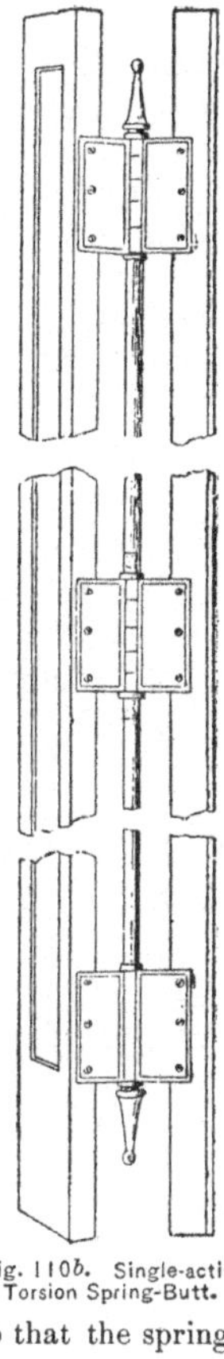

Fig. 110b. Single-acting Torsion Spring-Butt.

is complicated in its appearance and also that the springs are

Chapter V.

held in place by a moveable push-bar, which, under some circumstances might be knocked out of position and lost, in which case the hinge would be practically useless. On the whole, however, it is a most excellent hinge.

Union Hinge.

The "Union" spring-hinge, Figure 109, is a form in which the resistance is derived from flat band springs. The peg shown above the spring is moveable, and by turning up the hinge-pin the peg can be set in any of the holes in the pin, as may be necessary to retain the desired resistance. One end of the band spring is fastened to the hinge-pin and the other hooks into a slot on the adjoining hinge-plate. The hinge illustrated is intended for light doors. For heavier work a hinge is made with four such springs, two on each hinge-pin.

Torsion Spring-butt.

Figure 110 shows a form of spring-butt which depends for its action upon the torsion or twisting strain in a steel rod,

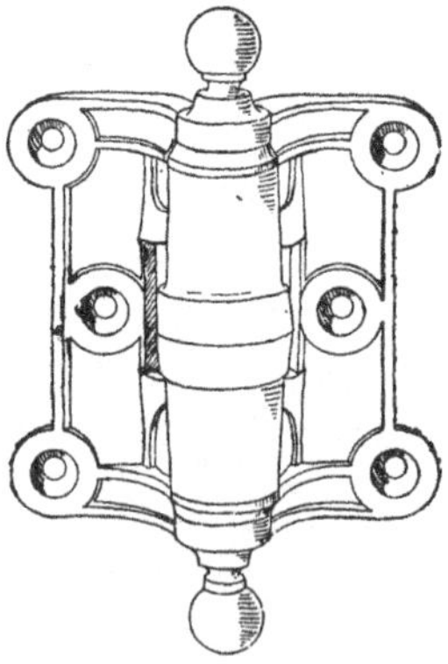

Fig. 111. Hero Spring-Hinge. Van Wagoner & Williams Co.

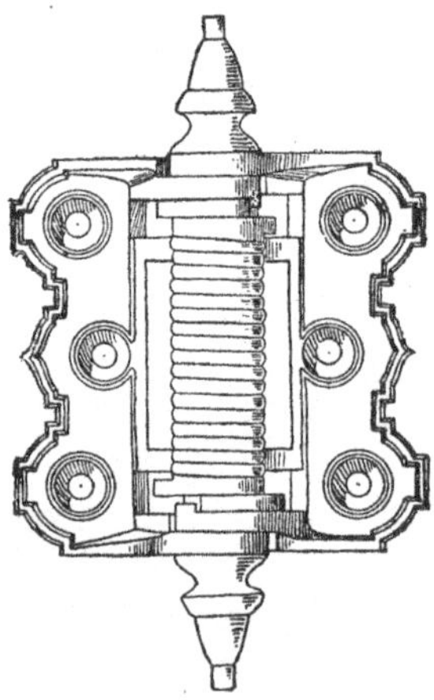

Fig. 112. Nickel Spring-Hinge. Coleman Hardware Co.

which is carried from the bottom to the top of the door.

The disadvantages of nearly all of the forms of spring-hinges previously considered is that they are apt to cause the door to slam, and the door cannot be kept open except by placing something against it. Spring-hinges which will keep the door open or cause it to close are peculiarly an American invention and one of the most ingenious of its kind. Chapter V.

Hold-back Spring-hinges

There are four leading styles of hinges which have a hold-back feature. They are the Hero, Figure 111; the Nickel, Figure 112; the Devore, Figure 113, and the Wiles, Figure 114. All of these hinges are necessarily single action. They are alike in that they are planted on the face of the door and door frame, and are delivered ready to be set, with the tension fully applied. The principle on which they work is simply this: The spring is coiled about a shank entirely disconnected from the pin of the butt and either united to the butt plates by top and bottom pieces which hook on the opposite sides of the

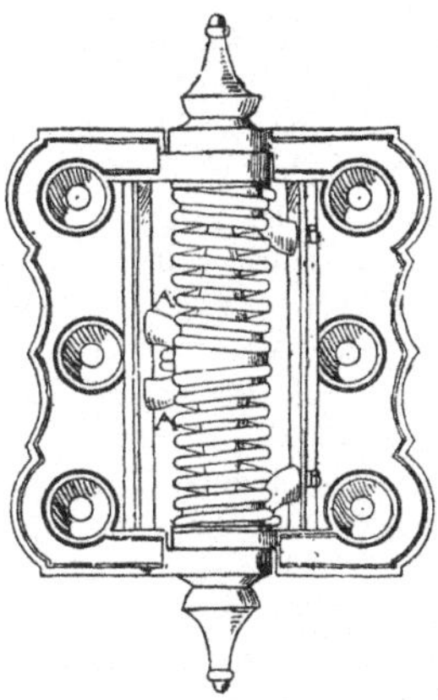

Fig 113. Devore Spring-Hinge. Freeport Hardware Mfg. Co.

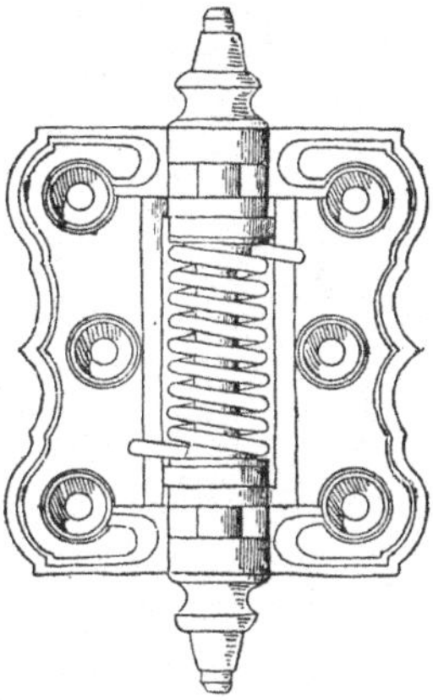

Fig. 114. Wiles Spring-Hinge. Freeport Hardware Mfg. Co.

hinge, as in the Nickel and the Wiles hinges, or with the springs themselves directly hooked onto the hinge-plates. In the Devore hinge two springs are used, the ends caught at the

Chapter V.

centre on one leaf *A,* while the outer ends catch on hooks at *B, B.* In this way, it will readily be seen that as the hinges open, the hooks to which the spring is attached are separated from each other, and consequently the spring is drawn tighter. But as soon as the hinges have passed through about three-quarters of the distance they are to swing, the horizontal distance between the points of attachment, at top and bottom of the spring becomes less, and the tendency is to draw the door open and hold it so. It is very difficult to show this action by a diagram, but Figure 115 may help to make it understood. Let *A* represent the jamb and *B* the door; *C*, a double turn of spring wire hooking on to the two arms, *D, D,* fastened respectively to the jamb and the door. When the door is opened, the ends of the spring are forced apart, but at the same time the spring forces itself out of centre, turning on the hooks of *D, D,* until, when the door has moved through 180 degrees, it is evident that the spring has both moved and turned so that the ends tend to draw the arms *D, D* together, rather than to push them apart.

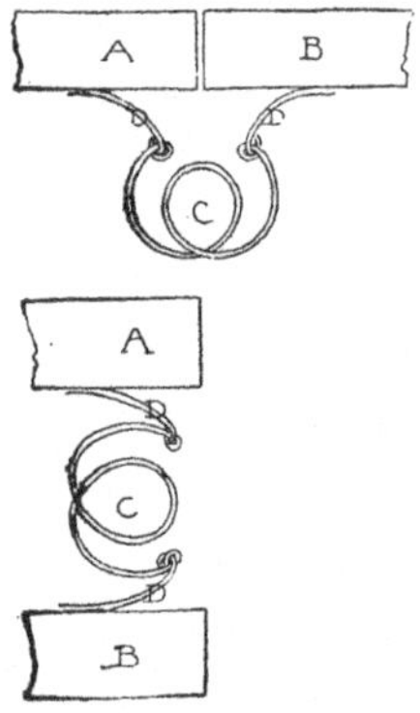

Fig. 115.

In order that the springs should be free to move, as just explained, the hinge-pins cannot extend through the butt, and the strength depends entirely on the flanges of the plates to which the pins are attached. As these cannot be made very large without rendering the hinge clumsy in appearance, it follows that the hinges can be used only for comparatively light doors. The "Hero" hinge is rather neater and apparently simpler in construction, and also has the advantage of having the hinge cased, though all the hinges are on essentially the same principle.

The metal used for the springs in connection with double-acting butts, is usually steel, in which case it is advisable that the spring should be nickel-plated, to guard against rust. Phospor-bronze is the best and most durable material to use, all things considered, though we are unable to state any particular hinge in which it is employed. Some compositions of brass, bronze, etc., are used with varying success. The different hinges are generally made with but one kind of spring throughout, so that a choice in the metals is implied in a choice of a hinge. The metal is, however, nearly always steel, as just stated. Chapter V.

TABLE OF SPRING–HINGES.—RETAIL PRICE PER PAIR.

Fig.	Name.	Bronzed or Japanned Iron.	Nickel Plated.	Brass or Bronze.
99	Garden City single-acting spring-butt....	$ .60	$1.60	$ 4.80
100	Garden City single-acting spring-butt....	.15	.75	1.80
101	Keene's saloon-door hinge..............	.60	1.20	4.00
102	Garden City double-acting spring-butt...	.80	2.16	6.40
103	Empire spring-hinge.................	.25	——	——
104	Star spring-hinge	1.36	3.00	3.00
105	Crown spring-hinge.......	1.20	——	——
106	Chicago double-acting spring-butt........	3.20	6.00	16.00
107	Champion spring-hinge..................	1.40	3.50	4.25
108	Jewett spring-butt......................	3.50	——	12.00
109	Union spring-hinge......................	2.50	——	7.00
110	Torsion spring-butt....	4.50	——	about 12.00
111	Hero spring-hinge...	.18	.33	
112	Nickel spring-hinge[1]....................			
113	Devore spring-hinge[1].....................			
114	Wiles spring-hinge[1]......................			

[1] We are unable to state any prices for these hinges, as they are not found in the Boston market. They would probably sell at the same prices as the Garden City Butts, Figure 100.

The foregoing table gives the retail prices of the spring butts and hinges previously described. For purposes of comparison, the figures represent the prices in each case of the size of hinge necessary for an ordinary door, from $\frac{7}{8}$ to $1\frac{1}{8}$ inches thick; excepting, however, that the price for Figure 101 is for a light screen-door hinge. The hinges are in general made in a number of sizes from those for the lightest kind of screen-doors, to those required to move doors weighing several hundred pounds.

CHAPTER VI.

Door Springs, Checks and Hangers.

DOOR-SPRINGS.

Door-Springs.

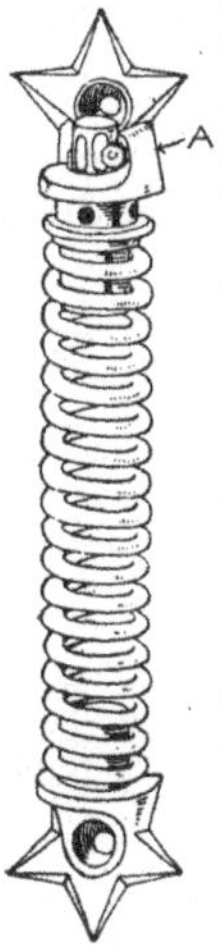

Fig 116.
Star Door-spring.
Van Wagoner & Williams Co.

THE simplest form of door-spring is a straight spiral coil of wire attached to the door and to the jamb, and drawing the door shut by a direct pull.

Such springs, of course, are used only on common work, though occasionally a spiral spring is used for gates in store-counters and railings, the spring being quite fine and long, and attached to the outside of the gate so that when the gate is closed the spring lies flat against it and does not show. The plain spiral spring is also a feature of many of the so-called door-checks ; but in the line of springs which simply draw the door to there are several forms which are more convenient to use. Figure 116 illustrates the "Star" spring, manufactured by the Van Wagoner & Williams Company. In application one end is screwed onto the door near the jamb, and the other onto the jamb itself, the spring being at an angle rather than strictly vertical, and inclined towards the latch of the door so that when the door opens the spring acts both by resistance to compression lengthwise and by the uncoiling effect of the wire. The

Star Spring.

Chapter VI.

spring can be tightened in the same manner as the spring butts previously described, by turning the upper spindle to which the spring is attached, the pin *A* holding the spindle in position. Figure 117 illustrates another form of door-spring not unlike the foregoing in principle, though in this the spring acts entirely by its resistance to a twisting strain. In the cut, the lower screw-plate and hubs are shown drawn slightly away from the spring spindle, so as to expose the ratchets which hold the spring at any desired tension.

Engine-house Spring.

Figure 118 is a very strong form of direct-acting spring, intended to be used on fire-engine-house doors. As shown by

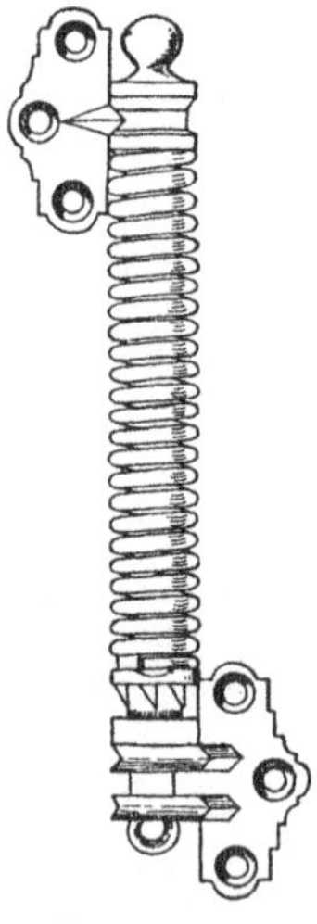

Fig. 117. Reliance Door-spring. Chicago Spring-Butt Co.

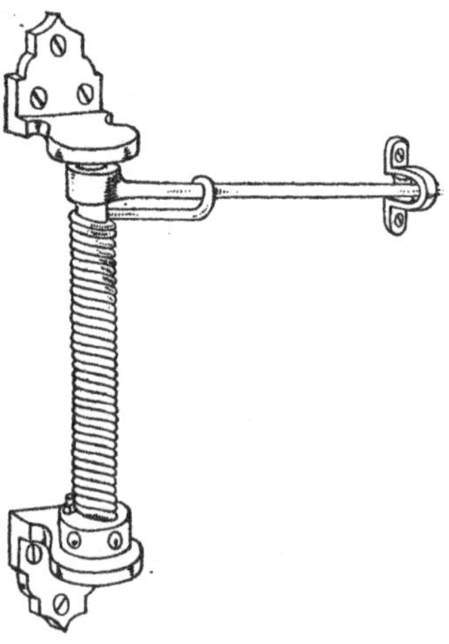

Fig. 118. Engine-house Spring. J. B. Shannon & Sons.

the cut, the spring would force the door open, which, of course, is the intention in an engine-house; but the same principle could be applied to springs which are to close a door.

The spring is tightened at the bottom, and the upper lever-arm works through a staple and pulley-wheel on the door. Chapter VI.

Screen Door-Spring.

For light screen-doors a spring is sometimes used which acts by the twisting strain or torsion of a single steel rod, Figure 119. The two side pieces, *A A*, are screwed to the jamb. The upper flange is fastened to the door and has a catch fitting into the ratchets of a drum attached to the rod. As usually applied, the rod is not strictly vertical, but is at an angle with the door jamb, so that when the door is opened the rod is subjected to both a torsion and a bending strain. If the rod is well tempered, the bending strain, of course, gives it an added efficiency. Figure 120 is a form of torsion door-spring which is attached to the door, and is operated by means of a bent, hinged-lever fastened to the door-head. It is asserted that this spring has its greatest power just as the door is opened, and that the strain in the rod diminishes as the door is swung around.

Torsion Door-Spring.

The following table gives the average prices of the door-springs described:

TABLE OF DOOR-SPRINGS. — PRICES FOR A SINGLE SPRING.

Fig.	Name.	Laminated Spring.			Nickeled Spring.		
		Light.	Medium.	Heavy.	Light.	Medium.	Heavy.
116	Star door-spring	$ .15					
117	Reliance door-spring	1.20	$1.80	$2.80	$2.00	$2.80	$4.00
118	Engine-house door-spring, 24, 30 and 36-inch	3.00	4.00	5.50	4.75	6.00	7.75
119	Torry door-spring	.20		.40			
120	Peabody door-spring	.25	.25	.35			
121	Devore door-spring[1]						
122	Warner door-spring[1]						

[1] Not found in Boston market.

Reverse-Action Spring.

The principle involved in a reverse-acting spring butt, which

Chapter VI.

has been explained in a previous chapter, can be applied to

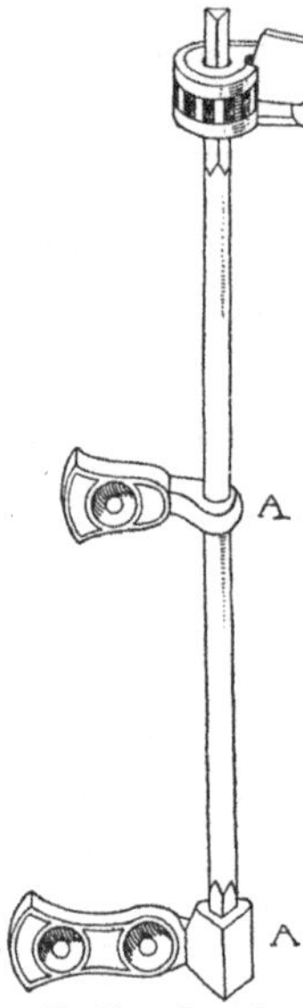

Fig. 119. Torry Door-spring.
Van Wagoner & Williams Co.

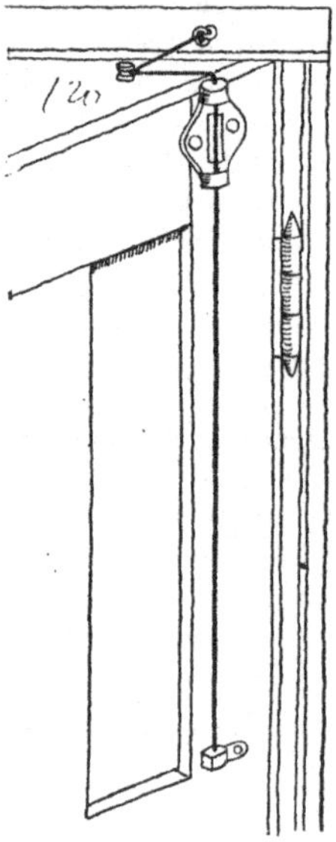

Fig. 120. Peabody Door-spring.
A. W. Paine.

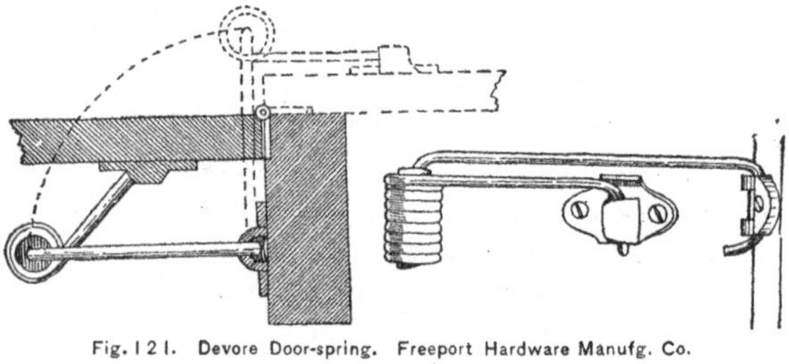

Fig. 121. Devore Door-spring. Freeport Hardware Manufg. Co.

simple door-springs. Figure 121 shows a form which is made by the Freeport Hardware Manufacturing Company. This

Chapter VI.

spring has a uniform tension, holds the door firmly when closed, and when the door is open about 120 degrees, the force of the spring is reversed and will hold the door open. Another advantage of this form is that the spring is easily unhooked and rehooked from the door, and also from the jamb without removing the screws. The action of the spring will be understood from the illustration. Figure 122 shows a different form of spring; though on the same principle as the Devore. It has all the advantages of the one described and acts in precisely the same manner, though it is slightly larger. Both of these forms are exceedingly ingenious and efficient, and are sold a great deal in some parts of the country.

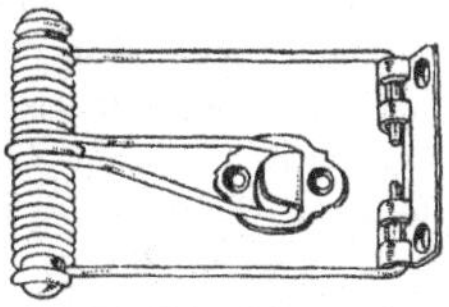

Fig. 122. Warner Door-spring, Warner Manuf'g. Co.

DOOR–CHECKS.

Spring Door-Checks.

A door-check is understood to be anything which will hold the door either open or shut, or which will keep the door from slamming when closed. All the door-checks in the market are combined, directly or indirectly, with some form of door-spring. In some cases the spring is a part of the check, but more often the spring is a separate fixture, and is used as an auxiliary to the action of the check. One of the simplest forms of door-check is that shown by Figure 123. This consists of a strong band or spring of flexible steel which is attached by a proper holder to the head of the door-frame, so as to project about half an inch below the soffit. The door, in closing, catches on the shoulder at the end of the spring, forcing it up against the soffit of the door-frame, while the pressure of the spring on the head of the door keeps it from slamming, and in a measure also

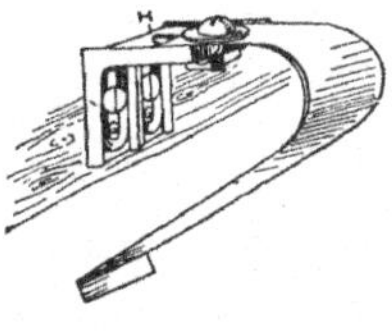

Fig. 123. Barlow Door-check. W. S. Barlow.

Chapter VI.

from opening too easily. The holder for the spring has slots instead of screw-holes, so that the spring can be moved up or down to any desired tension; while the upper slot *H* permits a further adjustment by moving the spring in or out. The retail price of this check is twenty-five cents each, japanned, or fifty cents, nickel-plated. For all ordinary practical purposes this door-check is as good a device as can be found, and is used a great deal on railroad cars, where there is constant liability to violent slamming. It will be noticed that the spriug has a double power; first, by the friction of the door as it closes, and secondly, by the door coming in contact with the shoulder at the end of the spring. There is a special form of spring used to close this door consisting of a straight coil, with a hook on the jamb and a shoulder on the door.

There are a few variations of the "Barlow" door-checks, but they all act on essentially the same principle, and this one will be sufficient for illustration. A very different kind of check is that which acts on the principle of a piston-pump, of which, perhaps, the best known is the "Norton" door-check. This article has been on the market a long time, and is used very extensively in some parts of the country. Figure 124 will give an idea of how it appears when set. The check consists simply of a plunger or piston working in a cylinder. Between the piston and the cylinder head is coiled a strong spiral spring, and the piston, as well as the cylinder, is pierced with a small hole to permit the air to escape.

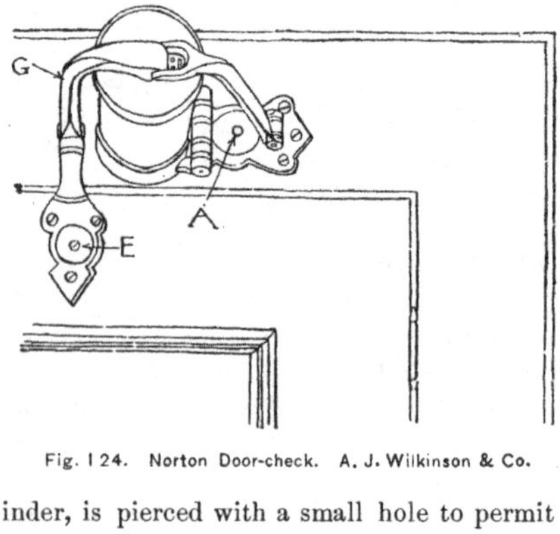

Fig. 124. Norton Door-check. A. J. Wilkinson & Co.

The cylinder is attached to the head of the door-frame, and is hinged at *A*. The piston-rod is connected by a hinged-joint with a lever hinged to the frame, and a lever *G* attached to the door. When the door is opened, the piston is drawn out, the internal spring compressed, and the air enters through the holes in the piston and the cylinder head, filling the space beyond the piston. When the door is released, the spring tends to close it, but the air behind the piston acting as a cushion, prevents the door from closing too quickly or from slamming. The orifice in the cylinder can be made larger or smaller, thus regulating the speed at which the door shall close. The spring is made sufficiently strong to both close and latch the door after the air has escaped from the cylinder. Chapter VI.

It will be seen that this door-check permits the door to be opened only about 120 degrees. This is usually more than enough for any doors requiring the use of a check, but a stop is always needed, otherwise the arm *G* may be broken.

A form of "Norton" door-check is also made to close the door from the outside. The action is exactly the same, except that the arm *G* is bent up and attached to the soffit, while the cylinder is attached to the door.

The "Norton" door-check is usually sold nickel-plated. The prices are as follows:

For screen-doors and doors not exceeding 2 feet 8 inches	by 1½ inches	$4.00
Doors not exceeding 2 feet 8 inches	by 2 inches	5.00
Doors not exceeding 3 feet	by 2½ inches	6.00
Doors not exceeding 4 feet	by 3 inches	8.00

A rather more cumbersome form of door spring and check has recently been put on the market by the Russell & Erwin Manufacturing Company. The check acts in practically the same manner as the Norton Door-check, but the spring is exposed and distinct from the cylinder, being acted upon by a bent lever. This door-check has hardly been before the public long enough to judge fairly of it.

A form of door-check, which has met with a great deal of approval, is the "Eclipse," manufactured by Sargent & Com-

Chapter VI.

pany. Figure 125 shows this check in position. It consists of piston secured to the head of the door-frame and working in a cylinder attached to the top of the door. The piston-rod is kept from lateral motion by a set-screw at one side of the foot, and a spring on the other, as shown by the figure, so that it can be accurately adjusted to meet the cylinder. When the door is opened, the cylinder is drawn entirely away from the

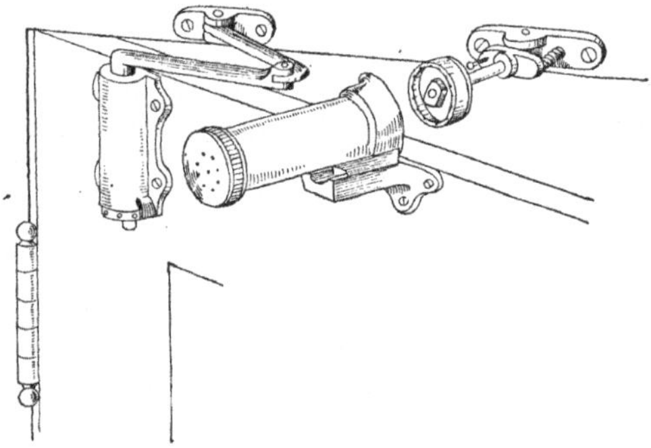

Fig. 125. Eclipse Door Check and Spring. Sargent & Co.

piston, while the compression of the air in the cylinder when the door closes, prevents any slamming. The air escapes through openings in the end of the cylinder, so arranged as to be easily regulated. The piston in both the "Norton" and the "Eclipse" door-check has leather washers.

The "Eclipse" door-check can be used with any suitable form of spring butt, though the door-spring, shown by Figure 125, is especially made for this purpose by Sargent & Company. It consists of a strong spiral spring, cased in a cylinder and connected with the door-frame by a hinged bent lever.

The "Eclipse" checks and springs are finished either Tuscan bronzed, bronze-plated or nickel-plated. The prices are as follows:

Chapter VI.

Door-checks.	Ordinary.	Heavy.	Very heavy.
Bronzed.	$2.10	$2.76	$4.00
Bronze.	2.40	3.30	4.60
Nickel.	4.20	5.00	6.30

Door-springs.	Ordinary.	Heavy inside.	Heavy outside.	Very heavy.
Bronzed.	$ .90	$1.25	$1.80	$2.52
Bronze.	1.10	1.50	2.10	3.00
Nickel.	3.00	3.36	4.00	5.00

The "Eclipse" check is applied to the outside of a door, if desired, a different form of holder securing the cylinder to the soffit of the door-opening, while the piston is fastened to the door.

Liquid Door-Checks.

It has been claimed that after being used for a certain time, the bearings in both the "Norton" and the "Sargent" door-check will wear loose, so that the air will escape too freely from the cylinders to form a reliable cushion; and several attempts have accordingly been made to produce a door-check in which the action should be regulated by the flow of some liquid, which would permit of metal instead of leather washers. Figure 126 illustrates one device on this principle. It consists of a piston and cylinder attached to the door. The cylinder is pivoted so as to admit of a slight rocking motion, and the piston is hinged to a bent-arm, also pivoted just above *C*, and provided with a spring which serves to keep the piston drawn out. The cylinder is filled with oil, which flows back and forth through a small tube at the back, the rate of flow being regulated by a screw at *C*. When the door closes, the shoulder *A* on the door-head strikes against the bent arm and forces the

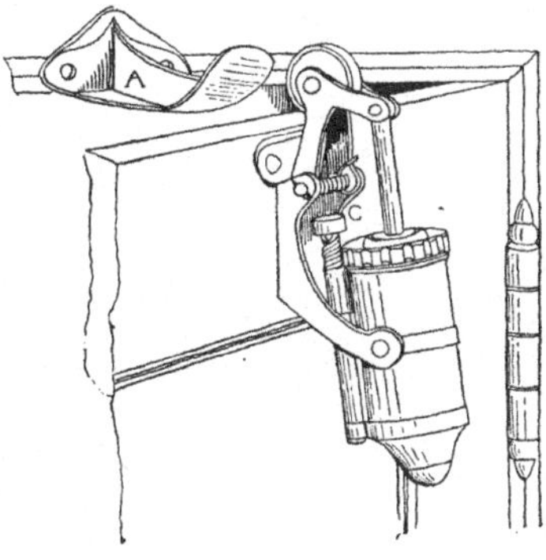

Fig. 126. House's Liquid Door-check Nimick & Brittan.

Chapter VI.

piston down, the oil preventing any sudden slamming, while the rate of the flow through the tube below *C* determines the rate at which the door will close. This fixture retails at about $1 per set.

Such a form of check necessitates a spring-hinge, in order that the door shall be self-closing. Figure 127 illustrates a form of door-check manufactured by the same parties, which has a coiled spring inside the cylinder acting by means of the

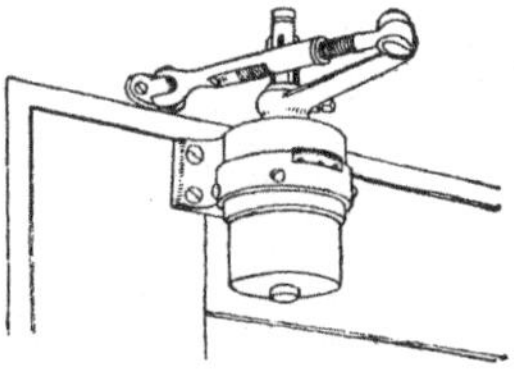

Fig. 127. House's Automatic Door-check. Nimick & Brittan.

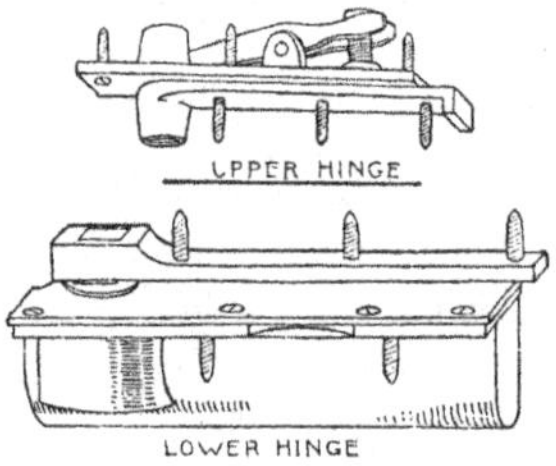

Fig. 128. Bardsley's Checking Spring-hinge. J. Bardsley.

hinged bent lever on the door-head. The interior of the cylinder is filled with oil, which checks the action of the door by flowing from one compartment to the other of the cylinder. The retail price is from $3.30 to $4.50 each, according to the finish.

Figure 128 is another form of combined door check and spring, which is intended especially for double-acting doors, though it can be used for any door. No regular hinges are required with this fixture. The top of the door is held by a species of pivot, fitting into a socket mortised into the top of the door, while the door-check answers for the lower hinge. The lever *A* is mortised into the bottom of the door, acting as a crank to turn the post *B*. The checking apparatus is encased in a box *C*, which is sunk into the door-sill and covered with a brass plate. Figure 129 shows a section of the box, which will illustrate more clearly its arrangement. The post *B*, when turned either way, moves a piston which travels in a cylinder completely filled with oil, in which is also a very heavy coiled spring whose action tends to close the door, while the oil pre- Chapter VI.

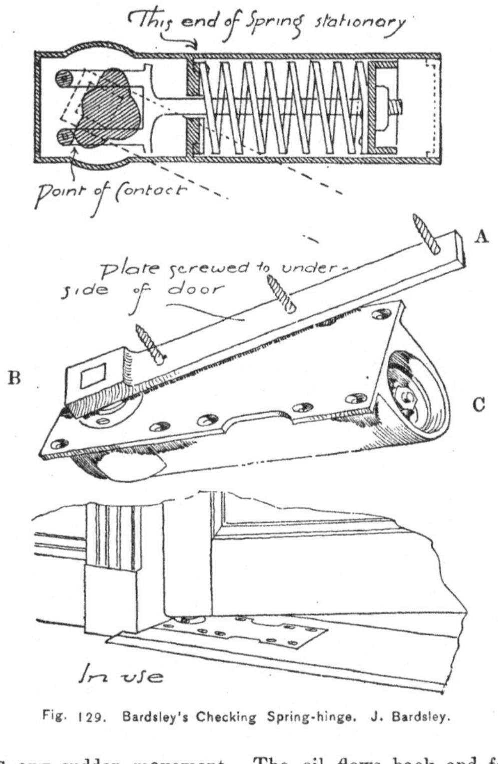

Fig. 129. Bardsley's Checking Spring-hinge. J. Bardsley.

vents any sudden movement. The oil flows back and forth through a narrow aperture, the size of which can be regulated by a screw extending up through the covering of the box.

Chapter VI. This check is listed at $17 per door complete for a door of ordinary size.

Door-Catches. Besides the door-checks which are automatic in their action, there are a number of devices for holding the door open or in

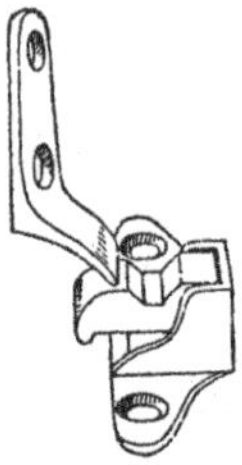

Fig. 130.
Fray's Door-holder.
J. B. Shannon & Sons.

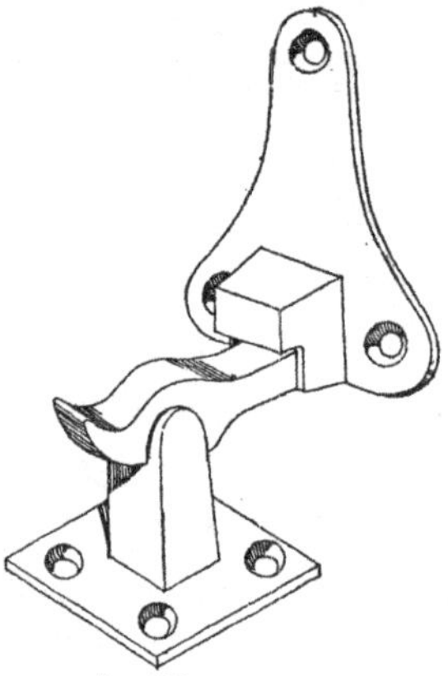

Fig. 131.

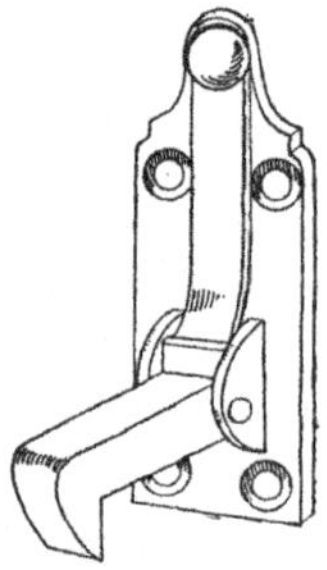

Fig. 132.

Door-holder. A. G. Newman.

any one position. Some of them are self-locking, but are arranged so that the door can be easily drawn to by slight pressure. Figures 130 and 131 are two varieties of this style of door-check. Another kind is made to absolutely hold the door fast when it is opened, so that in order to close the door the check must be released by hand: Figure 132 shows one of the many varieties of this form. It is varied by having a lever attached to the catch by which it can be more easily raised,

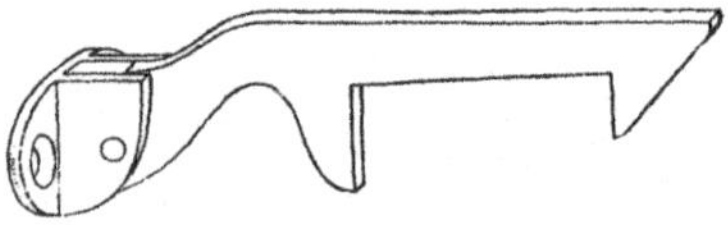

Fig. 133. Top Door-catch. P. & F. Corbin.

and also by the catch being placed so as to act sidewise instead of vertically. Figure 133 is a form of catch which is intended to be attached to the jamb and to work over the top of the door, nearly all the other forms being attached to the door and working on a striker which is screwed to the floor. Figure

Fig. 134. Drop Door-check.
A. G. Newman.

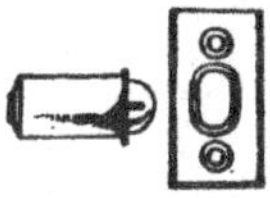

Fig. 135. Fray's Door-catch.
J. B. Shannon & Sons.

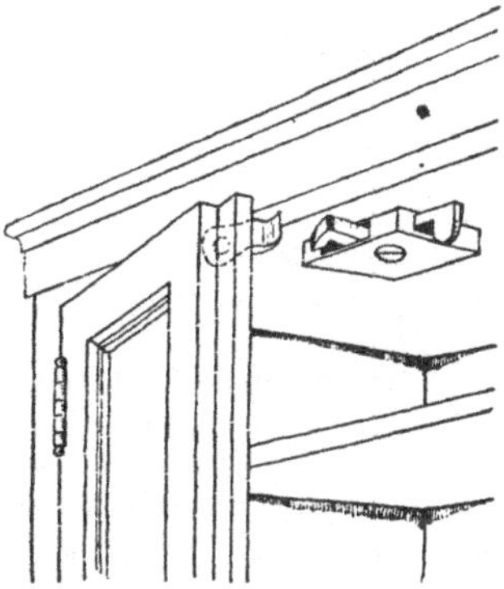

Fig. 136. Ross Inside Catch.
Stoddard Lock & Manufg. Co.

134 is a self-locking form of door-check which is screwed to the door, catching onto a hook projecting from the wall. Figure 135 is a form of door-catch which is used for show-cases and closets, acting by means of a spring coiled inside the cylinder. Figure 136 is a spring-catch used only for light work or for cupboards.

DOOR-STOPS.

Some form of stop is always desirable in order to keep the door from striking the finish when swung open, or breaking the

Chapter VI.

Door-Stops.

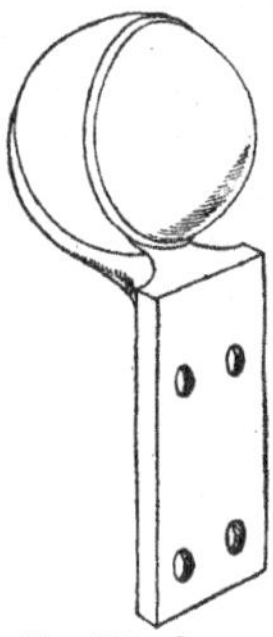

Fig. 137. Door-bumper. J. B. Shannon & Sons.

plastering. The commonest form consists of a wooden knob screwed straight into the base and tipped with rubber. The variation from this is a wooden-knob, which is screwed into the floor and has the rubber-tip on the side, to be used when the door does not swing against the wall, but has to be stopped at some point. These stops are made in birch, maple, ash, oak, chestnut, cherry, walnut and mahogany and are listed at $5.00 per gross for birch to $6.50 for mahogany. They are made in two sizes, 2½ and 3 inches long; the wood is turned and a gimlet-pointed screw is firmly attached to the stop, so that it can readily be put in place by hand. The prices are the same whether the rubber tip is on the side or the end. They are also made with a rubber ring entirely encircling the knob, the list price being $12.00 per gross.

The Meriden Malleable Iron Company manufactures elastic-headed screws which are used more for furniture than for doors, but which might be desirable in some cases. They consist simply of a half-round head which is covered in upholstery of some kind and fitted with a gimlet-pointed screw. These cost from $11.00 to $17.00 per gross, list price, depending upon the material with which they are covered. J. B. Shannon & Sons, manufacture a door-bumper shown by Figure 137, which is intended to prevent jar and noise in shutting the door: it consists of a cup with a brace attached to be screwed to the face of the door; a rubber ball is pressed into the cup so that it will not fall out. A piece of rubber made fast to the jamb casing over the door, for the ball to strike against, completes the contrivance. The rubber over the door may be increased or lessened in thickness so that when the rubber ball comes in contact with it the compression will allow the door to latch. This device effectually prevents any slamming of the door. The list price is $1.00 each. The only possible objection to its use would be that the head might prevent the door from being latched properly.

DOOR HANGERS AND ROLLERS.

Barn-door Hangers.

The commonest form of door-rollers are those used for barn-doors. Usually a barn-door slides on rollers or sheaves which are applied to the inner face of the door and run over a metal track secured to the floor. There are many kinds of large sheaves and rollers manufactured for barn-doors, which are too simple to require any illustration. The commoner kind consists of a large wheel with a steel or metal axle. The better kind of barn-door rollers are provided with anti-friction axle bearings ; that is to say, the axle of the wheel revolves in a cycle of small pins or rollers by which the friction is considerably reduced, and the wear on the bearings very materially diminished. Figure 138 is an ingenious device for a barn-door roller, the working of which will be readily apparent from the drawing. The inner plate, *A*, is screwed directly to the door, through the openings in the wheel, *B*, which revolves on the anti-frictional bearings.

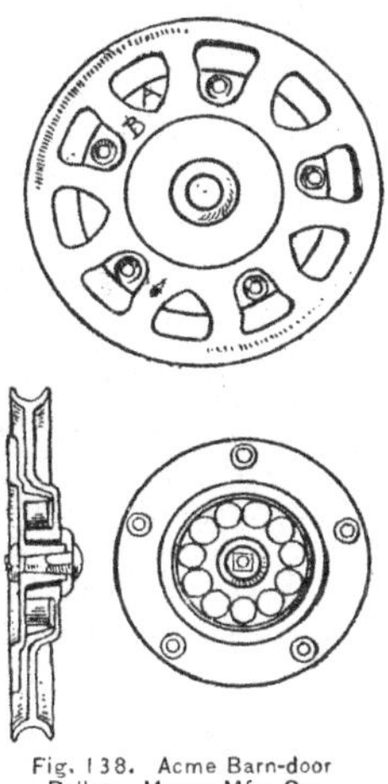

Fig. 138. Acme Barn-door Roller. Moore Mfg. Co.

Stay-roller.

Figure 139 shows a form of barn-door rail, intended to be used with a wheel which shall rest on the flanges and not bear at all on the upright portion. In

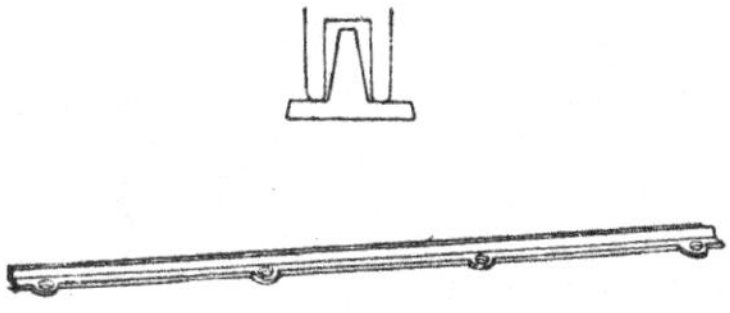

Fig. 139. Nickel Barn-door Rail. Coleman Hardware Co.

Chapter VI. this way the wheel will clear away any collection of snow or ice by its own action and enable the door to roll easily.

The standing objection to barn-door rollers which are applied to the bottom of the door, is that they are too easily thrown off the track by obstructions and also that the track itself is apt to get in the way and be a bother in driving over it. The greatest amount of ingenuity has been expended upon door-hangers in which the door is suspended from a track at the top. The only objection which is to be urged against this manner of arranging a sliding-door, is that in case of a violent wind the door would be forced inward. This difficulty can be in a measure obviated by the use of some form of stay-roller, such as Figure 140, which can be attached so as to prevent any lateral motion of the door. This form of stay-roller is also used for doors which slide on sheaves at the bottom, taking the place of an overhead groove.

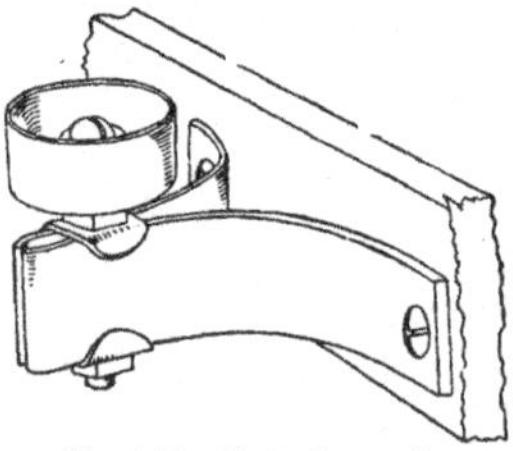

Fig. 140. Victor Stay-roller. Victor Mfg. Co.

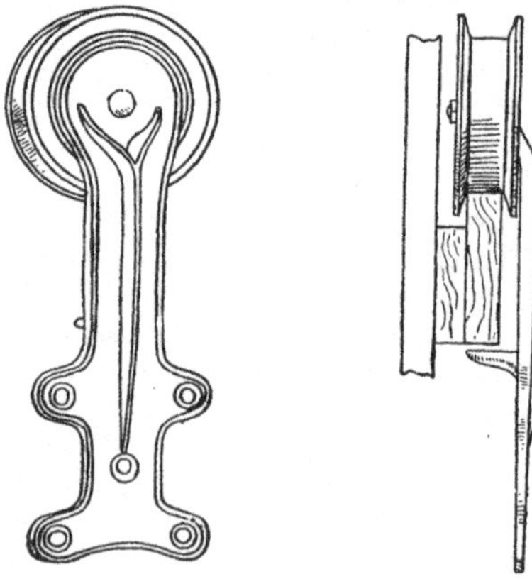

Fig. 141. Climax Barn-door Hanger. Moore Mfg. Co.

There are really but two distinct varieties of barn-door hangers; the first is represented by Figure 141, and consists of a single wheel running on an overhead-track and attached to a hanger which is screwed on to the inner face of the door. The same form of hanger is made to be used with an iron rail. This form is rather old but is very good and we should imagine would give little trouble. It is made with anti-friction bearings. The second variety of hanger is one in which the axle is not fixed but travels along a single bearing beam, as in the "Moody Hanger," Figure 142, which is one of the

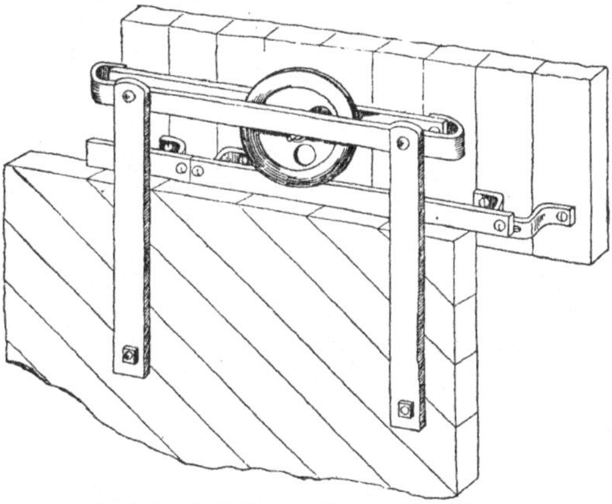

Fig. 142. Moody Barn-door Hanger. Victor Mfg. Co.

simplest of this kind. In this the axle bears on two bars, and the uprights to which the bearing bars are secured, are placed sufficiently far apart to admit of the axle having enough play for the opening of a single door. The rail and brackets are made of steel and the wheel has a steel axle.

The "Victor Hanger," Figure 143, is a slight improvement over the "Moody" in that the bearing is on a single bar instead of on two; and that the wheels work on each side of a high, ridged track which prevents them from slipping off or becoming interfered with.

The "Lane Hanger," Figure 144, is very similar to the

Chapter VI. "Moody" though the arrangements of the supports is somewhat different.

All of the foregoing hangers are made of wrought-iron or steel, with steel or chilled-iron bearings. Several kinds of hangers are made with malleable iron, such as the "Nickel," Figure 145, which follows the pattern of the "Moody Hanger." The "Nickel Hanger" is also made in steel, with a slightly different shape. The "Hatfield Hanger," Figure 146, also in malleable iron, is a form after the pattern of the "Lane," on which the patent seems to have run out; at any rate, a similar form is made by several of the manufacturers, and the principle embodied in the "Hatfield" and the "Moody Hangers" is the one which is usually consid-

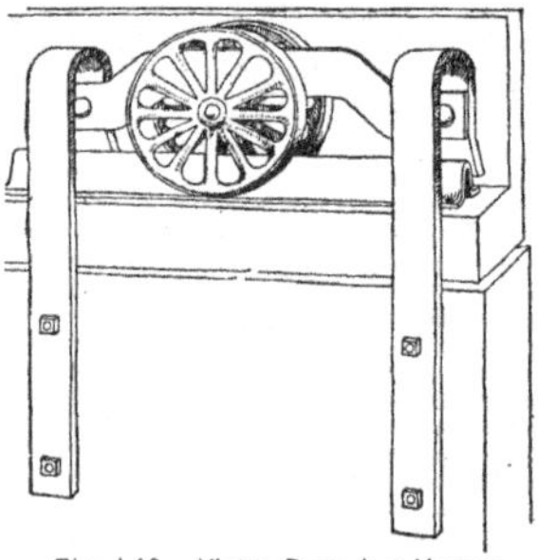

Fig. 143. Victor Barn-door Hanger. Victor Mfg. Co.

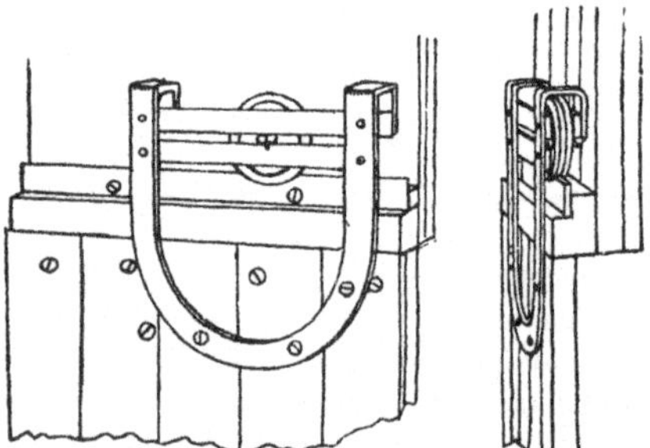

Fig. 144. Lane Barn-door Hanger. Lane Bros.

ered to be the most satisfactory; that is to say, one in which the axle bears on two parallel plates and works in slots; indeed,

Chapter VI.

this principle is applied to nearly all the most successful hangers, both for barn-doors and parlor-doors.

For parlor doors many builders and architects still prefer sheaves mortised into the bottom of the door. The rail, which is a necessary part of this arrangement, is admitted to be a trouble, but the rollers are so easily taken out that the rail is retained. There seems to be an idea with many people that the overhead hanger more easily gets out of order, and that it is more complicated. This is a mistake, for there is hardly a hanger in the market that will not give satisfaction if properly applied, while any one who has had experience with sheaves and rail at the bottom can testify to the trouble which will sometimes occur, with the settlement of the woodwork throwing the door out of plumb or obstructions getting on the track and throwing the wheels off. Some of the overhead hangers require more care in setting, and others have some special adjustment which must be understood; but the principle on which they all work is so simple that, to the uninitiated, there seems to be but little choice between the various kinds.

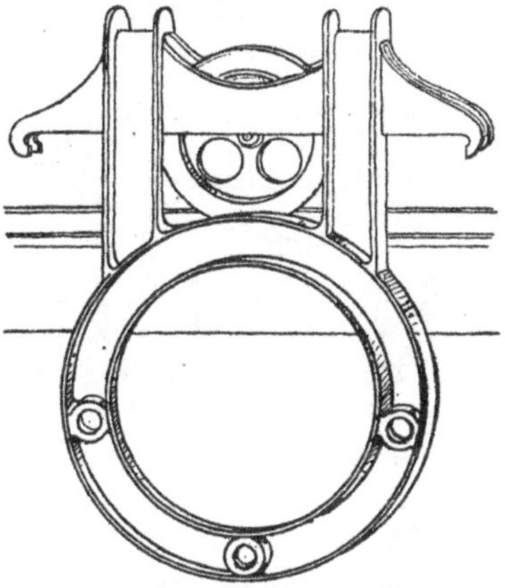

Fig. 145. Nickel Barn-door Hanger. Coleman Hardware Co.

Sliding-door Rails.

The ordinary mortised sheave, of which Figure 147 is a type, runs on a brass rail, which is generally made with a raised section, though a form is sometimes used which is channelled instead of being raised. The former will be something

Chapter VI.

to stub the foot against; the latter will collect dust. The only form of rail in the market, which presents neither of these difficulties, is the "Climax," Figure 148. This consists of a double brass track with a central strip, which is held flush with the two sides by springs inserted at intervals in the track. A special form of wheel is manufactured to go with this rail. The wheel, in passing along over the rail, presses down the central strip, forming a groove for the wheel to run in. When the door is opened, the springs force the flexible central strip up again, so that when the doorway is entirely clear the appearance is of a single brass plate perfectly flush with the floor. The list price of this rail is sixty cents per foot in brass.

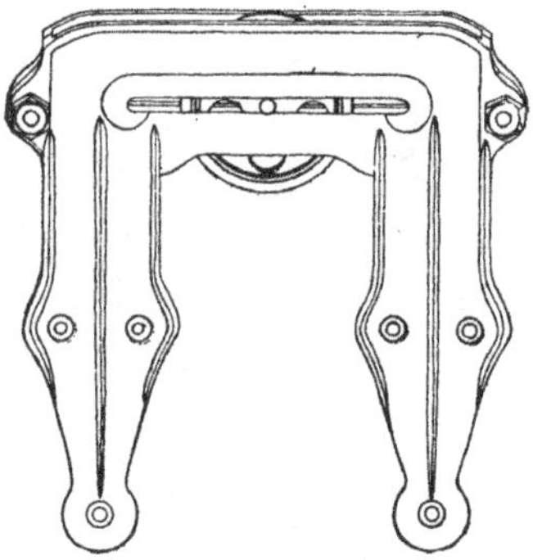

Fig. 146. Hatfield Barn-door Hanger.

Parlor-door Sheaves.

Besides the common pattern of mortised sheaves, shown by Figure 147, there is another form which works more easily, made on the same principle as the "Hatfield" barn-door hanger. This is shown by Figure 149. This sheave is made in five sizes, from two-and-one-half inches to six inches in diameter of wheel and costs from $1.50 to $4 per set of four sheaves.

Parlor-door Hangers.

Parlor-door hangers are usually arranged to run on a wooden or metal track which is bolted to the side-studding. There are one or two points which should be considered in judging of any door-hanger as ordinarily applied. In most houses the studs which form one side of the sliding-door pocket are made to rest

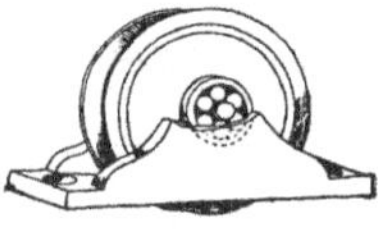

Fig. 147. Sliding-door Sneave. Russell & Erwin.

on something pretty solid, a foundation wall, or, at least, a heavy timber, while on the other side of the pocket the studs are supported on the floor-joist, and are left to settle with the shrinkage of the timbers, thus bringing about a difference in

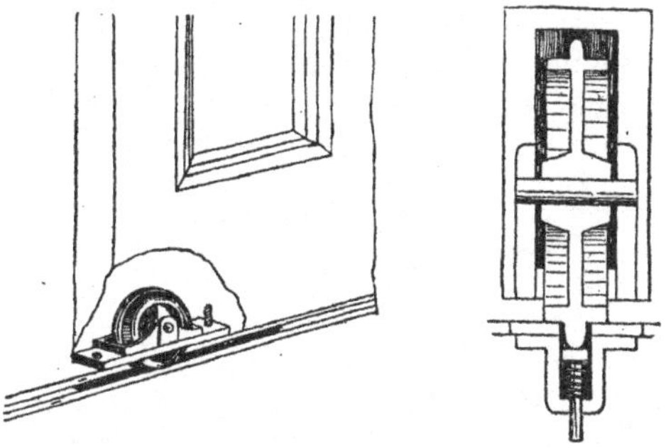

Fig. 148. Climax Rail. Climax Rail Co.

level of the two sides of the pocket. It may, then, be stated as a general rule that the best form of hanger would be that which is supported on one side only, since if any inequality of settlement takes place, it does not affect the hanger. Another consideration is, that it would be well to have the door-hangers so arranged that in case the door should not hang perfectly plumb, there would be no inequality of bearing on the axles of the wheels. It will be seen that this is perfectly

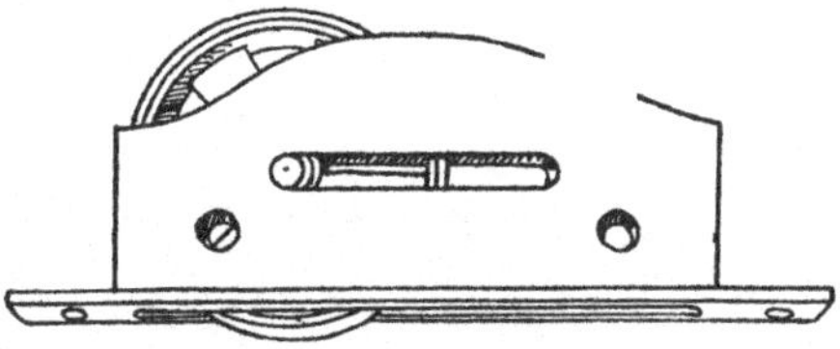

Fig. 149. Hatfield Anti-friction Sheave.

Chapter VI.

possible, and that it has been considered in some of the forms of door-hangers.

One of the earlier patents is the "Moore" parlor-door hanger, Figure 150. This is a very good form in the main, being hung by a single rod which is mortised into the top of the door. The adjustment may be obtained by turning up the nut at the bottom of the rod, through a hole cut in the edge of the door in the same manner as a stair-rail bolt is turned up. The difficulty is, that the hanger cannot readily be readjusted when once set. Another of the early forms which has since been but little improved upon, is the "Warner" hanger, Figure 151. This consists of two sets of double wheels connected by a rod, and working directly on the double track secured to each side of the door-pocket. The manner of supporting the door is much the same as with the "Moore" hanger, except that in the "Warner" the supporting rods can be got at after the door is finished by means of a face-plate on the edge of the door. The wheels are made perfectly flat, and it is claimed that under no combination of circumstances can they run off the track. The axles of the wheels are attached to the connecting rod by means of a universal bearing, thus enabling the weight of the doors to bear equally upon both tracks, no matter how much out of plumb or level they may be. A somewhat similar door-hanger is that shown by Figure 152, manufactured by the Reading Hardware Company. In this variety, however, the adjustment is entirely from the top of the door, and no

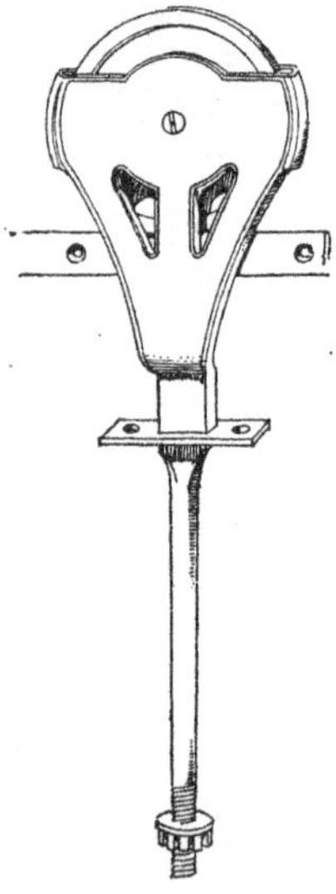

Fig. 150. Moore's Anti-friction Parlor-door Hanger. S. H. & E. Y. Moore.

mortise is required. The axles are not attached to any part of the hanger, but work in the slot somewhat on the principle of a "Victor" hanger.

Figure 153 shows a door-hanger which is very popular, and which for simplicity and perfection of construction is one of the best yet produced. The wheels run on two flat wooden tracks, one secured to each side of the studding. The axles bear on the short

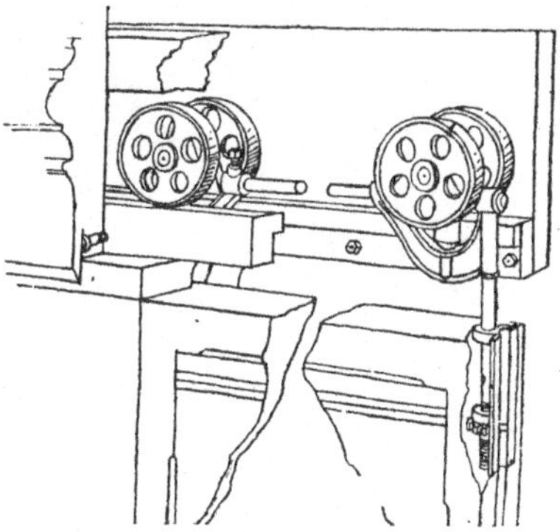

Fig. 151. Warner Parlor-door Hanger. E. C. Stearns & Co.

connecting rod which is made sufficiently long to allow for the run of an eight-foot door. The rod being round there will

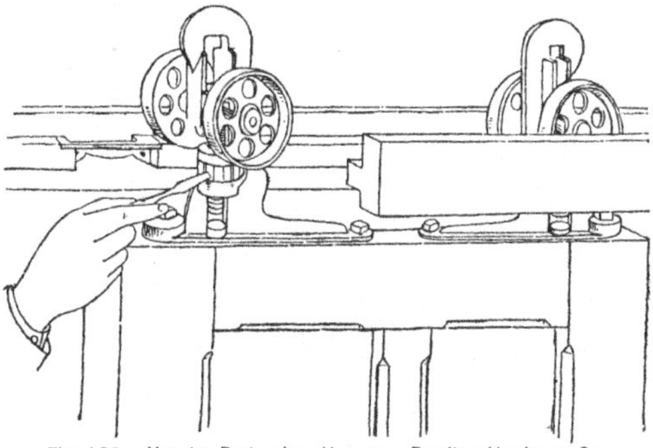

Fig. 152. Novelty Parlor-door Hanger. Reading Hardware Co.

never be an uneven bearing. The hangers are adjusted by means of a sliding screw-joint which is operated from the edge

Chapter VI. of the door, and which, by forcing the hanger away or drawing it towards the edge of the door, raises or lowers the bearing rod.

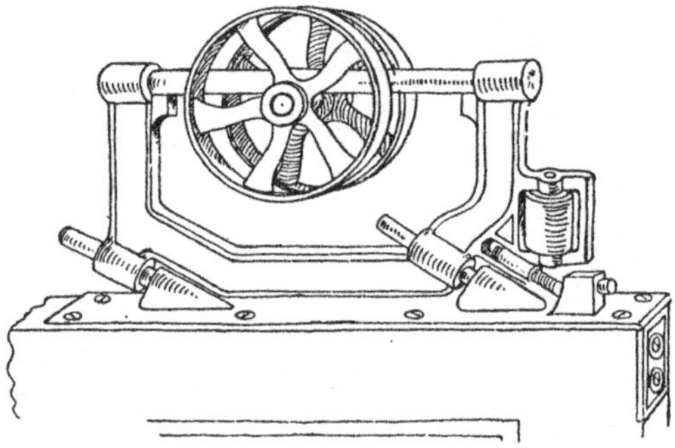

Fig. 153. Prindle Parlor-door Hanger. Prindle Mfg. Co.

A very simple application of the same principle is embodied in the "Nickel" parlor-door hanger, Figure 154. This con-

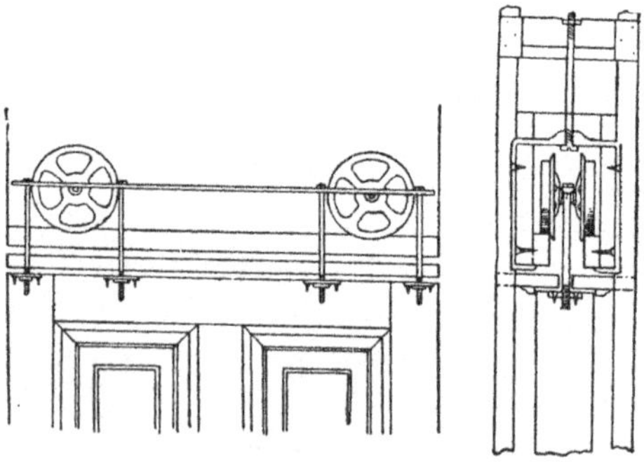

Fig. 154. Nickel Parlor-door Hanger. Coleman Hardware Co.

sists of a double set of flanged wheels, which run on a double track suspended by iron hanger-rods attached at intervals to a cross-piece at the top of the door-pocket. The axles of the wheels bear against a half-round bar, which is secured by

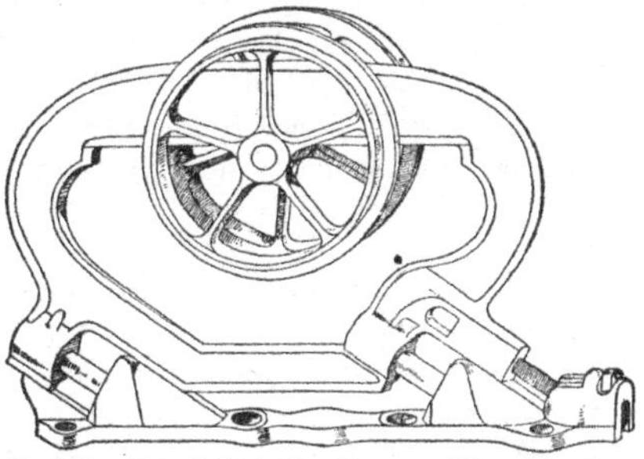

Fig. 155. Richards Parlor-door Hanger. Wilcox Mfg. Co.

upright bars to the top of the door. The hangers are adjusted by turning up the hanger-rods in the top of the pocket, thus

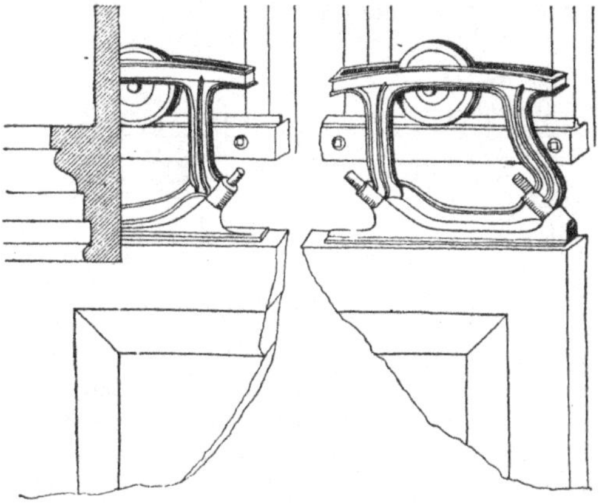

Fig. 156. Paragon Parlor-door Hanger. Dunham Mfg. Co.

lifting the track bodily. One objection to this form is that it requires considerable width of pocket—four-and-one-half inches.

Chapter VI.

Figure 155 illustrates the "Richards" hanger, which, with the "Prindle," rather leads the market just at present. The principle is almost exactly the same with both forms, except that in the "Richards" the axle has a flat instead of a round bearing, and the wheels are grooved. The "Prindle" manufacturers claim that the flat wheel is preferable; the "Richards," on the other hand, maintain that the flanged wheel is more desirable. There is really very little to choose between the two kinds.

The American Manufacturing Company has a parlor-door hanger on the market which is essentially the same as the "Richards" hanger. The "Paragon" door-hanger, Figure 156, is on the principle of the "Moody" barn-door hanger previously described. It consists of a single grooved wheel running on a rail secured to one side of the pocket, the axles bearing against two flat surfaces. It would seem as though this fulfilled the conditions of a perfect door-hanger more fully than anything else in the market. It can be adjusted with

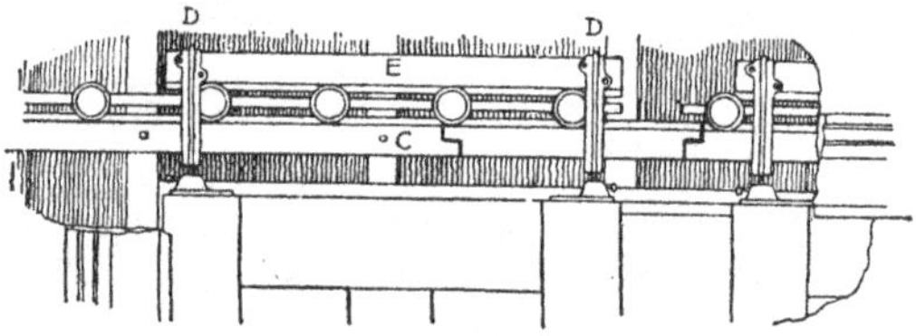

Fig. 157. Emerson Parlor-door Hanger. B. D. Washburn, Agent.

very little trouble; and as the centre of support is directly over the centre of the door, there is no tendency to bind; while as the track is secured to only one side of the door-pocket, the possible effects of shrinkages and settlements are reduced to a minimum.

A form of door-hanger which is essentially the same as this, but in which the axle of the wheels work in a slot on the principle of the "Hatfield" sheave, has been manufactured by

Burditt & Williams, for one of the Boston builders, but has received no patent, and is not really in the market.

The "Emerson" door-hanger, Figure 157, is yet another variety, and represents in some respects a different principle from any of the former, in that the rollers are entirely separate and distinct from each other, being connected merely by a thin strip of wood notched over the axles. The rollers bear on the wooden rail *C*. The hangers *D*, to which the door is directly attached, are fastened to a rider bar *E*, which bears directly on

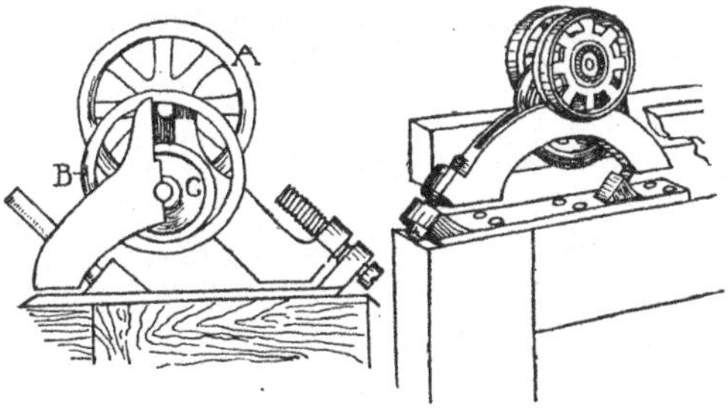

Fig. 158. Endless Anti-friction Parlor-door Hanger. Reading Hardware Co.

the rollers. This form of hanger is very effective, does not get out of order, and works very smoothly. Adjustment is obtained by a small set-screw in the attachment of the hanger. The track is fastened to one partition only.

Figure 158 is a very ingenious combination of the principles of the "Warner" and the "Prindle" hangers. It consists of a set of double, flanged wheels *A*, bearing on two tracks, which are bolted to each side of the pocket. The axle of the double wheel supports a ring *B*, which in turn supports a plain, grooved, pulley-wheel. The axle of the pulley-wheel turns in iron flange-plates which are fastened to the top of the door. In this way the friction is greatly reduced; and the principle of the continuous run of the axle, which is embodied in the "Warner" hanger, is here provided for by means of the ring

Chapter VI. connected with the flanged wheels, and with the lower pulley. The hanger is adjusted by a turn-screw near the edge of the door. The point of support of the hangers is always in the same relative position to the door, thus overcoming one of the weak points of the horizontal, anti-friction hangers in which the wheels travel back and forth from one end to the other, continually changing the strain on the screws.

The following table gives the relative prices of the various door-hangers that have been described:

TABLE OF DOOR-HANGERS.

Fig.		Per set of four.
138	Acme barn-door roller, 8 inch wheel	$3.00
141	Climax barn-door hanger	3.00
142	Moody barn-door hanger	3.20
143	Victor barn-door hanger	4.40
144	Lane barn-door hanger	2.30
145	Nickel barn-door hanger	3.50
146	Hatfield barn-door hanger	4.50
147	Parlor-door sheave, anti-friction	2.00
149	Hatfield parlor-door sheave	2.50
150	Moore parlor-door hanger	4.25
151	Warner parlor-door hanger	5.50
152	Novelty parlor-door hanger	3.50
153	Prindle parlor-door hanger	5.00
154	Nickel parlor-door hanger, (approximately)	3.50
155	Richards parlor-door hanger	5.00
156	Paragon parlor-door hanger	4.50
157	Emerson parlor-door hanger	5.00
158	Endless parlor-door hanger	3.75

There is still another distinct type of door-hanger, one which is unique of its kind, and for certain purposes is decidedly better than anything else in the market, though not always

Chapter VI.

applicable or always desirable: this is the "Prescott" hanger. It is difficult to illustrate this hanger properly; it should be seen in order to appreciate fully its workings. Figure 159 is the common form of hanger, consisting essentially of two flat bars joined, scissors fashion, in the centre. The lower end of one bar is fastened to a pin on the jamb of the pocket. The lower end of the other bar is fastened to a pin on the back of the door. The upper end of the bar which is fastened to the jamb, works with a roller in a slot on the back of the upper part of the door, while the upper end of the other bar works in a small slot let into the upper part of the jamb-pocket. A

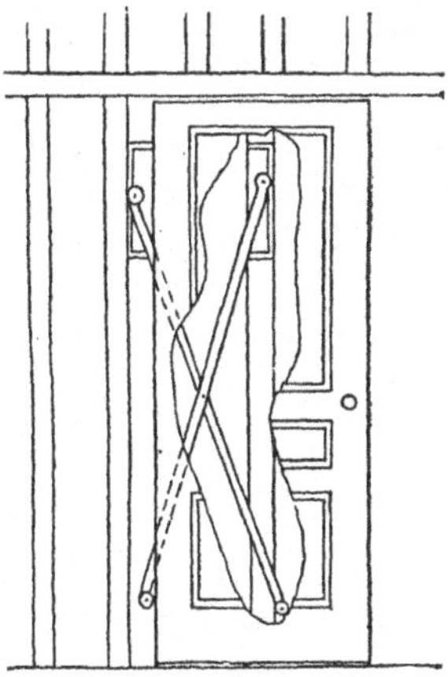

Fig. 159. Prescott Hanger. Prescott Mfg. Co.

little reasoning will show one that the door in this manner is held absolutely free from either the top or the bottom of the door-opening, and can be easily moved forward or backward. It cannot settle without the scissors part spreading out, and as the movable roller ends are on opposite sides, that is, one on the joint and one on the door; any tendency to spreading out of the rollers is counteracted by the opposite ends of the bars, which are always on the same level. In practice, the hanger, when properly set, works to perfection. The door never can bind, but can be operated by the slightest pressure in one direction or the other. Figure 160 shows a compound hanger on the same principle for use in very wide doors. The difference is simply that there are two hangers joined by bolts instead of one. Figure 161 shows a trussed hanger, which is used for doors that are wider than they are high. It may be said, incidentally, that these door-hangers can be exactly re-

Chapter VI.

versed; that is to say, the fixed ends may be at the top instead of the bottom.

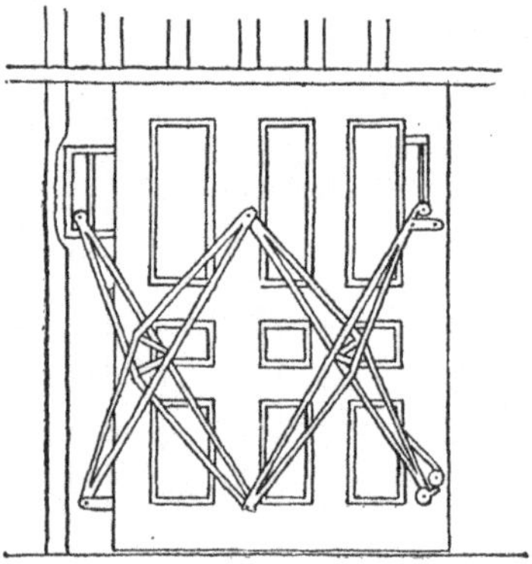

Fig. 160. Prescott Hanger. Prescott Mfg. Co.

The "Prescott" hanger is used to great advantage for elevator doors, as it permits of an opening the entire width of the car, if desired, while the ordinary width would be half that size. In such a case, a door across half of the opening is hung with ordinary butts, and the rest of the space is closed with a door hung by "Prescott" hangers to the first. The whole, or a part of the opening, can then be left unobstructed for the removal of boxes or trunks. Hangers for this purpose can be made of bronze, so as to present a neat appearance. These hangers are also used to advantage for barn-doors, car-doors, etc., and for any places where the hangers are exposed. The only objection to their use for parlor-doors, is that they have to be pretty carefully set by a mechanic who thoroughly understands the

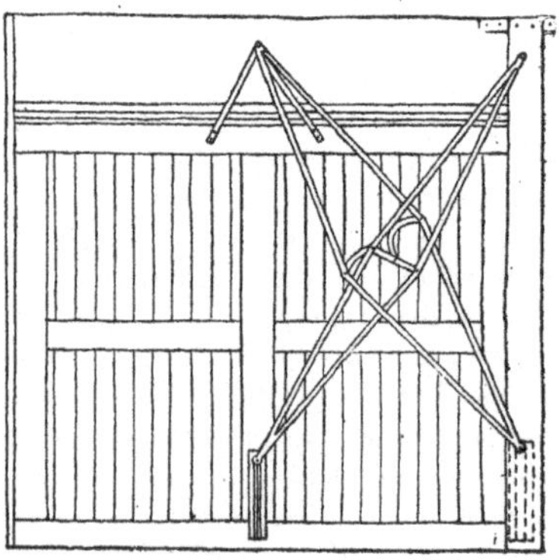

Fig. 161. Prescott Hanger. Prescott Mfg. Co.

workings, otherwise they are apt to rattle. It is not easy to adjust the hanger in case of settlements. They also take up considerable width in thickness of the pocket; still, they work so beautifully that they deserve all the popularity that they have enjoyed.

The prices of the ordinary form of Prescott hangers for inside doors, are as follows:

SIZE OF DOOR.	PRICE.
2½ x 8½ x 1⅛ inches	$3.25
3 x 9 x 1¼ inches	4.25
3½ x 9 x 1¼ inches	5.25
4 x 9 x 1⅜ inches	6.25
4 x 10 x 1½ inches	6.75
4½ x 10 x 1⅝ inches	7.50
5 x 10 x 1¾ inches	8.50
5½ x 12 x 2 inches	9.50
6 x 12 x 2¼ inches	10.50

CHAPTER VII.

PULLEYS.

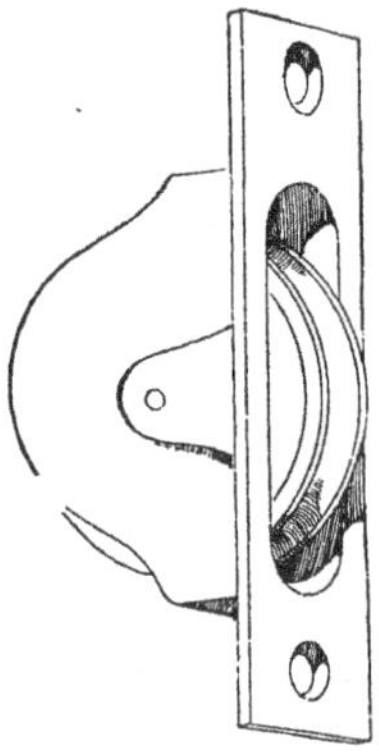

Fig. 162. Ordinary Axle-pulley.

ALMOST the only forms of pulley used by builders are those which are employed for double-hanging windows. These are made with cast-iron frames for the cheaper styles of work, or frames of malleable iron for a better class of goods; while some manufacturers use wrought-iron entirely. The wheels are usually made of cast-iron, with a groove shaped to receive the sash-cord or chain. The pulley is fitted in a mortise cut into the face of the hanging-style of the window-frame, and the part visible, or the face, is made of almost any material, but most often of bronzed, nickel-plated, painted or japanned iron. The commonest form is bronzed iron. The axles of the wheels are of steel or gun-metal, and the wheels themselves, in the better class of goods, are turned to accurate dimensions, though some cheaper grades are simply cast and polished. Some manufacturers finish pulleys with plain or ornamented bronze faces, in which case the face is made of a separate piece of metal, riveted to the iron frame of the pulley. There are also in the market, a few fine grades of pulleys made with brass wheels and cast-brass frames. This is, however, a needless expense, and such pulleys are used more in connection with furniture than with building.

The essential qualities of a good pulley-wheel, are simply that it shall run lightly, smoothly and easily. There should be a broad hub on the axle in order to prevent the flanges from jarring or rattling against the pulley frame, and the wheels selected should be of such a size that when the face-plate is mortised-in flush with the face of the hanging-style, the inner edge of the wheel will be over the centre of the box, so that sash-weights will not strike against the frame when raised or lowered.

Sash Pulleys.

Sash-pulleys are usually made in five sizes: $1\frac{3}{4}$ inch, 2, $2\frac{1}{4}$, $2\frac{1}{2}$ and 3 inches, the size referring to the diameter of the wheel. The two-inch wheel is sufficiently large for most cases, but for heavy, plate-glass windows larger sizes are used, though the chief advantage of a large pulley is not so much that it will wear better, but that it will throw the sash-cord farther away from the hanging-style, and so permit of larger sash-weights. When the expense is not an item to be considered in the selection, it is well to employ some form of anti-friction, ball or pin bearing pulley for all sash weighing over fifty pounds. A poor pulley will soon wear loose so as to rattle on the axle. If anti-friction wheels are not advisable, the next best form is one with a large gun-metal axle. Some compositions of phosphor-bronze would seem to be peculiarly well suited for pulley axles, though not at present in the market to any extent.

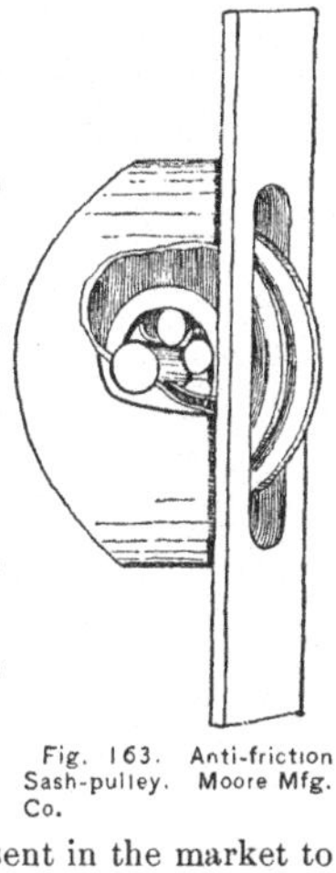

Fig. 163. Anti-friction Sash-pulley. Moore Mfg. Co.

There are a great many varieties of sash-pulleys, though the

Chapter VII.

differences are so slight that a few examples will serve to illustrate the whole. Figure 162 is a fair type of the ordinary axle-pulley, and Figure 163 is a type of the best form of anti-friction sash-pulley. All of the most commonly used forms of sash-pulleys are on essentially one or the other of these lines.

The only important deviations from the common types of sash-pulleys have been made with a view to reducing the amount of labor required to properly set the fixtures in the window frames. It should be said that none of the patent forms have thus far met with either very wide sales or general approval, which would seem to indicate that the common form answers pretty fully all the requirements of the case. There are, however, a few styles which have met with considerable favor in the market, and which will serve to give an idea of the lines the attempted improvements have followed.

Figure 164 illustrates a form known as the "Empire" sash-pulley, in which the case is corrugated horizontally so that it will exactly fit into a series of holes bored into the frame with an auger or bit of standard size, a great reduction in the labor of mortising thus being effected. The advantages claimed for it are that it cuts away less of the frame than any other pulley, is held more securely, does not require any screws, and can be inserted much more readily and quickly than any other kind. It is claimed that these pulleys can be fitted to the window frames at the rate of sixty per hour.

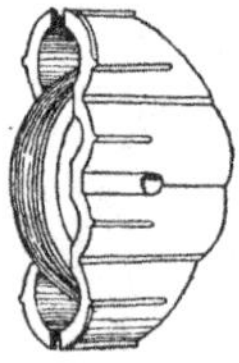

Fig. 164. Empire Sash-pulley. Empire Portable Forge Company.

A pulley requiring even less work in setting, though somewhat more complicated in construction, is shown by Figure 165. This consists of two small wheels set in a cylindrical case, and requiring no more labor to fit in place than is involved in the boring of a single hole. Shoulders or flanges at top and bottom of the case serve to plumb the pulley properly, and kept it from twisting. It is claimed that this pulley will hold its position quite as well as any other form, though it would seem more apt to work loose by reason of the leverage of the weight over the inner wheel, than the ordinary form. Chapter VII.

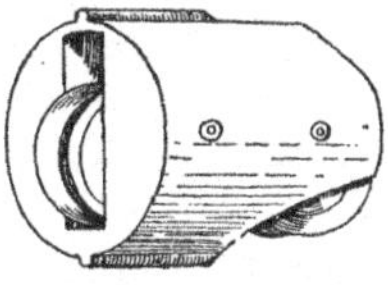

Fig 165. Corey's Two-wheel Sash-pulley. J. B. Johnston.

Another style of pulley which does not avoid cutting the mortise on the frame, but saves somewhat in the screws, and has a finer appearance than either the "Empire" or the "Corey," is shown by Figure 166. In this pulley the face-plate and frame are cast together, and the frame is made with a wide shoulder or flange at the bottom, which is cast on a bevel, so that when placed in position in the rebate, the pulley cannot slip down or out, by reason of the bevel wedging into the mortise. A single screw at the top of the pulley holds it securely in place; but it will be seen that it does not depend

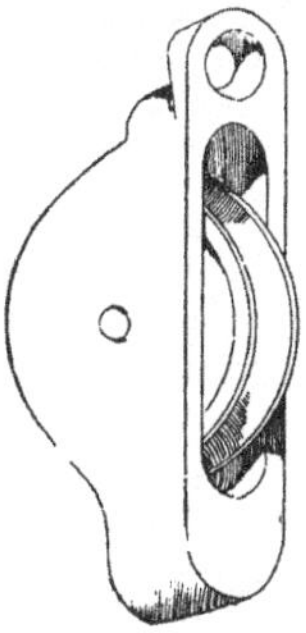

Fig. 166. Norris Pulley. C. Sidney Norris & Co.

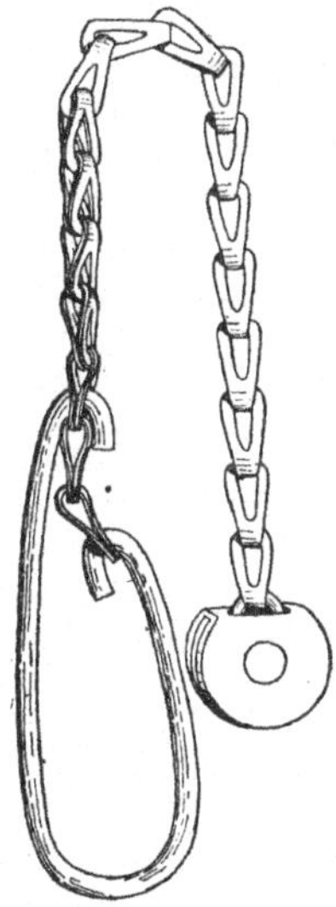

Fig. 167. Single Sash-chain. Smith & Egge Mfg. Co.

upon the screw for its stability. With an ordinary pulley, the heavier the sash the greater is the possibility of the wheel being forced out from its mortise, whereas, with the "Norris" pulley, the greater the load, the more securely is it wedged in

Chapter VII.

place. Another obvious advantage is that it requires just half the quantity of screws and amount of labor to set this pulley as it does the ordinary pulley; and it is said that the carpenters who have used this, have liked it very much. The labor of mortising is slightly more than for the ordinary form, but the company controlling the patent also manufactures a mortising-machine specially adapted to this kind of work, by which the labor is greatly reduced. Aside from the labor of mortising, the only possible objection to this sash-pulley seems to be that it would require a pretty heavy hanging-style to the window frame, and would cut away the wood a good deal, the bevelled flange being three-quarters of an inch through for an ordinary sized pulley. This patent is manufactured in the same sizes and styles as the ordinary pulleys, including those with flat-grooved wheels for sash-chains.

The following table gives the average retail prices of the principal marketable varieties of ordinary sash-pulley wheels.

TABLE OF SASH-PULLEYS.—PRICES PER DOZEN.

Description.	1¾ in.	2 in.	2¼ in.	2½ in.
Painted iron, cast wheel	$ .25	$ .30		
Bronzed iron, steel axle, cast wheel	.50	.65	$ .70	
Bronzed iron, steel axle, turned wheel	—	.90	1.10	$1.35
Bronzed iron, anti-friction steel axle, turned wheel			1.75	2.00
Polished brass face, anti-friction steel axle, turned brass wheel			5.50	6.25
Brass or bronze face, steel axle, polished iron wheel		1.35	1.75	
Corey's fine bronzed iron, steel axle, turned wheel	.60	.75		
Empire fine bronzed iron, polished face and wheel		.45		
Norris's fine bronzed iron, polished face and wheel	.49	.50	1.10	
Smith & Egge, polished iron, flat grooved, turned wheel		2.25		2.65
Smith & Egge, polished iron, 3-inch double grooved wheel $ 8.00				
Smith & Egge, polished iron, 4-inch double grooved wheel 13.50				

SASH-CHAINS AND WEIGHTS.

Sash-chains.

In the better-class of buildings it is usual to hang all sashes weighing over forty pounds with some form of sash-chain; indeed, except for the expense, it would often be well to use nothing but chains, especially in buildings of a public character where the windows are apt to be moved with little care. The ordinary cords used for windows are liable to wear out and break, and experience has often shown that a good sash-chain will outwear enough of the ordinary sash-cord to make it more than worth while to use the stronger material.

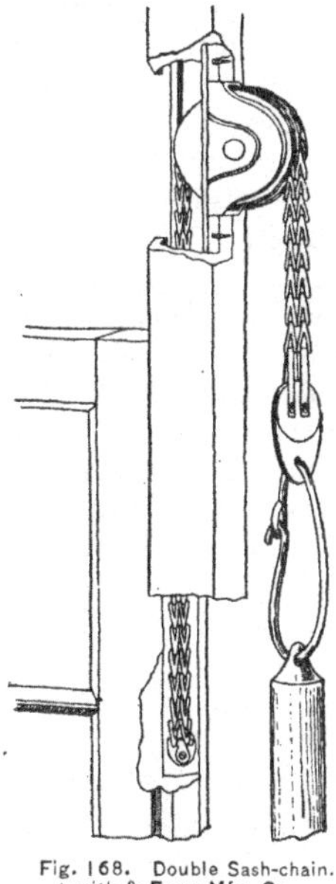

Fig. 168. Double Sash-chain. Smith & Egge Mfg. Co.

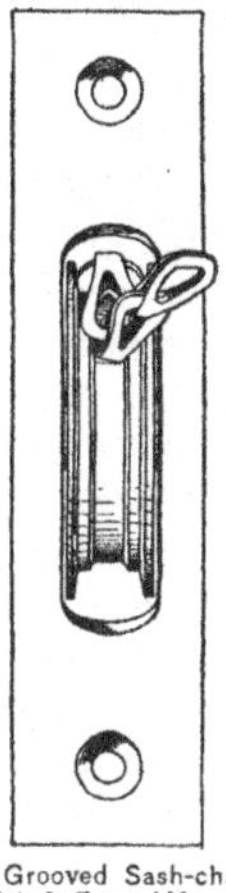

Fig. 169. Grooved Sash-chain Pulley. Smith & Egge Mfg. Co.

The sash-chain which appears to meet with the greatest favor in the Eastern market, is that which is made by the Smith & Egge Manufacturing Company. The form of this chain is illustrated by Figure 167. It is a species of flat-link

Chapter VII.

Smith & Egge Chains.

chain, the form being the same as the well-known plumbers' safety-link, which has been in use for various purposes for a great many years. Smith & Egge adopted this form as best adapted for window-chains, and they have perfected special machinery which does away almost entirely with handwork, and enables them to produce the chain at marketable prices and of a superior quality. The chain is made with a great deal of care, each link being automatically tested as it leaves the machine. The metal preferably employed for this purpose is a bronze composition specially prepared by the manufacturers, designated as "giant metal," which is, in appearance, very much like pure copper, but is tougher and harder. A cheaper composition is also used, which is known as "red metal," and steel chains are manufactured to a certain extent, though the giant-metal chains are the best in every respect. The best giant-metal chains will sustain loads as high as 700 pounds. The red-metal chains are manufactured to sustain from 380 to 500 pounds. Steel chains are made in three grades: one capable of sustaining from 125 to 175 pounds; another from 400 to 450; and the strongest from 600 to 700 pounds.

Double Chains.

Instead of one chain, it is often more desirable to use a double chain for very heavy windows, as shown by Figure 168. This figure, and Figure 167, also show the manner in which the chain is attached to the sash and to the weights.

Chain Pulleys.

Pulleys intended for use with sash-chains, require a different groove from that usual where rope is employed. The Smith & Egge Manufacturing Company has a special form of pulley intended to go with their goods, Figure 169. For convenience this pulley has been listed with the others in the preceding table of prices, together with one or two different makes of flat grooved-pulleys which would answer for the purpose equally well.

Morton Chains.

What has been said of the Smith & Egge chains applies equally well to the "Champion" sash-chains, manufactured by Thomas Morton, which have the same shape of link, though the sash and weight fastenings are slightly different. The "champion" metal is a bronze composition probably not differing essentially from the giant or the red metal. Thomas Morton, however, manufactures another form of sash-chain with quite a different link, which is shown by Figure 170. This is known as the cable-chain, and is a very strong, durable form, never twisting or kinking. The sash attachment used with this chain is very simple and efficient, consisting of a short half cylinder

Chapter VII.

Cable-chains.

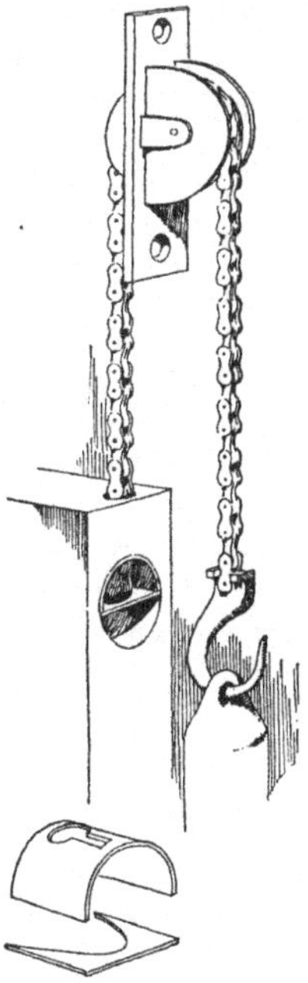

Fig. 170. Cable Sash-chain. Thomas Morton.

Fig. 171. Solid-link Sash-chain. Solid-link Sash-chain Co.

with a slot cut down from the top, wide enough at the bottom to admit a link of the chain, but narrowing at the top so as to prevent the swelling at the link-joint from passing through. The same sort of slot is cut in the weight-hook to hold the other end of the chain.

The cable-chains are usually made with alternately two and three pieces to each link, joined by a pin passing through the

Chapter VII.

five thicknesses. For the lightest work the pieces are arranged two and two. The cables are made of either copper or steel, and vary in strength from a size for a thirty-pound sash to one capable of sustaining a door weighing 1,500 pounds.

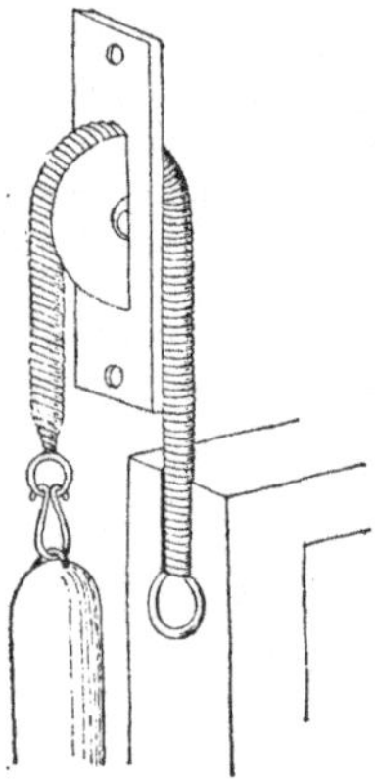

Fig. 172. Acme Sash-cord. Coiled-wire Belting Co.

Solid-link Chain.

There seems to be but one other form of sash-chain at present in the market. It is known as the "Solid Link" chain. It consists of a compound link on much the same principle as the Smith & Egge chain, but made double and with rather finer brass or bronze wire, so that the sash-chain is nearly as flexible as ordinary sash-cord, and can be bent or twisted in all directions without knotting or kinking, a quality which the Smith & Egge chain does not possess. The "Solid Link" chain can even be tied into a knot without kinking. Figure 171 will give an idea of the construction of the link. This form of sash-chain requires no special pulley, but will run over an ordinary grooved-wheel.

Spring Sash-cord.

An entirely different kind of sash-cord is shown by Figure 172. This consists of a steel wire spring so closely and strongly coiled as to have the resistance necessary to sustain any sash weighing sixty pounds or less. It has a stretching-capacity of only about five per cent. It enables the sash to rest easily and lightly on the pulleys, and enables it to be raised or lowered with half the effort required with rope sash-cord. It is fastened at one end to the sash by an eye or ring, and to the weight in the manner shown by the figure.

The following table gives the average, comparative, retail prices per foot of the various makes and sizes of sash-chains:

TABLE OF SASH-CHAINS.—PRICES IN CENTS.

Description of chain.	For sashes or doors weighing											
	15 lbs.	30 lbs.	40 lbs.	75 lbs.	100 lbs.	125 lbs.	150 lbs.	250 lbs.	400 lbs.	[illegible]00 lbs.	1000 lbs.	1500 lbs.
Smith & Egge's plumbers' link, Giant-metal.... ...		5	6.5	8		10						
Smith & Egge's plumbers' link, Red-metal....	3.5	5	6									
Smith & Egge's plumbers' link, steel........	2.5	3.5		5								
Smith & Egge's plumbers' link, steel, black enamelled....................	3	4		5.5								
Morton's plumbers' link, Champion-metal........	3.5	3.5		4.1	4.9	6						
Morton's plumbers' link, steel....................				3.5	4.1							
Morton's cable-chain, copper.....................		8		7.8	9.6		11.2	13.6	18.4			
Morton's cable-chain, steel.....................				7.2	8		9.6	11.2	13.6	17.6	20	28
Solid-link chain[1], brass....					14		22					
Solid-link chain[1], silver-plated...................					16		24					
Acme sash-cord...........	3	4	5	6								

SASH-BALANCES.

The inconveniences attending the use of weights and pulleys for raising sashes, while in most cases due to bad workmanship in setting the pulleys and imperfect arrangement of the boxes rather than to any intrinsic deficiencies in the system, have given rise to several so-called sash-balances, which are intended to permit of weights, boxes and ropes being entirely dispensed with. Indeed, the natural outgrowth from the idea involved in the "Acme" sash-cord would be that a plain spring could be made to answer the purpose of both weight and

[1] There are but two sizes of solid-link chain. They are tested by the makers at 200 lbs. and 300 lbs. before leaving the factory.

Chapter VII.

cord. This has been done with the "Anderson" sash-balance. Figure 173, which consists simply of a steel, spiral spring for each sash, from $\frac{3}{8}$ inch to $\frac{1}{2}$ inch in diameter. The springs are fastened to the hanging-style of the window-frame and to the sashes, and are made of just sufficient strength to sustain the sash in any position, so that a very slight exertion is sufficient to move it either up or down. The springs are made the same

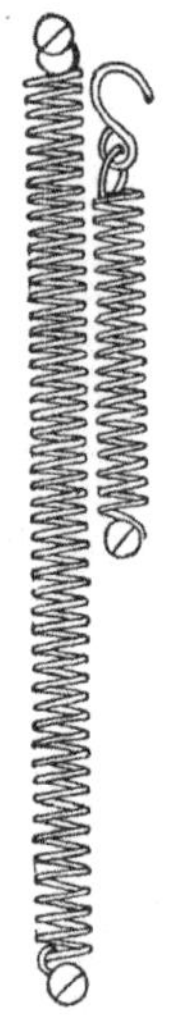

Fig. 173. Anderson Sash-balance. Wm. G. Anderson.

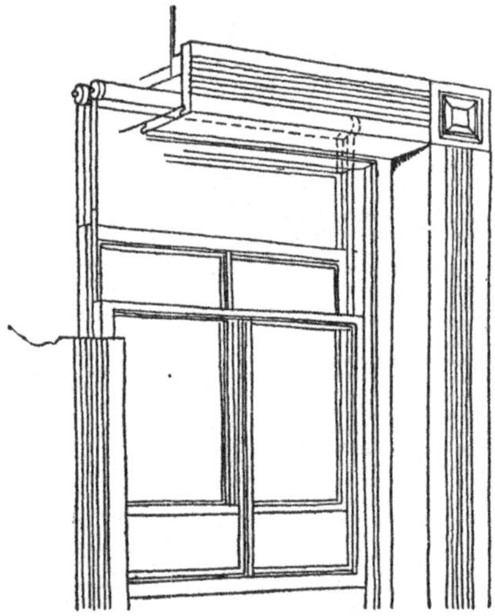

Fig. 174. Ormsby Sash-balance. Ormsby Sash-holder Co.

length as the sash; and, being secured near the bottom of the sash, are drawn out to twice their length when the sash is down. When a sash is hung with this kind of balance, it has to be fitted with some form of self-catching sash-fast, as otherwise the window might fly open as soon as the hand was withdrawn. Anderson uses the "Attwell" sash-fast for this purpose, which will be described in a subsequent chapter, though any other self-locking form would answer equally well. The retail price for the four springs necessary for two sashes weighing fifteen pounds each, is $1.35. For forty-pound sashes the price is $2.50, and other sizes in proportion. When the sashes weigh over fifty pounds, the size of the spring required becomes so large as to render it rather too conspicuous for ordinary use.

The "Ormsby" sash-balance, Figure 174, is on exactly the same principle as the ordinary shade-roller, consisting of two strong spring-rollers which are concealed in a pocket over the window-head, the sashes being suspended therefrom by thin brass bands which coil around the roller. The price of this sash-balance varies from 75 cents for a window with fifteen-pound sashes, to $1.75 for fifty-five pound sashes.

A third type of sash-balance is illustrated by Figure 175. This has the general appearance of an ordinary sash-pulley, being mortised into the hanging-style in the same manner. Inside of the pulley, however, is coiled a strong band-spring of steel, attached to the axle, which is fixed, and to the outer edge of the wheel, in the groove of which is wound a narrow brass ribbon serving instead of a sash-cord, so that when the sash is drawn down, the ribbon is uncoiled and the spring acted upon. The spring can be set to any desired tension, and its action can further be regulated by a brake on top of the wheel, which is tightened or loosened by turning a screw in the face-plate. The "Shumard" sash-balance is made for runs of from thirty-two to forty-six inches, and for sashes of from eight to forty pounds' weight. The price per pair, for a sixteen-pound sash, is $1.40; for a forty-pound sash, $3.25 per pair, and other weights in proportion.

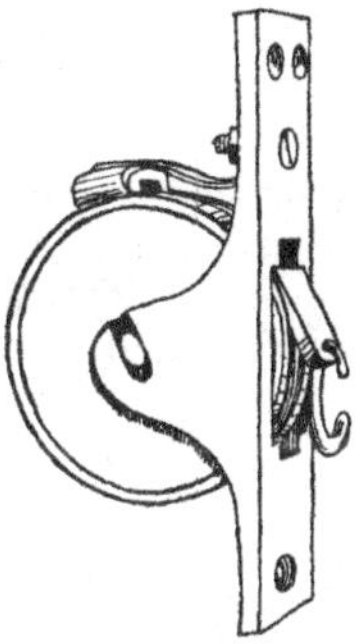
Fig. 175. Shumard Sash-balance. Coleman Hardware Co.

Chapter VII.

One advantage which all of these spring-balances possess is, that they act most strongly when the sash is down, enabling one to move a binding window more readily than if it were hung with ordinary weights and cords, while, when the sash is up, the springs barely suffice to hold it in position and do not offer resistance to drawing down, as is the case with weights. The objection, of course, is, that the springs are in constant tension, and will, in time, loose their elasticity. They can be replaced quite as easily as worn-out ropes; still, most people seem to prefer the old-fashioned weights and pulleys.

SASH-CORD ATTACHMENTS AND WEIGHTS.

Sash-cord Irons.

There are several devices for attaching the cords to the sashes. The commonest method is to cut a groove on the side of the sash with an enlargement towards the bottom, and then simply knot the end of the cord, the knot holding in position.

Fig. 176. Sash-cord Iron.

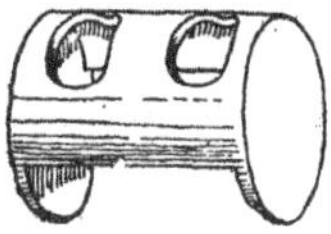

Fig. 177. Double Sash-cord Iron.

It is better to use some form of sash-cord iron. Figure 176 is one of the simplest forms. It is mortised into the side of the sash and held in position by a screw, the sash-cord being knotted under the hook. This form retails at 35 cts. per gross. Figure 177 illustrates an iron used when the sash is hung with two cords on each side. This retails at 52 cents per gross. Figure 178 is a form quite

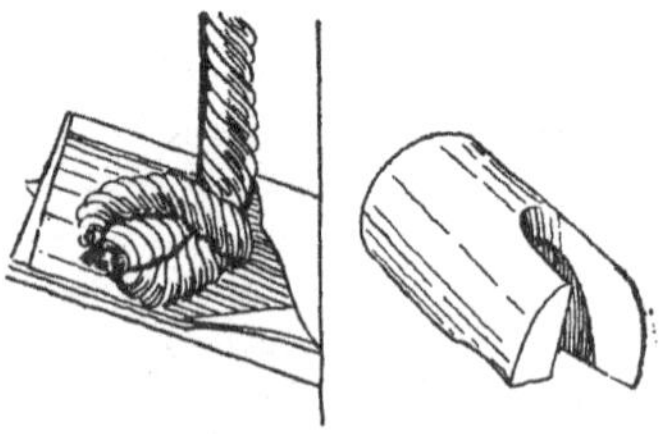

Fig. 178. Jackson's Sash-cord Iron. Ireland Mfg. Co.

Chapter VII.

similar to Figure 176, though requiring a deeper mortise and being driven in on a slant, so that it cannot work loose. It retails at $1 per gross. Figure 179 is different from either of the preceding forms, consisting of a cartridge-shaped cylinder, closed at both ends, but with an opening at the top and the bottom, through which the cord is passed and wedged by the eccentric cam shown by the figure. The cord is released by inserting a wire, as shown. This fixture retails at $1.60 per gross.

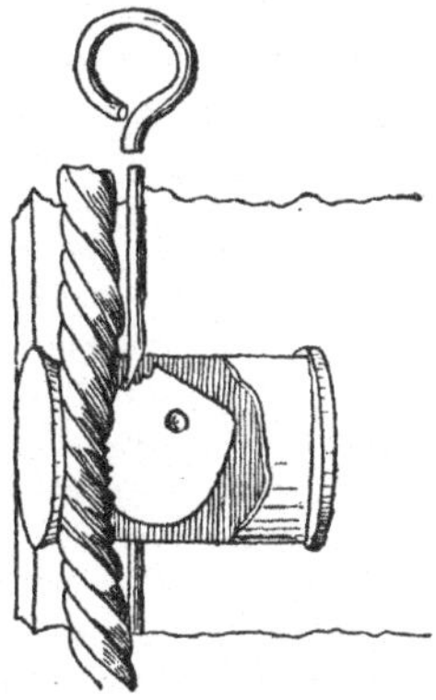

Fig. 179. Rodgers Sash-cord or Chain-fastener. Nimick & Brittan.

Sash-weights.

Sash-weights are usually made of cast-iron, to order. They are cast in plain round-bars with an eye at the top, substantially as has been shown in Figure 170. When they are very heavy, or the space for the sash-boxes is restricted, lead weights are used, as they occupy less space than iron. They are usually made to order, and can be had in any desired shape, but are manufactured in regular weights by a few of the lead-works. Figure 180 is the form adopted by the Raymond Lead Company. Each section is cast on an iron rod extending through the weight, with a hook at one end and an eye at the other, so that as many sections may be hung to each other as may be necessary. Iron sash-weights cost $1\frac{1}{2}$ cents, and lead-weights from 6 to 7 cents per pound.

Fig. 180. Raymond's Sash-weight. Raymond Lead Co.

CHAPTER VIII.

Sash–Fastenings.

SASH-LOCKS may generally be said to be devised for the moral encouragement of the faint-hearted, who cherish a fond belief that when the lock is turned no intruder can possibly enter through the window. Most of the forms in the market are sufficient protection against a sneak-thief, but while nearly every sash-lock in existence is claimed to be strictly burglar-proof, and advertised as such, the burglar must be a novice, indeed, who would let even the best of them keep him out of a house. When the window is secured with a fast which cannot be opened by slipping a knife-blade between the meeting-rails and pushing back the bolt, an enterprising burglar would simply break out a pane of glass, which can be accomplished with less noise than is made in picking a lock; so that, after all, the protection afforded by a sash-lock is more in sentiment than in fact. Still, that the timid be cheered and the stray tramp kept out of the silver closet, some form of sash-lock is always considered a necessity for all windows.

The requirements of a perfect sash-fast or lock, are that it shall fulfil the following conditions:

First, it must be so constructed that it cannot be opened from without by a knife, or by jarring the window. Second, it should, in closing, draw the two sashes tightly together, and, at the same time, should not be affected by any small inequalities of adjustment. Third, it should always remain either open or shut, with some sort of spring-contrivance to hold the lever in position, so that it will not be possible to leave the lock partially turned, thereby running the risk of breaking the muntins when the window is raised. Fourth, it should have no projections which could possibly tear the sash; and fifth, and perhaps most important of all, it should be simple in construction and in its operation.

It is not essential, though it is generally very well that the fast should be self-locking. It is not well to trust to anything which acts by gravity, or which depends on any perfectly fitted sash, as such appliances are apt to get out of order. It also is well that the lock should be as inconspicuous as possible, though neither is this essential. Some sash-fasts are provided with rebated appliances which fit down between the two sashes; this is not necessary, though it is perhaps a greater security, as in this way the fast can be more securely screwed onto the sash. In selecting any form of rebated sash-fast, however, it must be borne in mind, that sashes are made differently in the East and in the West. In the vicinity of Boston, it is customary to rebate the meeting-rails where they come together, but elsewhere, we believe the meeting-rails are usually simply bevelled.

Much ingenuity is to be observed in the line of patents for sash fasts and locks. This might be interpreted as an indication that either the sash-fastenings at present in the market are quite insufficient for their purpose, or that there is an extraordinary necessity for the species of protection which such contrivances can afford. This chapter, however, will but faintly indicate the variety of devices having in view the securing of sashes. One must wade through the list of Patent Office reports in order to fairly appreciate what has been done in this direction. A great many of these inventions never get beyond the Patent Office. Still, there are all sorts and kinds of sash locks and fasts in the market. Each one appears to be covered by a patent, yet somehow or other, the best locks and fasts are to be found in the catalogues of nearly all the manufacturers, so slightly disguised it is easy to see that such ideas are in a measure common property.

Chapter VIII.

The terms sash-lock and sash-fast have been used synonymously, though a distinction should be observed between them. A sash-lock is understood to be some contrivance which actually locks a sash by means of some form of key. All of the other numerous devices which, by means of levers, catches or springs hold the sash either open or shut, are technically termed sash-fasts. The distinction, however, cannot be rigidly adhered to. Nearly all of the self-locking fasts might be classed with sash-fasts, while, with equal propriety, the sash-locks can be said to possess the essential qualities of sash-fasts; though, with very few exceptions, all are designated by the hardware dealers, as sash-fasts.

SASH–LOCKS.

Figure 181 illustrates a form of sash-lock which is secured to the face of the sash, the bolt working into staples at intervals in the window-frame or stop-bead. The bolt works with a spring, so as to be self-acting, and by means of the key it can be locked, thus permitting the sash to be left partly open and secured against intrusion. This lock is light and strong, and well adapted to be used on screens as well as sashes.

Figure 182 is a type of several varieties of sash-locks which mortise into the sash and throw out a bolt in the same manner as an ordinary door-lock. In this example the bolt works with a spring, which

Fig. 181. Sash-lock. Yale & Towne Mfg. Co.

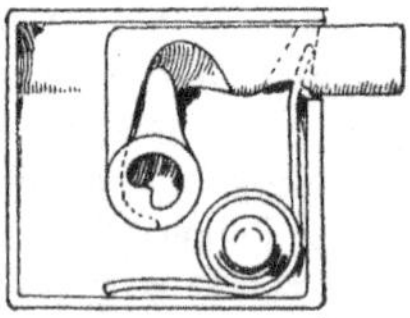

Fig. 182. King Sash-lock. Burditt & Williams.

makes the lock self-acting after the bolt has been thrown; that is to say, if the bolt were pushed back it would not remain

Chapter VIII.

closed until the knob or handle were turned. This lock may be operated with either a movable key or a flat knob or button.

Of course, any mortise drawer-lock can be applied to a sash, if desired, or any other lock sufficiently narrow to fit the thickness of the sash, though the two forms just described are about the only ones especially made for windows, which can properly be termed sash-locks.

SASH—FASTS.

There are so many varieties of sash-fasts to be considered, that, in order to make the descriptions less confusing, it will be well to classify the sash-fasts according to their most prominent characteristics.

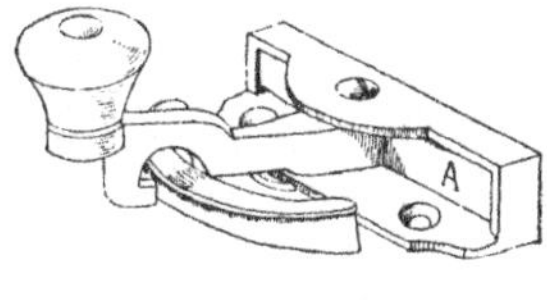

Fig. 183. Sash-fast. P. & F. Corbin.

Plain Lever Sash-fasts.

I. *Old Style of plain lever-fast.* This consists of a plain, pivoted bar, attached to the upper sash, with a hook on the end, which works over a catch or raised plate on the lower sash. A knob, often of porcelain, is fastened to the end of the bar. Figure 183 illustrates this type. In the oldest makes of this kind, the lever was simply pivoted on the upper sash, and a knife-blade, slipped between the meeting-rails, could push the lever to one side and unlock the window without the slightest difficulty. In the sash-fast shown by the figure, the lever is made with a broad, flat end, which presses against a strong spring, *A*, at the back. The spring serves to stiffen the action of the lever, which is further protected against intrusion, in some cases, by dovetailing the bottom plates, so that the lever cannot be got at through the joint between the meeting-rails. Varieties of this same form are made with

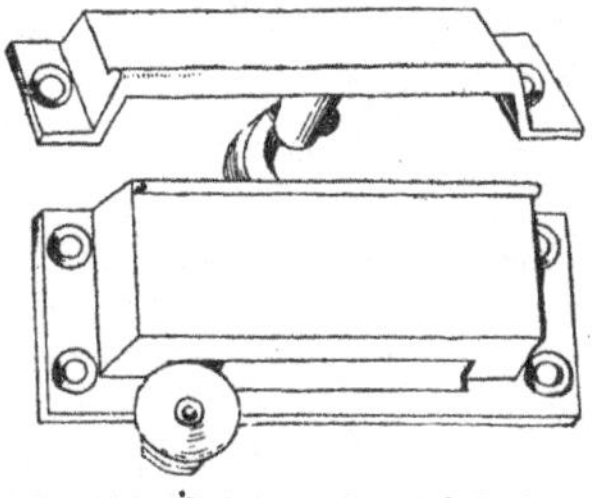
Fig. 184. Sash-fast. Russell & Erwin.

Chapter VIII. the lever swinging only half around in one direction, the gain thereby being that the back spring can be made longer and stiffer, without increasing the size of the sash-fast.

Figure 184 represents a variety of sash-fast, in which the lever is on the lower sash and hooks over an inverted peg on the upper sash. The "Judd" sash-fast, Figure 185, has a

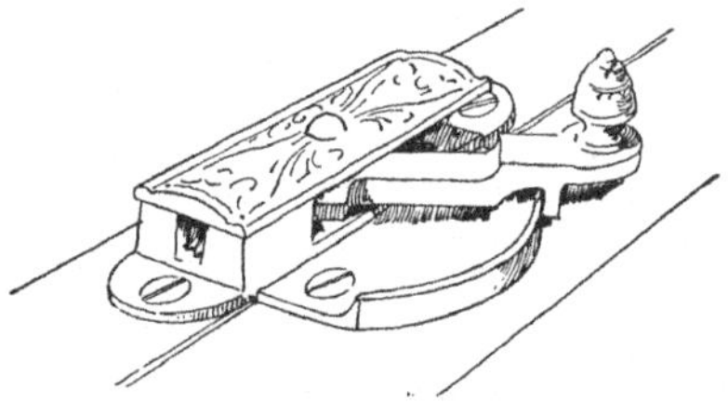

Fig. 185. Judd Sash-fast.

shoulder on the side of the lever so arranged that a knife-blade would catch on it and be broken before the lever could be moved sufficiently to open the window. Figure 186 shows a

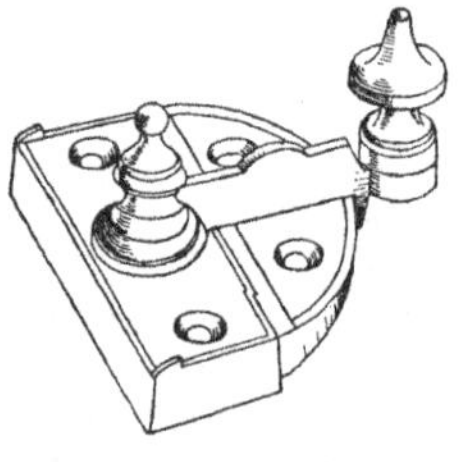

Fig. 186. Sash-fast. Norwich Lock Mfg. Co.

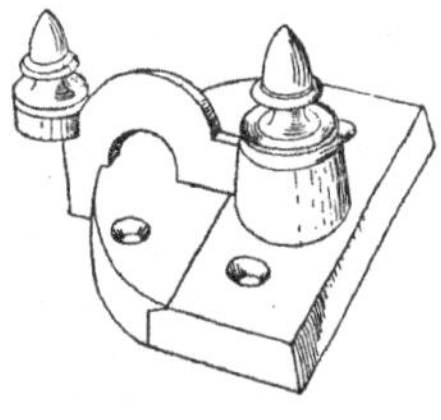

Fig. 187. Sash-fast. P. & F. Corbin.

strong form of sash-fast with no spring of any sort, but with a protection for the lever by dovetailing the plates together, as described in the previous paragraph; while the sash-fast indicated by Figure 187 works in exactly the same manner, but the plates are rebated together. Figure 188 is yet another variety, the plates here being not only dovetailed together, but also lipped down into the joint between the meeting-rails. The lever works in the same manner as the first sash-fast noticed. Chapter VIII.

II. *Spring lever sash-fasts.* While some of the forms just described might be classed under this category, none of them actually have spring levers, as the springs are not so arranged as to force the lever open or shut from any position. Figure 189 shows what is known as the ordinary "Boston" sash-fast, which is used a great deal in Boston, and is much liked for its simplicity and sureness of action. This form is, apparently, made by almost every manufacturer of builders' hardware in the country. There is a coil spring around the hub of the lever tending to throw it back, and a simpler spring which bears against the ratchet connection, so as to lock the lever when it is closed. Hopkins & Dickinson manufacture a variety of this form, in which the spring-catch, has several cuts or ratchets on its edge, and the catch on the lever is bevelled, so that it will be held by any one of the ratchets. The advantage claimed for this is, that if the lever is drawn around hastily, it will be more likely to catch on the ratchets and be locked, than the ordinary pattern, which has but a single ratchet. This is known as the "Ladd" sash-fast. Spring Lever Sash-fasts.

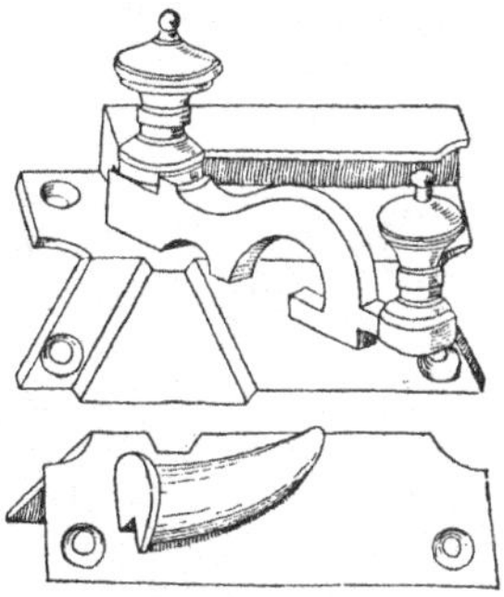

Fig. 188. Metropolitan Sash-fast. A. G. Newman.

The "Boston" sash-fast has to be set pretty carefully in order to be effective, and in the case of very excessive shrinkage, the space between the plates might be so reduced that the ratchet would not hold the lever. Such difficulties would, of course, arise only in a third-rate building. The form is believed by many of the dealers to be the best in the market, and it surely answers very fully the requirements of a perfect sash-fast. "Boston" Pattern.

Chapter VIII.

A scarcely less admirable sash-fast is shown by Figure 190. The lever is on the lower sash, hooking under a tooth on the upper sash, which is bevelled so as to draw the sashes more closely together. A strong spring about the axle of the lever tends to throw it open, while a small bolt, inside the lever, locks into a concealed catch on the post or axle of the lever. The bolt ends in a knob, *A*, and

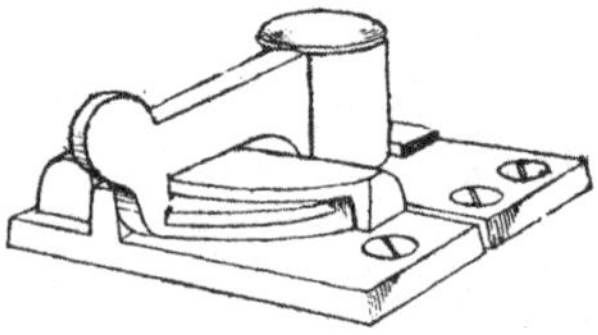

Fig. 189. Boston Pattern Sash-fast.

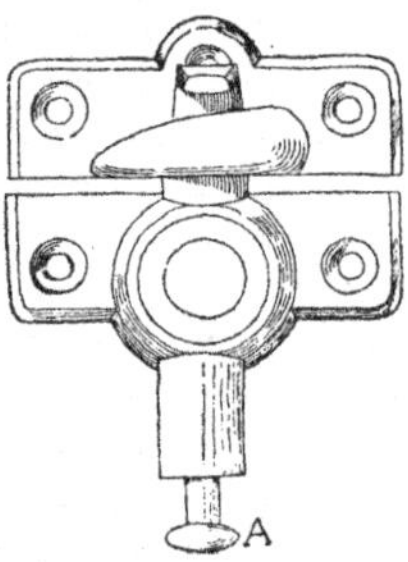

Fig. 190. Sash-fast. Hopkins & Dickinson Mfg. Co.

is fitted with a spring which keeps it pressed tightly towards the centre, so that the lever is caught and held when it is turned clear around through 90 degrees. On pulling the knob, *A*, the catch is released and the lever flies open.

Figure 191 illustrates a sash-fast which works in very nearly the same manner as the preceding, except that in order to release the lever the knob is pushed in. The Hopkins &

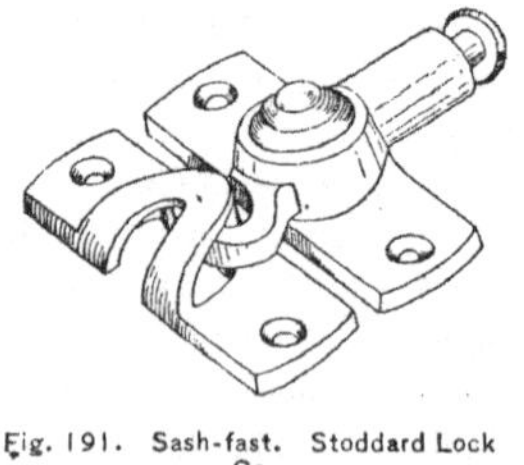

Fig. 191. Sash-fast. Stoddard Lock Co.

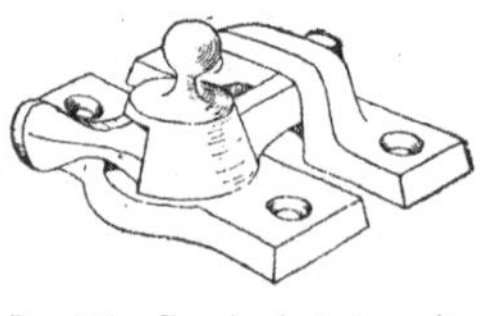

Fig. 192. Favorite Sash-fast. Reading Hardware Co.

Dickinson Manufacturing Company also has a sash-fast which hooks around a pin, like Figure 191, but is otherwise the same

Chapter VIII.

as Figure 190. P. & F. Corbin manufacture two forms of sash-fasts which are essentially the same as Figure 190.

Lever Sash-fasts with locking lever.

III. ***Lever sash-fasts with locking lever.*** This includes lever sash-fasts with locking lever, those in which the lever works without a spring, but is held either closed or open by means of an auxiliary lever. Figure 192 gives one variety of this kind, the lever being pulled down in order to permit the bolt to turn. The locking lever here is held in place by a

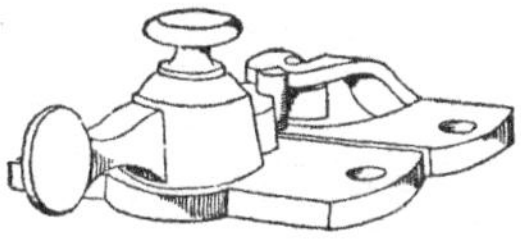

Fig. 193. Sash-fast. Yale & Towne Mfg. Co.

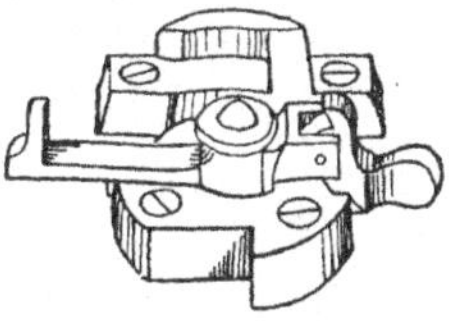

Fig. 194. Morris Sash-fast. Ireland Mfg. Co.

spring, and catches into a slot in the bed-plate, preventing the lever from being turned. Figure 193 has a lever which works in the same manner as that of Figure 191, except that it has no spring. It is locked in place when turned, by a catch which is released by pressing the knob on top of the hub.

Figure 194 represents a form of sash-fast which has met with considerable favor, as being one of the first which had any right to the qualification of being burglar-proof. The action

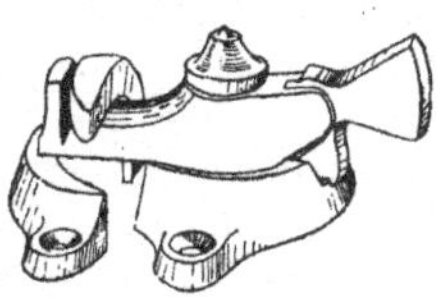

Fig. 195. Triumph Sash-fast. Ireland Mfg. Co.

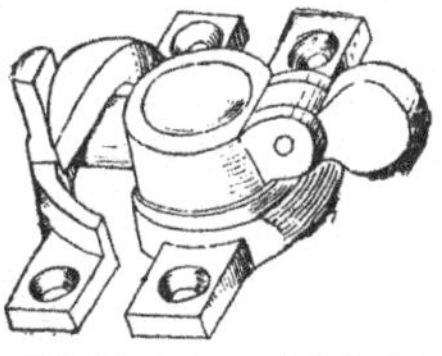

Fig. 196. Sash-fast. P. & F. Corbin.

is perfectly simple. The lever is on the lower sash, and is held either open or shut by a smaller hinged lever which drops

Chapter VIII.

by gravity into the rebates of the bed-plate. Figure 195 shows a fast which operates in exactly the same manner, the smaller, gravity lever being raised to release the main lever; and Figure 196 is a type of a number of similar forms manufactured by P. & F. Corbin. The lever in this example is released by raising the secondary lever at the rear.

Locking Lever Sash-fasts.

IV. *Locking lever sash-fasts.* This class includes those sash-fasts in which the lever locks itself when turned. Figure 197 is a form which has been on the market for some time, and

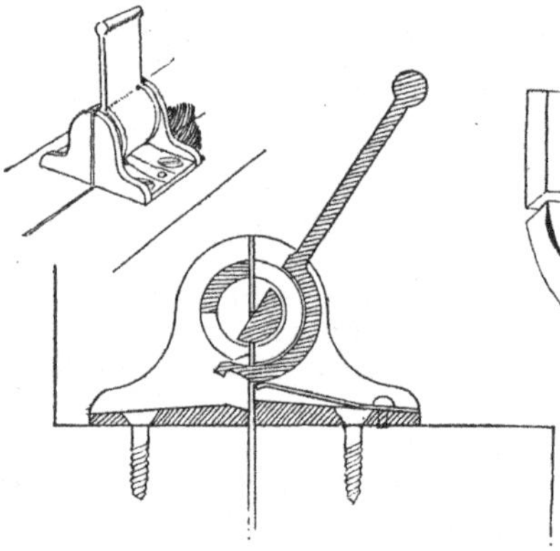

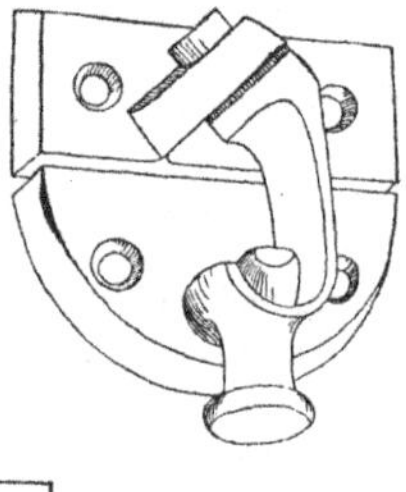

Fig. 198. Payson's Sash-fast. J. B. Johnston.

Fig. 197. Mathes' Sash-fast. Nimick & Brittan.

is now comparatively little used, though the chief objection to it is only in reference to its size. It is as near to being absolutely burglar-proof as any form of sash-fast which has been devised. Half the axis, about which the lever is rotated, forms a part of the upper and of the lower sash portion of the fast. The figure shows the position when the sashes are locked, the lever hooking down under both half-axles, and not only drawing the two sashes together, but binding and holding them so they cannot be moved. To unlock the sash, the lever is thrown up to a vertical position.

The "Payson" sash-fast, Figure 198, is very simple and effective, ranking as one of the best in the market. The lever is on the top sash, and locks itself over the opposite post with the help of a small spring-bolt, the knob of which shows at the end of the lever. The attachment of the lever is such as to permit it to fold back on the upper sash. This sash-fast is not liable to get out of order, it draws the sashes together, and is as burglar-proof as would ordinarily be desirable. The only objection to it is that the connection between the lever and the locking-post does not allow for much shrinkage in the sash.

Chapter VIII.

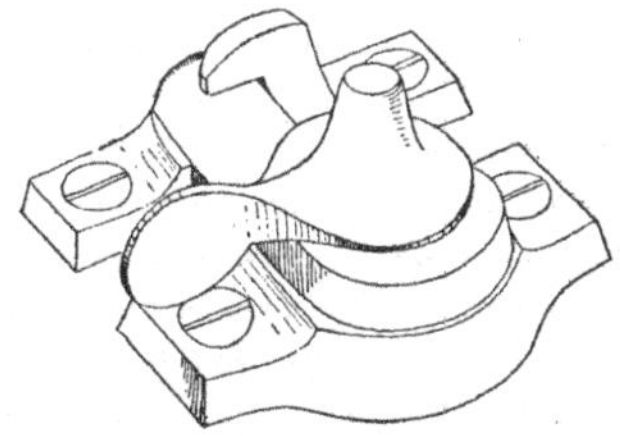
Fig. 199. Ives Sash-fast. H. B. Ives & Co.

Cam Sash-fasts.

V. "*Cam*" *sash-fasts.* The sash-fasts of which Figure 199 is a type, are quite difficult to represent by a drawing. The action of the fast is as follows: The levers are fastened to the lower sash. When the upper lever is turned, the lower, or locking lever, is first thrown out until released from the hook on the upper sash, and then drawn around and in toward the hub, until both levers are on a line with the edge of the sash, the upper lever moving through 180 degrees, while the lower lever is moved only 90 degrees. Though the action sounds complicated, the sash-fast is perfectly simple in its construction, and there is nothing about it that can get out of order, or even wear loose, except by such excessive use as would practi-

Fig. 200. Plan of inside of Figure 199.

Chapter VIII.

cally never be given it in a house. There are no springs about it, consequently it has no automatic action, and in unlocking, care must be taken that the locking lever is turned clear around, as otherwise it will project beyond the meeting-rail and catch on the sash-muntins. This is really the only objection to it. Figure 200 illustrates the internal construction of this sash-fast.

Figure 201 is a very similar sash-fast manufactured by P. & F. Corbin. It differs only in the internal, eccentric arrangement. The Reading Hardware Company also manufactures a sash-fast very much like the "Ives."

Self-locking Sash-fasts.

VI. *Self-locking sash-fasts.* The "Boston" sash-fast flies open of itself, unless properly locked. Many people believe that a sash-fast should lock itself the moment the sash is drawn down, so as to leave no chances of the windows being unlocked, and, accordingly, there are in the market several varieties of self-closing sash-fasts. For gen-

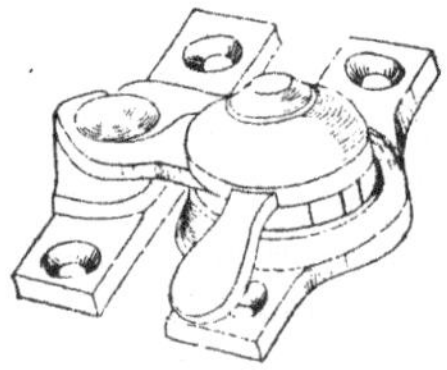

Fig. 201. Sash-fast.
P. & F. Corbin.

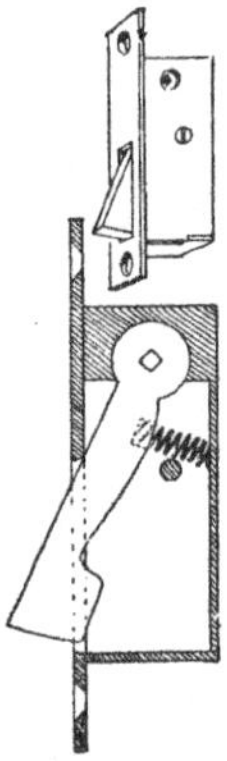

Fig. 202. Ticket-office Sash-lock.
Enoch Robinson.

eral house use, it is believed that such fasts would prove a great nuisance, as the window would, of course, have to be unlocked every time it was to be opened; besides, nearly all of these self-locking appliances are much more liable to get out of order, either through rust or neglect, than the ordinary

Chapter VIII.

sash-fasts; still, in some cases, there seems to be a necessity for them.

Ticket-office Sash-fasts.

Perhaps the simplest form of self-acting sash-fast is that illustrated by Figure 202. This consists of a spring bolt, acting not unlike the latch-bolt of an ordinary lock, which flies out whenever the sash is closed. It is mortised into one of the styles of the upper sash, or into the hanging-style, and the bolt bears on a plate on top of the meeting-rail of the lower sash. The figure shows one of the case-

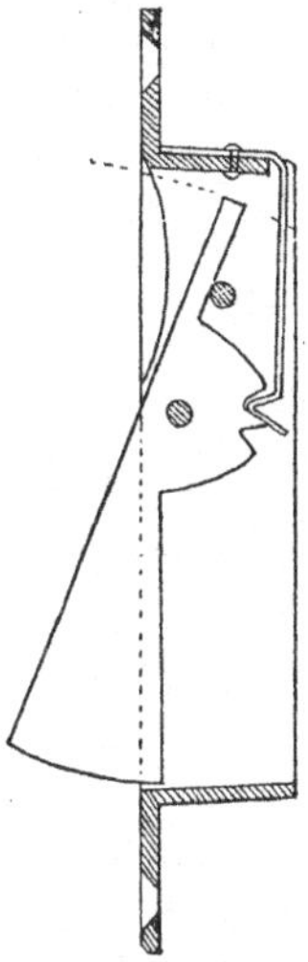

Fig. 203. Sash-fast. E. Robinson.

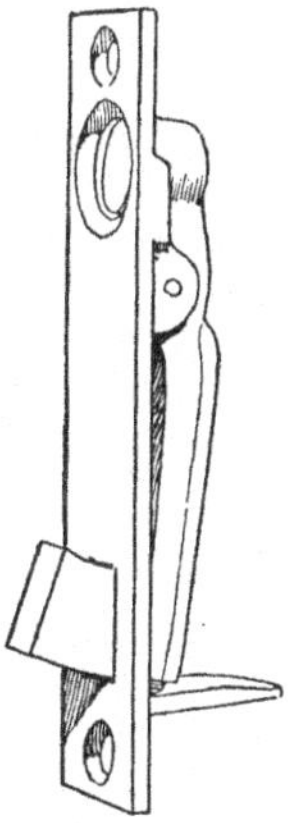

Fig. 204. Sash-lock. Enoch Robinson.

plates removed, to illustrate the construction. A very similar fast is shown by Figure 203. This, however, is not self-locking, as the top of the lever must be pressed in to force out the lower portion. Both of these appliances might almost be classed as sash-locks, instead of sash-fasts.

Figure 204 shows a variation of the last form of sash-fast, working without springs of any sort. This is a neater looking form than Figure 203, and perhaps a trifle stronger in princi-

Chapter VIII.

ple. Variations of these forms are manufactured under several different patents.

A self-locking fast of the description of Figure 202 has the disadvantage that the bolt must be held back when the window is to be raised, and, if the sash should bind, it is rather awkward to attempt to hold back the bolt with one hand and move the sash with the other. There is a device, the Security Self-locking Sash-fast, Figure 205, which obviates this difficulty. This consists of a bolt similar to that of Figure 202, but with a locking-lever which falls out when the bolt is pushed back. The locking-lever holds the bolt flush with the sash and allows the sashes to be raised or lowered, but when the meeting-rails pass each other, the locking-lever is raised, releasing the bolt, which flies out as soon as the window is closed, thus locking the sashes.

Figure 206 is a self-locking sash-fast on a different principle. The cut shows the lower sash partly raised, the locking portion being attached to the upper sash. When the window is closed the hook, *D*, strikes against the catch, *C*, forcing it away from *B*, until, when the meeting-rails are on a level, *D*, is hooked in between *C* and *B*. A spring at *A* keeps the two parts pressed against each other.

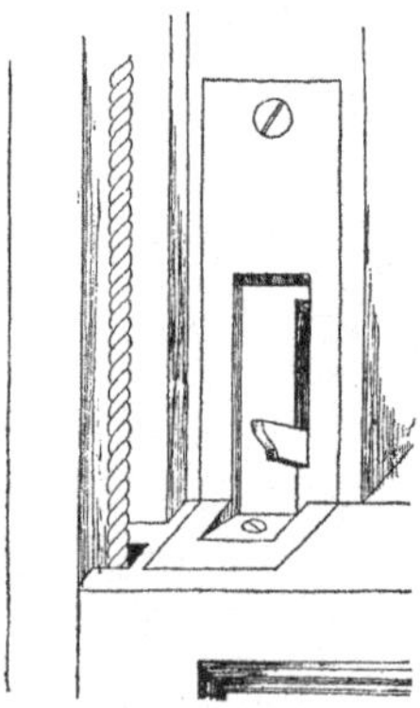

Fig. 205. Security Sash-fast.

Gravity Sash-fasts.

A very ingenious sash-fast, which works almost entirely by gravity, is shown by Figure 207. The cut shows it in the position it takes when the window is locked. The mechanism is attached to the upper sash. *A* is hung on each side to *B*, which hooks over the post on the lower sash. To open the window, the thumb is placed under *B* and the forefinger on top of *A*. Both pieces are lifted together until *B* assumes a vertical position, and *A* catches over the hook *C*. The sash can then be opened freely. On closing the window, however, the lower rail strikes against a hidden lever or cam at the back of *A*, lifting it from its hold on *C*, so the piece *B* can descend to hook in the position shown by the figure. The only spring used is one which pushes out *B* when *A* is released. This sash-fast is very

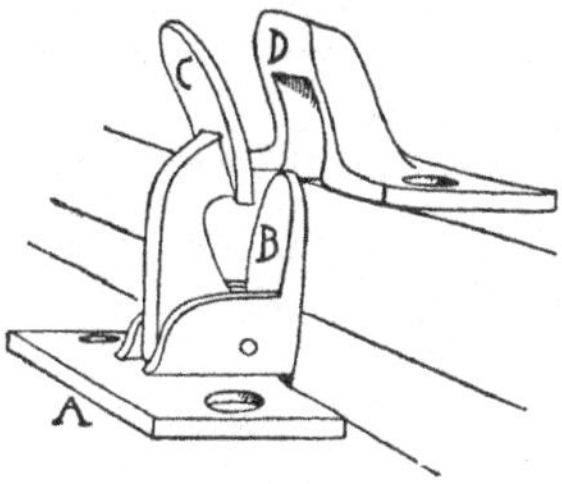

Fig. 206. Shaw's Sash-fast. Nichols & Bellamy, Agents.

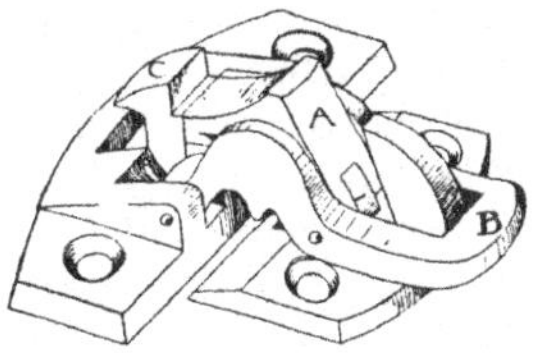

Fig. 207. Davis Sash-fast. Sise, Gibson & Co.

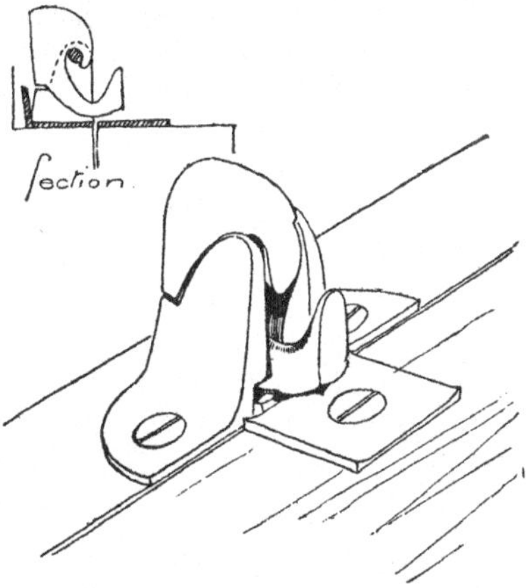

Fig. 208. Byam's Sash-fast. Byam, Stewart & Baker.

nicely made, and is about as good a self-locking form as is in the market. It has an added advantage in that it locks

Chapter VIII.

itself before the sash is entirely closed, the post, *C*, being double notched on the face so that *A* will slip down and wedge the hook *B* when the window is within about a quarter of an inch of being closed. It will be observed that the locking is effected entirely by the piece *A*. *B* is brought over the hook on the lower sash merely in order to draw the meeting-rails more closely together.

The "Byam" sash-fast, Figure 208, is a very simple device acting entirely by gravity, the central bent lever being so counterbalanced that the lower arm will always project over the lower sash. The section will show how this lever is hung.

Sash-fasts which lock in different positions.

VII. *Sash-fasts which lock in different positions.* The difficulty with all sash-fasts of this description is that, of necessity, they operate on one side of the window, instead of in the centre of the meeting-rail, and that, consequently, every attempt to open the window when it is locked, wrenches the sash so as to, in time, make it loose in the joints. Also, with nearly all the forms, the mechanism is concealed, so that the sashes are liable to many unnecessary wrenches. The advantages are that the window can be left partly open and still be secured from intrusion, and that, in most cases, either sash can be locked independently of the other.

Fig. 209. Sash-fastening. J. B. Shannon & Sons.

Figure 209 illustrates a very primitive appliance, consisting simply of a ratchet rail, with a spring-catch on the bottom sash. Figure 210 shows a more complete form, which is mortised into the face of the hanging-style, the levers working into holes in each sash. The sashes are fitted with other holes on the edges, at intervals, so that they can be locked at various heights. In the cut the section shows more clearly the working. A single spring, coiled about each lever, serves to throw them both out. Pushing up the knob on the inside bead draws back the upper lever, releasing the upper sash. Pushing the knob down releases in the same manner the lower sash. Chapter VIII.

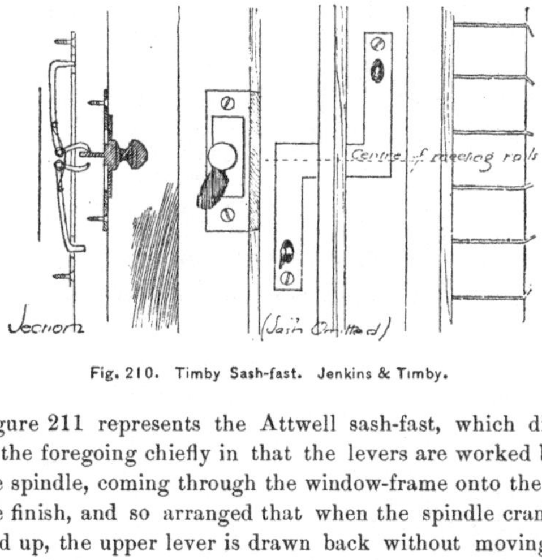

Fig. 210. Timby Sash-fast. Jenkins & Timby.

Figure 211 represents the Attwell sash-fast, which differs from the foregoing chiefly in that the levers are worked by a single spindle, coming through the window-frame onto the face of the finish, and so arranged that when the spindle crank is turned up, the upper lever is drawn back without moving the lower, and *vice versa*.

Excentric Cam-fast.

There are a few forms of sash-fasts which will hold the window in any position. Figure 212 is a very common form, consisting of an excentric cam which screws to the face of the sash and wedges against the stop-bead, holding the sash by friction. Figure 213 and Figure 214 are used chiefly for car work, the former acting in the same manner as Figure 212, while the latter works into slots in the jamb.

The sash-fasts thus far considered include all the principal

Chapter VIII. forms commonly known to the hardware trade, as well as types of many styles which have only limited sales. It would be

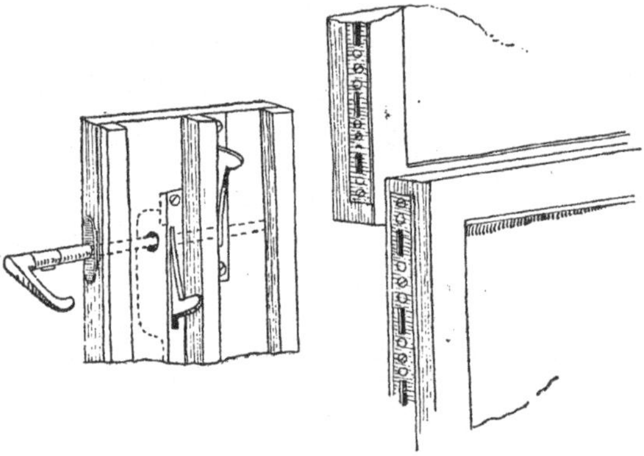

Fig. 211. Atwell Sash-fast.

impracticable to attempt an enumeration of all the sash-fasts which have been put on the market throughout the country, nor would any such list be of great value. The forms discussed and illustrated, will, it is believed, serve

Fig. 212. Brown's Window-lock. S. A. Brown.

Fig. 213. Eccentric Sash-fast.

Fig. 214. Sash-lock.

every purpose of comparison, and will enable the retail buyer to select to advantage, and to know the worth of what he is choosing from.

SASH-HOLDERS AND SPRINGS.

Chapter VIII.

The sash-fasts which have been described up to this point, are those intended to be used in connection with double-hung windows; and, while any of them would answer for a window having double sashes which are simply set in the frame, without weights or balances of any description, there are a few appliances especially intended for windows without weights, which can best be considered in this connection. "Hammond's" window-springs are used so extensively for this purpose that they are almost standard. The springs are simply flat bands secured to the hanging-style and catching in ratchet cuts on the edge of the sash, each spring having a bent handle or thumb-piece coming out in front of the sash. One spring is used for each sash. A different form of spring is shown by Figure 215. This is mortised into the hanging-style and catches into square cuts on the edge of the sash, the thumb-piece showing in front. Figures 216 and 217 show two forms of window-springs on an entirely

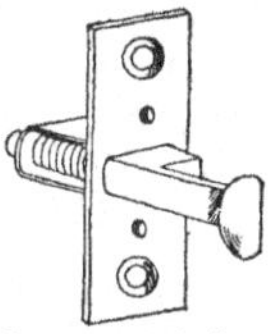

Fig. 215. Window-spring. P. & F. Corbin.

Window-springs.

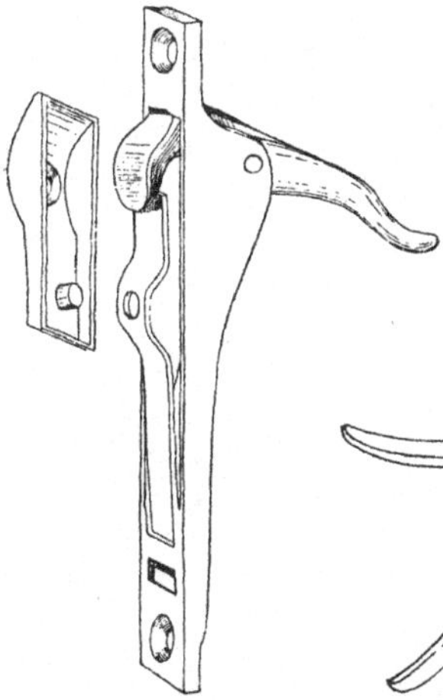

Fig. 216. Sweet's Window-spring, Stanley Works.

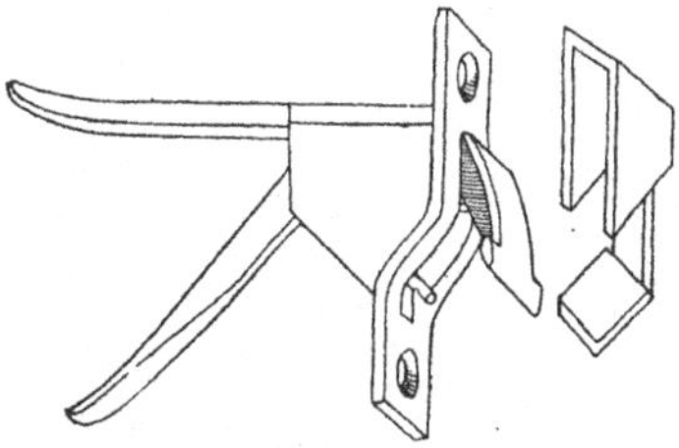

Fig. 217. Northrup's Window-spring. Stanley Works.

Chapter VIII.

different principle. The portion containing the spring is mortised into the sash, the lever showing in front of the glass. In the first instance the spring is exposed, and catches on posts set in mortised plates at intervals on the hanging-style. In latter example, the spring is concealed, the end of the lever hooking into plates of the shape shown by the figure, which are mortised into the hanging-style. These springs also serve as holders, the ends of the levers pressing strongly against the hanging-style and holding the sash from slipping.

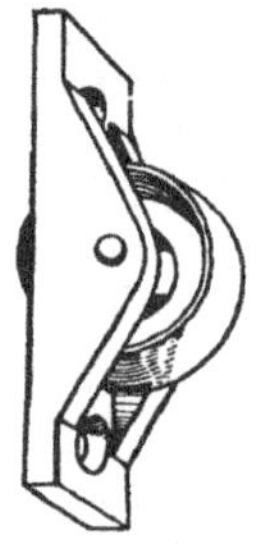

Fig. 218. Sash-roller. P. & F. Corbin.

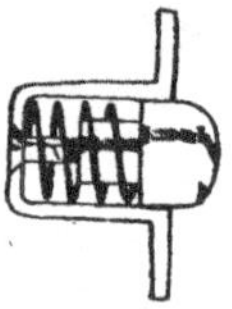

Fig. 219. Ayer's Sash-holder. Ayer's Sash-holder Co.

Sash-holders.

A sash-holder is something intended to keep unweighted sashes in any given position by friction on the jambs of the frame. Besides the two previous forms, there is a very simple device consisting of a wooden or vulcanized-rubber roller, Figure 218, mortised into the sash and bearing against the window-frame. There is also the "Ayer" sash-holder, Figure 219, better adapted to its purpose than any wheel could be, consisting of a round peg mortised into the sash and pressed constantly against the hanging-style by means of a strong spiral spring. Still another form is shown by Figure 220. This is mortised into the hanging-style, and consists of a small wooden cylinder, or roll, laid loosely in a slot and against a heavily milled surface which prevents it from rolling down. The milled plate is backed by a heavy steel spring.

Fig. 220. Byam's Sash-balance. Byam, Stewart & Baker.

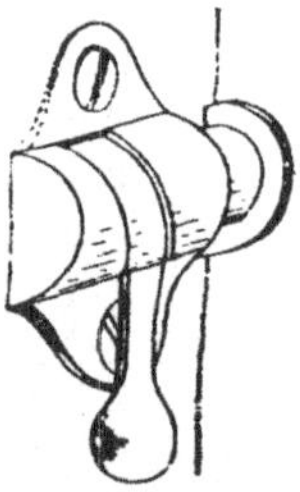

Fig. 221. Sash-fastener. Stoddard Lock & Mfg. Co.

When the sash is lifted, the roll turns easily in the slot; but, when the sash is being lowered, the wooden roll wedges between the milled-plate and the edge of the sash, offering a resistance in proportion to the stiffness of the spring. Chapter VIII.

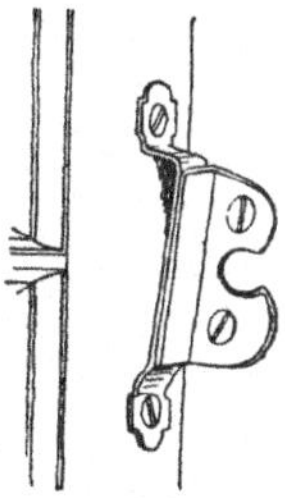

Fig. 222. Clauson's Anti-rattling Wedges. Sise, Gibson & Co.

Fig. 223. Ideal Anti-rattler. American Mfg. Co.

Figure 221 illustrates an appliance which can be used as a sash-fastener, but which is intended more especially to bind the sash so it will not rattle. It consists of a plain bolt sliding in a case attached to the window, with a lever working in an oblique slot, so that when it is drawn down, the bolt is pressed out with considerable force, and holds by friction against the jamb. Figure 222 shows another anti-rattling device, a metal surface on the sash which wedges against an inclined plate on the jamb; and Figure 223 is the simplest of all, a metal wedge, suspended by a small chain attached to the sash, which can **be inserted in the joint between the sash and the stop-bead.**

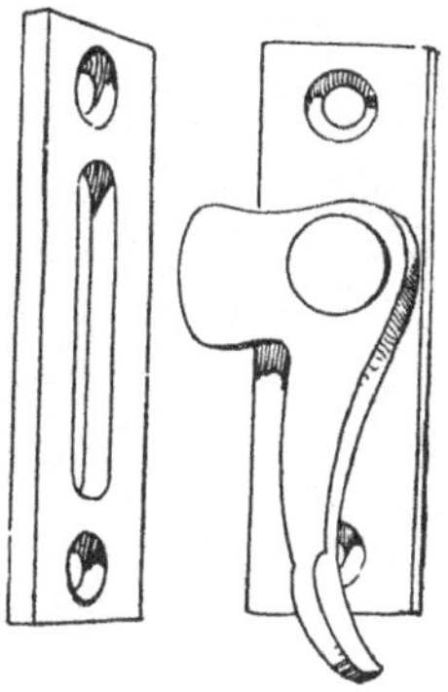

Fig. 224. Storm Sash-holder.

Chapter VIII.

The following table gives the average retail prices per dozen of the sash-fasts, etc., referred to in this chapter. The prices are for plain goods, when such are manufactured.

TABLE OF SASH-FASTS, ETC.

Fig.	Name.	Bronzed Iron	Brass or Bronze.
181	Yale & Towne sash-lock	—	$12.00
182	King sash-lock	$2.50	
183	Sash-fast, P. & F. Corbin	1.00	2.50
184	Sash-fast, Russell & Erwin	1.35	6.60
185	Judd sash-fast	1.38	4.00
186	Sash-fast, Norwich Co	1.15	3.30
187	Sash-fast, P. & F. Corbin	—	6.00
188	Metropolitan sash-fast	3.00	7.50
189	Boston pattern sash-fast	1.50	6.00[1]
—	Ladd sash-fast	2.30	7.00
190	Sash-fast, Hopkins & Dickinson	1.60	5.00
191	Sash-fast, Stoddard	1.00	4.50
192	Favorite sash-fast	1.00	3.50
193	Sash-fast, Yale & Towne	—	6.00
194	Morris sash-fast	1.68	4.20
195	Triumph sash-fast	1.68	4.20
196	Sash-fast, P. & F. Corbin	.88	—
197	Mathes's sash-fast	1.50	3.70
198	Payson's sash-fast	1.00	4.00
199	Ives' sash-fast	1.00	3.50
201	Sash-fast, P. & F. Corbin	.60	2.50
202	Ticket-office sash-fast	—	6.00
203	Sash-fast, E. Robinson	—	6.00
205	Security sash-fast	—	3.75
206	Shaw's sash-fast	—	5.00[2]
207	Davis sash-fast	3.60	5.85
208	Byam's sash-fast	1.00[5]	4.20
210	Timby sash-fast		1.00[6]
211	Atwell sash-fast	3.00	6.00[2]
212	Brown's window-lock	.60	.75[2]
213	Eccentric sash-fast	.12	—
214	Sash-lock	.33	—
—	Hammond's sash-springs	.70[4]	—
215	Window-spring, P. & F. Corbin		.58
216	Sweet's window-spring	1.25[3]	
217	Northrup's window-spring	1.25[3]	
220	Byam's sash-balance	7.20[3]	
222	Clauson's anti-rattler, per set	—	.18
223	Ideal anti-rattler	—	.60

Figure 224 shows a rude but effective storm sash-holder, often useful in holding in position the outside sashes of a window, which are intended to be removed in the summer time. The lever is screwed to the window-casing, and catches

[1] Price in Bower-Barffed iron, Yale & Towne Mfg. Co., same as in bronze.
[2] Nickel plated. [3] Plain iron only. [4] Japanned iron with silver-plated knobs.
[5] Japanned iron. [6] Face-plate only of bronze.

in the metal slot which is mortised into the edge of the sash. The upper edge of the lever is bevelled, so that when it is turned into the slot the sash is wedged tightly against the window-frame.

Window-Tightener.

There is also a very satisfactory and simple device consisting of a plain, hard-rubber button with a milled edge, which is screwed on to the bead and against the sash, so as to turn when the window is opened. This is known as "Patten's" window-tightener. It is manufactured by the Portsmouth Wrench Company, and retails at 4 and 5 cents per hundred. The same, or a much similar form, is manufactured by the Ayer's Patent Sash-Holder Company.

Stop-bead Fastener.

"Nelson's Perfect Fastener," is a name applied to a device for screwing the stop-bead to the window-frame; it consists of

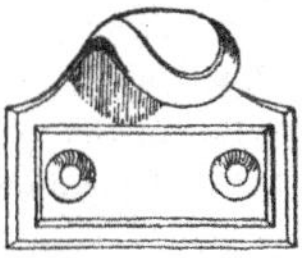

Fig. 225. Ordinary Sash-lift.

Fig. 226. Byam's Sash-lifter. Byam, Stewart & Baker.

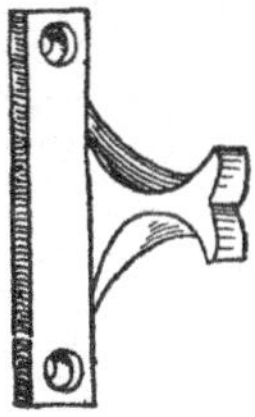

Fig. 227. Sweet's Reversible Sash-lift. Stanley Works.

metal eyelets which are sunk into the bead, with an elliptical instead of round hole to receive the screw. As the longest dimension of the screw hole is in a horizontal direction, the bead can be set so as to permit of a play of quite a quarter of an inch. Such a contrivance must prove a boon to those who are suffering with windows which bind in summer and rattle in winter. The fasteners cost from $2 to $3.25 per gross, including either round or flat headed screws.

Chapter VIII.

SASH–LIFTS.

Sash-lifts.

Sash-lifts are often omitted from architects' specifications, though they are usually very desirable, and when properly applied, will save a great deal of wear on the sash. The common form of lift is shown by Figure 225. A form which is not quite as convenient to use, though sometimes preferred, is the flush lift, the type of which is similar in the main to Figure 229. Both of these are intended to be attached to the bottom-rail of the lower sash. Figure 226 is a species of sash-lift which is applied over the mouldings of the sash-styles, in the shape of a concave strip of metal, with shoulders at intervals. Figure 227 is in the form of an angle-iron, likewise secured to the sash mouldings, and Figure 228 shows a form which can be applied either over the mouldings or on the face of the sash.

Fig. 228. Wigger's Sash-lifters. Brainerd & Co.

Sash Lifts and Locks.

Figures 229, 230, 231 and 232 show four styles of combined

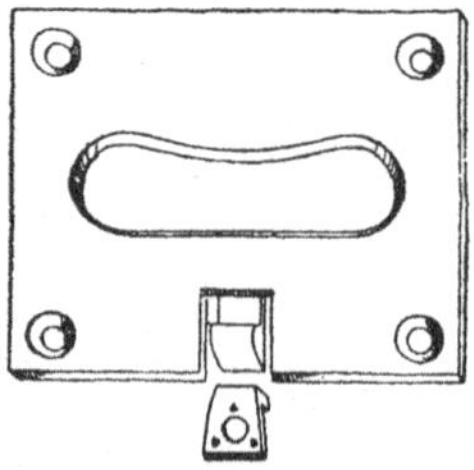

Fig. 229. Flush Sash Lift and Lock. Ireland Mfg. Co.

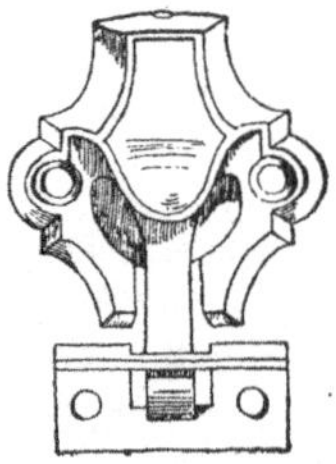

Fig. 230. Sash Lift and Lock. Ireland Mfg. Co.

sash lift and lock. No. 231 works by gravity. The others are each fitted with a spring which forces out the catch so that it will hold on the sill-plate or catch. Figure 233 shows a very practical form of sash-lift so arranged that by pressing down the lever handle a great lifting power can be ap-

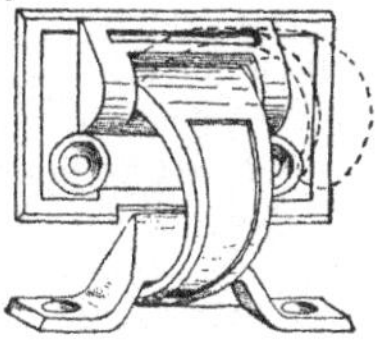

Fig. 231. Sash Lift and Lock. Russell & Erwin.

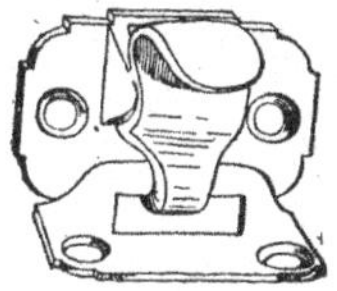

Fig. 232. Sash Lift and Lock. P. & F. Corbin.

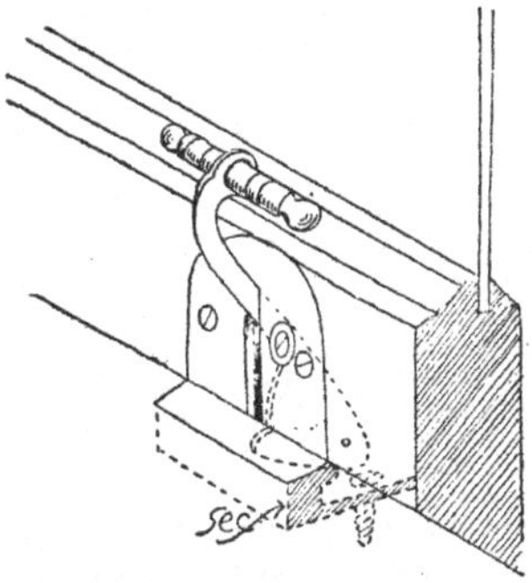

Fig. 233. Anderson Sash-starter.

piled, sufficient to start and lift a heavy sash, or to start a sash that has become wedged or frozen in. After the sash is

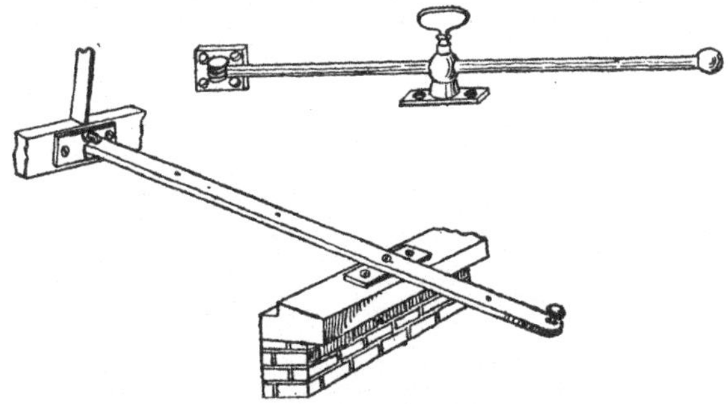

Fig. 234. Sash-openers.

started, the lift is in the right position to raise it with ease, and when the window is closed the lift can serve as a lock, by

Chapter VIII. turning a button which will prevent the lever handle from moving.

Figure 234 shows two of the common forms of casement or hinged sash-openers. These are best made to order, but can be had in stock, 12, 15 and 18 inches long, and cost from 50 cents to $1 each, depending on the metal and the finish.

TABLE OF SASH-LIFTS.—AVERAGE RETAIL PRICES PER DOZEN.

Fig.	Name.	Bronzed or Japanned.	Brass or Bronze.
225	Common sash-lift	$ .30	$1.35
	Flush sash-lift	—	2.00
226	Byam's sash-lift	.30	.90
227	Sweet's sash-lift	.25	—
228	Wiggers's sash-lifts	1.50	2.25
229	Sash lift and lock, Ireland Mfg. Co.	2.70	4.80
230	Sash lift and lock, Ireland Mfg. Co.	2.10	4.20
231	Sash lift and lock, Russell & Erwin	1.00	4.00
232	Sash lift and lock, P. & F. Corbin.	.80	3.20
233	Anderson's sash-starter	6.00	9.00

CHAPTER IX.

SHUTTER-FIXTURES.

SHUTTER FASTS AND LOCKS.

Chapter IX.

Outside Blind-fasts.

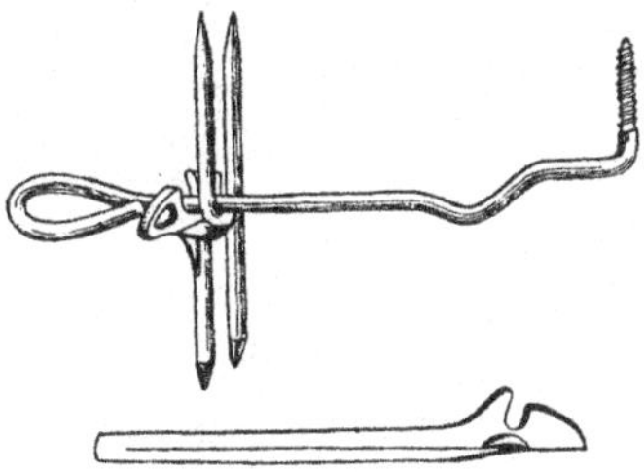

Fig. 235. Spring Wire Blind-fast.

THE appliances for securing outside blinds, though in some cases combined directly or indirectly with the blind hinges, are more often distinct fixtures, acting independently of the blind attachments. The usage in regard to shutter fasts and locks varies in different portions of the country. In the West there seems to be a willingness to accept considerable complication in the devices, whereas the standard Eastern goods are mostly very simple; though, of course, this distinction is not a rigid one, by any means. The West, however, is rapidly developing new ideas and fresh combinations, in hardware no less than in nearly every other department of mechanical industry, and special patent forms seem to be more naturally expected there than elsewhere. This does not imply that the Eastern cities are united in the usage of particular forms, for places as near to each other as New York, Providence and Boston employ different forms, as will be seen later on.

Chapter IX.

Wire Fasts.

Figure 235 will serve to illustrate one of the most common forms of shutter or blind fast, consisting of a tempered steel rod, or wire, one end of which is cut with a thread and screws into the under side of the blind, while the other end is held by a staple. The rod is bent so that the loop is kept away from the blind, and the elasticity of the metal enables it to spring

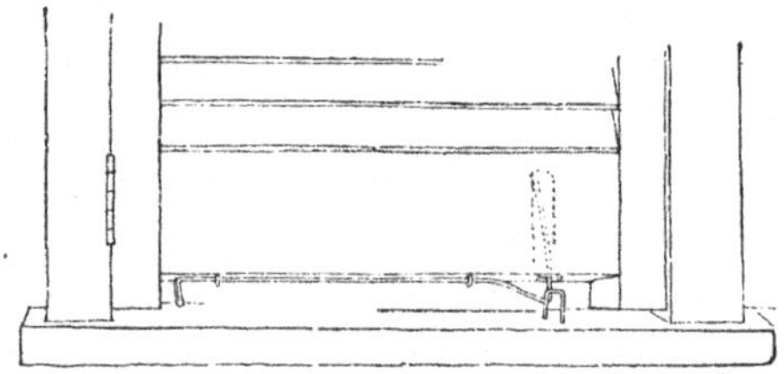

Fig. 236. Folsom's Blind-fast. J. H. Hiller.

into the malleable-iron catch on the sill, or on the outside of the wall. The well-known "Shedd" blind fastener is practically the same as this, except that the rod is bent in a complete twist to gain the elasticity, and a common screw takes the place of the threaded end. The same form is made, with slight variations, by several of the leading manufacturers.

Figure 236 shows the only form of wire blind-fast which allows one to close the blind without leaning out of the window, or in any way lifting the shutter to release it from the back catch. It consists of a steel wire, bent as shown by the figure, but carried as far back towards the hinge as the hanging-style of the blind will permit. To release the blind, the fastener is simply pulled inward. Any form of back catch may be used. For the sill-catch a wide staple is used, which is set on an angle to the blind, so as to force the spring back and permit it to catch behind the staple. This fastener has but very recently been put on the market.

Gravity Fast.

The blind-fast shown by Figure 237 works entirely by gravity. It consists of a bent lever, working in a mortise cut through

Chapter IX.

the bottom rail of the blind, pivoted so that one arm protrudes above the top of the rail, while the other catches over an ordinary hook on the sill or against the wall. Lugs on the end of the horizontal lever arm catch on a thin plate screwed to the under side of the rail and prevent the fast from dropping too low or being lifted too high. This fast is made of coppered malleable-iron, and seems like a very satisfactory article.

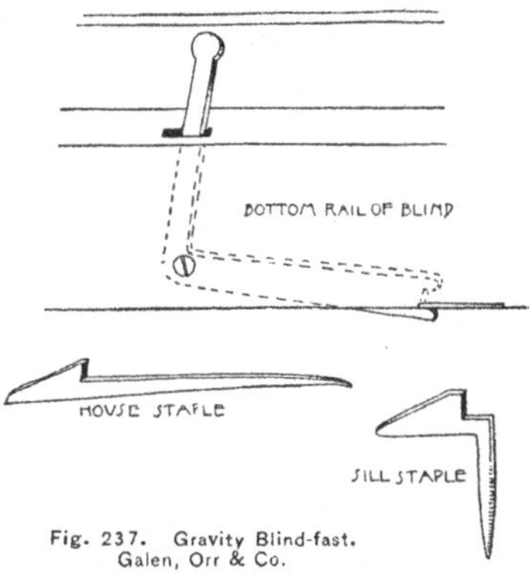

Fig. 237. Gravity Blind-fast. Galen, Orr & Co.

Figure 238 is an older style of blind-fast, on essentially the same principle as Figure 235; using, however, a flat bar instead of the spring wire. This form requires a little more work in adjustment. It is designated peculiarly as the "Boston" pattern blind-fast. The so called "New York" pattern is illustrated by Figure 239. The action of this fast will be better appreciated when it is remembered that in New York, the blinds are usually hung flush with the outer casing, and the sill is rebated so that the bottom of the blind strikes against the upper rebate. The latch is hinged on the inner plate, the weight of the long arm keeping the inner hook thrown up. The sill-staple is driven perpendicularly, while the back catch is screwed

Boston pattern.

New York pattern.

Fig. 238. Boston Pattern Blind-fast. Stanley Works.

Chapter IX.

horizontally into the wall. The Stanley Works also has what is designated as the "Providence" style of blind-fast. This is exactly the same as the "New York" pattern, except that the

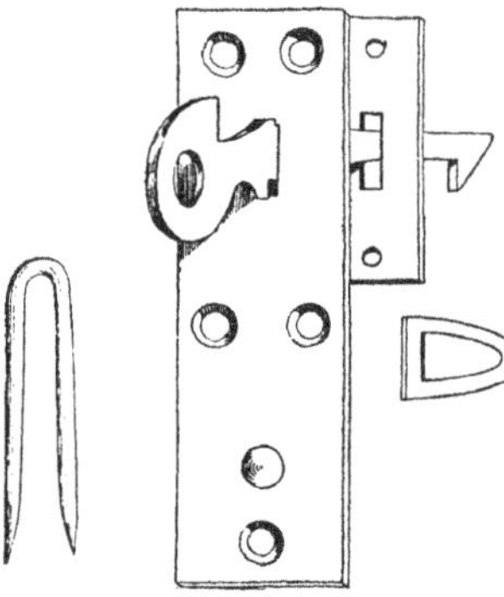

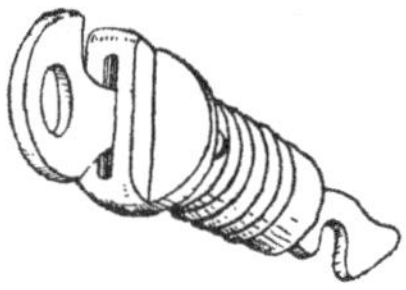

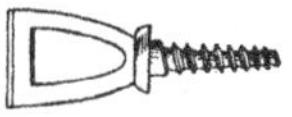

Fig. 240. Standard Screw Blind-fast. Stanley Works.

Fig. 239. New York Pattern Blind-fast. Stanley Works.

inner hook catches over instead of under the sill-staple, and is shaped like the back catch of Figure 235, inverted.

Screw Fasts.

Figure 240 shows a form of blind-fast which is screwed bodily through the blind, catching on sill and wall staples in the same manner as the preceding styles. A flat spring inside of the case keeps the inner hook constantly pressed up and against the sill-staple. A variation of this same pattern is made which acts by gravity, the catch working in an oblique slot in such a manner that the weight of the outer catch forces the inner catch always against the sill-staple. Figures 241 and 242 illustrate two forms of fasts which are screwed to the

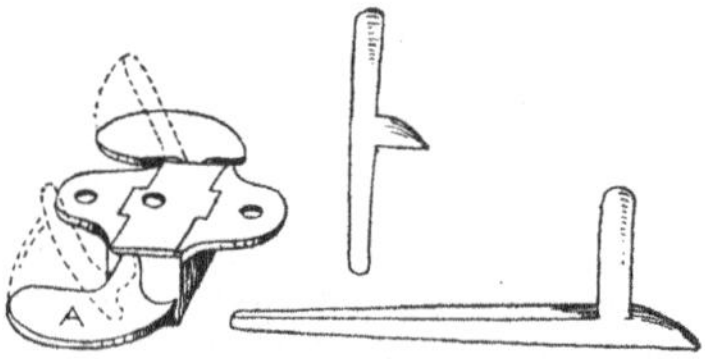

Fig. 241. Security Blind-fast. Stanley Works.

Chapter IX.

under side of the blind. The former acts entirely by gravity. The lobes, ***A A***, are connected through the case, and are counterbalanced so as to always drop to the position shown. When the blind is closed, the lobe strikes against the sill-pin

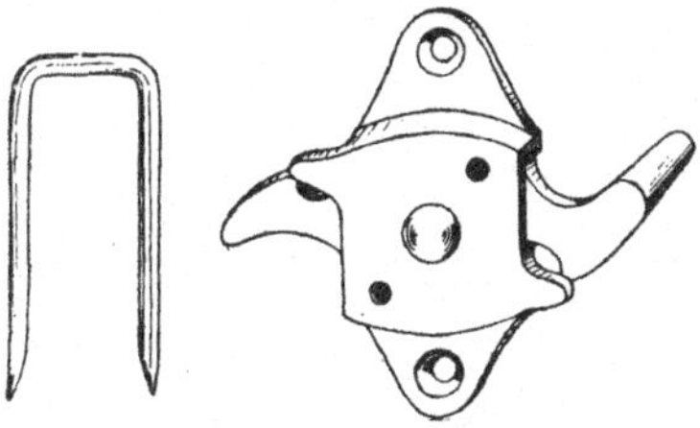

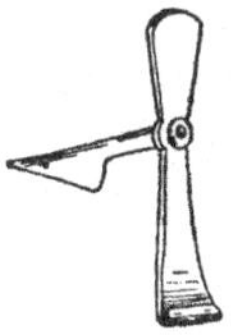

Fig. 243. Turn-buckle A. G. Newman.

Fig. 242. Lock Blind-fast. Stanley Works.

and is forced up as shown by the dotted lines, dropping so as to catch inside of the pin. Figure 242 has a concealed spring, to force the action of the lever.

The foregoing styles of blind-fasts are intended to be used on wooden buildings, but with some modifications in the sizes might also serve for brick buildings. In New York, it is

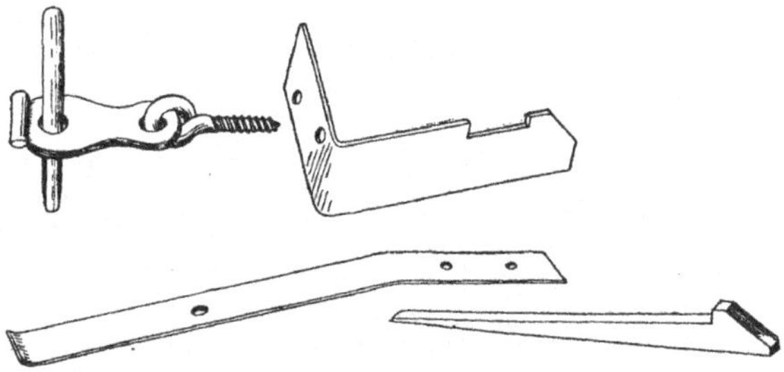

Fig. 244. Drop-and-Pin-fast. Stanley Works.

Turn-buckles.

customary to use some form of turn-buckle, Figure 243, which is driven into the joints of the brickwork, the cross-piece being

Chapter IX.

free to turn, but hanging naturally in a vertical position by reason of the greater weight of the longer arm. Turn-buckles of a slightly different shape are sometimes used, also, for wooden buildings.

All of the foregoing are, in a certain sense, automatic; that is to say, the blind, if flung open or shut will stay in position, requiring no special adjustment. Figure 244 is a form of drop-and-pin fast, much used in some cases, consisting simply of a plate secured to the blind by a screw-eye, perforated with a hole to fit over the pin driven into the sill. For holding the blind open, a back catch is made as shown by the figure, which locks with a plain, flat spring, screwed to the under side of the blind. The figure also shows the form of back catch used for brick buildings.

Fig. 245. Seymour's Blind-catch. P. & F. Corbin.

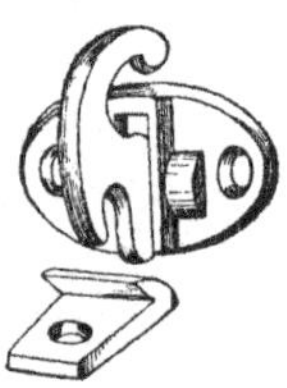

Fig. 246. Blind-catch. Shepard Hardware Co.

Fig. 247. Seymour's Blind catch and lock. P. & F. Corbin.

Figures 245 and 246 show two very simple forms of blind-catch serving only to keep the blind closed, and generally used with some form of turn-buckle to hold the blind open. Figure 245 works with the aid of a small spring, as shown; Figure 246 works entirely by gravity. There are several varieties of each of these forms in the market. The catch shown by Figure 247 acts in the same manner as Figure 245, but has, in addition, a locking-lever, operated by a key, which secures the catch so that the blind cannot be opened.

BLIND-FASTENERS.

There are a number of forms of blind-hinges, which have been previously described in the chapter on hinges, that in a measure serve as blind-fasteners, keeping the blind either open or shut. They are all perfectly simple in their operations, and it is difficult to discriminate between them. The common fault with them all is in the difficulty of opening and closing the blind. With most of the forms of patent self-locking blind-hinge, the blind must be raised from its seat in order to be swung around. With the blind-fasts previously described in this chapter, it is necessary to lean far out of the window to release the catch from underneath. Figure 248 shows a device intended to overcome the difficulties of both styles. It consists simply of a lever attached to the blind, and hooking into a plate screwed onto the jamb of the window.

Fig. 248. Rochester Blind-hinge. Byam, Stewart & Baker.

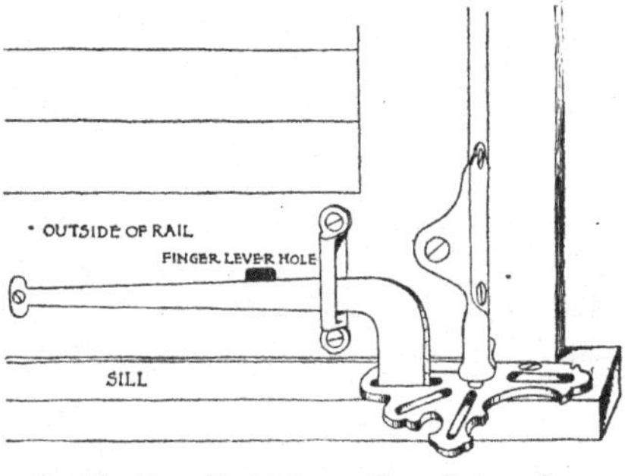

Fig. 249. Tenon Blind-fastener. Tenon Fastener Co.

It is only necessary to lift the end of the lever in order to swing the blind shut. The advantages are that in closing, no lifting of

Chapter IX.

the blind is necessary; there is no danger of throwing it off the hinges, and no chance of pinching the fingers or bumping the head.

Blind-adjuster.

There are several other devices intended to hold the blind, either shut or open. Figure 249 illustrates the "Tenon" blind-fastener, which consists of a bent, flat bar, attached to the outside of the blind and catching in slots cut in a plate which is secured to the sill, so that the blind can be held either open or shut, or in either of two intermediate positions. The bar is lifted by means of a lever on the inside of the blind. This fixture does away with the ordinary bottom hinge, substituting therefor a pivot working in the locking sill-plate. A blind-fastener of this description is especially suitable for bay-windows, or any place where the blinds cannot open clear back. Being placed on the outside of the blind exposes it to the weather to an undesirable degree, though it is made of Bower-Barffed iron to prevent it from rusting.

Fig. 250. Excelsior Blind-adjuster. Russell & Erwin.

Fig. 251. Washburn's Blind-adjuster. B. D. Washburn.

Figure 250 is a very simple form of bar blind-adjuster, the bar being attached to the blind, and held in position by the action of the thumb-screw on the jamb; Figure 251 shows a variation of the same principle, consisting of a bar which fits into the sockets at several points on the sill, enabling the blind to be held in several different positions. The action of the adjuster will readily be understood by the figure. Zimmerman's Blind-fast is on practically the same principle as this. Chapter IX.

The difficulty with the two foregoing patterns is, that they do not hold the blind perfectly rigid, and the rods are likely to get in the way, specially as the rods and sockets take up considerable space on the sill. There is but little practical advantage in having a fixture which permits of the blind being open at various degrees, for, as a rule, most people prefer to have their blinds either entirely open or entirely shut.

SHUTTER-WORKERS.

The desire to open and operate blinds without opening the window has led to the invention of several devices which are worked by rods passing entirely through the frame of the house and attached to the blind. It is not altogether easy to under-

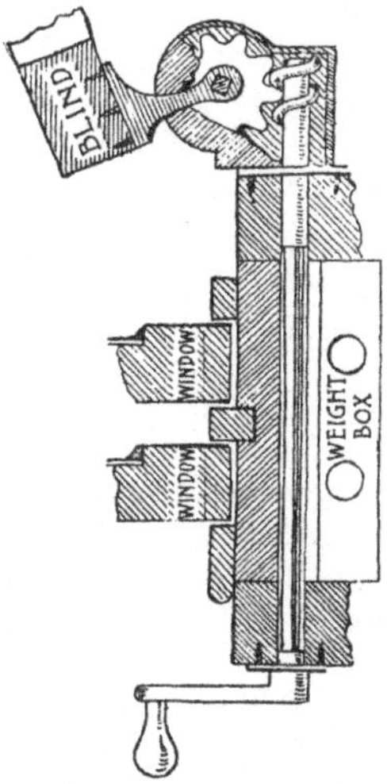

Fig. 252. Mallory's Shutter-worker. Frank B. Mallory.

Fig. 253. Brown's Shutter-worker. Ireland Mfg. Co.

stand why such devices are used so little, but it must be

Chapter IX.

admitted, that all of those now in the market are more or less clumsy. Still, the idea is an excellent one, and if there were greater demand for such appliances, undoubtedly better ones would be put before the public. The shutter-worker of this description that is the most natural in its adjustment is illustrated by Figure 252. This consists simply of a rod, at the

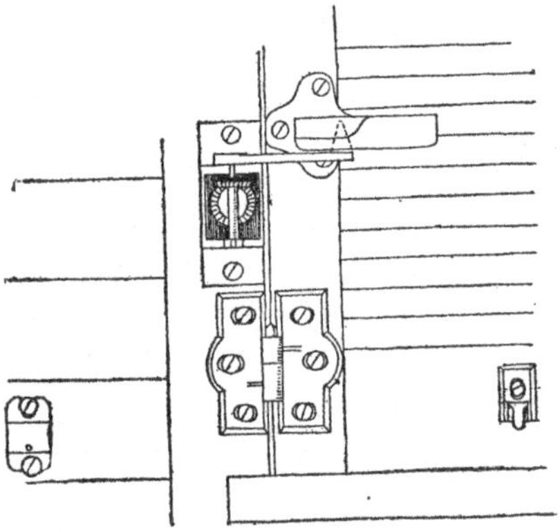

Fig. 254. Automatic Shutter-worker. Dudley Shutter-Worker Co.

end of which is a thread working against a cog-wheel forming a part of the bottom hinge of the blind. On account of the slowness of pitch of the thread, it is very difficult to move the blind from the outside, but the leverage is sufficiently strong to enable one to easily open the blind from within by turning the crank.

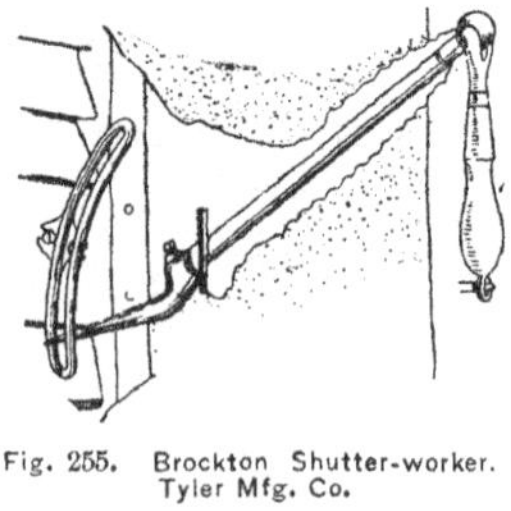

Fig. 255. Brockton Shutter-worker. Tyler Mfg. Co.

A very similar appliance to this is the Brown shutter-worker, Figure 253, in which the thread on the spindle works into teeth on the bottom of a plate forming a part of the lower shutter-hinge.

Chapter IX.

The Automatic Shutter-worker, Figure 254, combines the good points of several other devices, and is somewhat more complicated than either of the preceding. Two cog-wheels gear into each other. The shaft of one wheel is carried through the wall and can be operated by a crank or handle inside the house. The shaft of the other wheel turns a crank, or bent lever, the end of which works in a slide attached to the face of the blind. The cog-wheels are encased in an iron box, which is shown partly removed in the figure, in order to illustrate the workings. Aside from the number of parts, which is no very great objection, this shutter-worker has a great deal to recommend it. It is strong and compact, and can act on the shutter with such force that, it is asserted, a child can work the blind with it in a high wind. It has the advantage of permitting the blind to be removed without disturbing the fixtures.

One of the simplest acting shutter-workers, is illustrated by Figure 255. This is very ingenious in its idea, consisting of a straight rod set on an angle, with a bent lever on the end

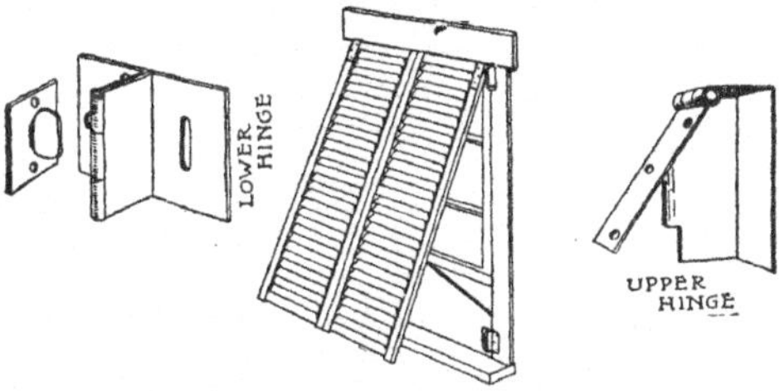

Fig. 256. Tucker Awning Blind-hinge. Hamblin & Russell Mfg. Co.

working in a curved slot or catch secured to the outer face of the blind. This shutter-worker will lock the blind as securely as any door can be locked, the handle of the rod being dropped down onto the pin as shown by the lock.

The company which manufactures the Brockton shutter-worker has bought up the patents of the Prescott shutter-worker, which was somewhat on the same principle.

Chapter IX.

There are some other shapes in the market; but practically very few which embody ideas essentially different from those described.

AWNING–HINGES.

Awning-hinges.

Awning-hinges might more properly be considered with common blind-hinges, but they are included in this connection, as they are in a measure blind-adjusters, permitting the blind to be opened part way. The writer has been able to find only two forms in the market. The simplest is shown

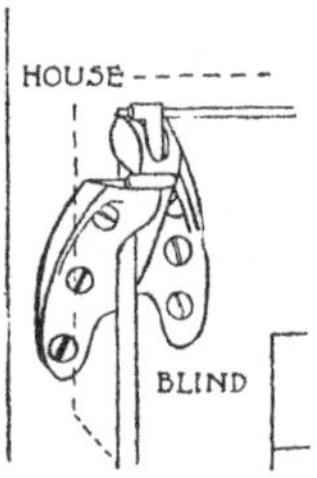

Fig. 258. Byam's Blind-slat Adjuster. Byam, Stewart & Baker.

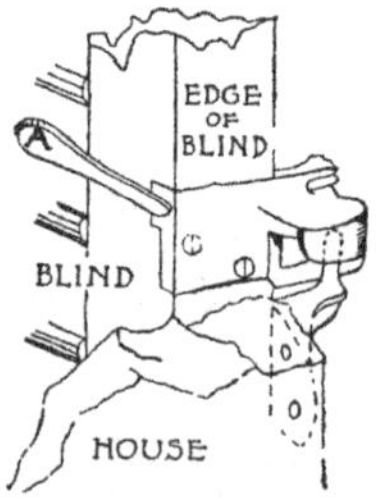

Fig. 257. Automatic Blind-awning Fixtures. F. O. North & Co.

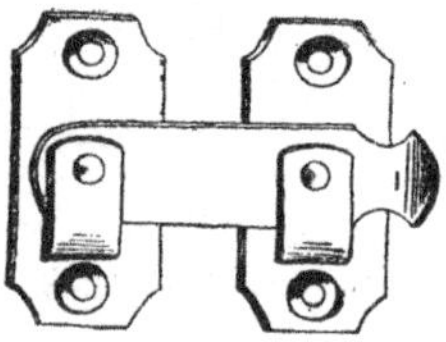

Fig. 259. Shutter-bar.

by Figure 256. This consists of a double-acting hinge for the upper portion of the blind, a lower hinge being screwed to the jamb and fastened to the blind only by a turn-button.

The other form of awning-fixture is more commonly used about Boston, Figure 257. The upper hinge is so made as to work in either direction, while the lower hinge consists of a cup fitting over a pin screwed to the jamb. A small catch

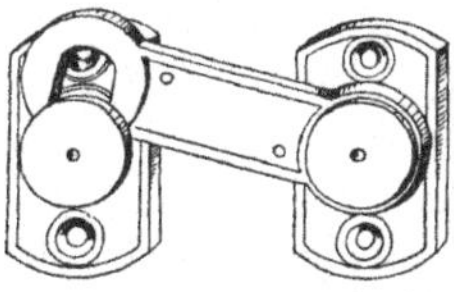

Fig. 260. Shutter-bar.

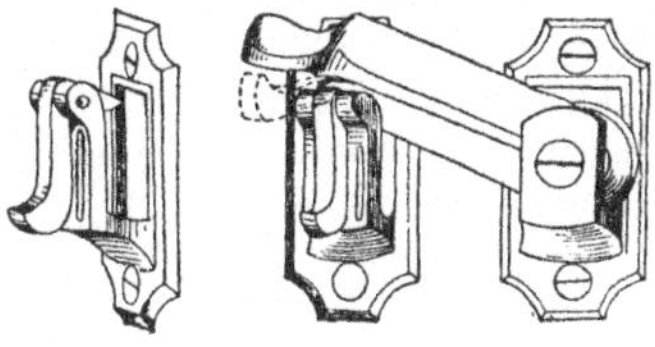

Fig. 261. Morris' Self-locking Shutter-bar. Ireland Mfg. Co.

keeps the blind from pushing out when the hinges are to be used in the ordinary manner, but is readily lifted when the blinds are to be pushed out from the bottom. The fixtures are sold with side-bars to hold the bottom of the blind away from the building, and with a centre cross-bar which permits the blinds to be opened part way in the ordinary manner, and secured. The description and the figure might seem to imply a somewhat complicated arrangement; but the fixtures work very simply, and seldom fail to give satisfaction.

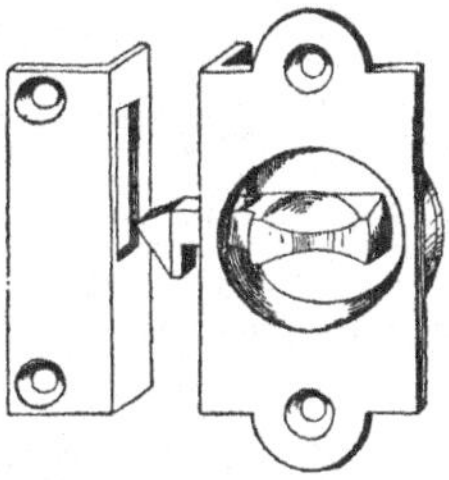

Fig. 262. Sliding Shutter-hook. P. & F. Corbin.

Slat-adjuster.

Figure 258 shows a form of slat-adjuster intended to be operated by a key from the inside of the house without opening the window. The slats are connected with an eccentric which is turned by the key, so that the slats can be either raised or lowered as desired.

HARDWARE FOR INSIDE SHUTTERS.

Shutter-bars.

There is little to be said as regards fasts or locks for inside shutters. The shutters themselves are usually provided with knobs of some description, with porcelain or metal heads secured in position by a screw. The shutters are also pro-

Chapter IX

vided with some form of latch or bar, of which Figure 259 is a very simple type. Figure 260 shows a more elaborate form, for inside work. There are, of course, many variations of these forms. A few of the hardware manufacturers have been making self-locking shutter-bars, in which the cross-bar is secured by some form of auxilliary lever or cam. Figure 261 illustrates one variety. There is, however, but little demand for such appliances.

For sliding shutters a bar like that shown by Figure 260 may be employed, but there are also several varieties of mortise hooks, Figure 262, which work with a spring, and are rather preferable for most cases.

The retail prices of the foregoing blind and shutter fixtures are as follows:

TABLE OF SHUTTER-FIXTURES.—PRICES PER WINDOW, WITH TWO SINGLE-FOLD BLINDS.

Fig.	Name.	Price.
235	Stanley's wire blind-fast	$ .07
236	Folsom's shutter-fastener	.08
237	Boston pattern blind-fast	.08
238	Gravity blind-fast	.13
239	New York pattern blind-fast	.09
240	Standard screw blind-fast	.09
241	Security blind-fast	.08
242	Lock blind-fast	.08
243	Turn-buckles or drop-buttons for brick	.10⅓
	Turn-buckles or drop-buttons for wood	.08⅓
244	Drop-and-pin fast	.08
245	Seymour's blind-catch	.12½
246	Shepard blind-fast	.04
247	Seymour's blind catch and lock	.21
248	Rochester blind-hinge	.17
249	Tenon blind-fastener	.45
250	Excelsior blind-adjuster, galvanized	.55
251	Washburn's blind-adjuster,[1] galvanized, 10-inch bar	.56
252	Mallory's shutter-worker, with hinges and handle	1.25
253	Brown's shutter-worker, japanned	.85
254	Automatic shutter-worker, with hinges and handle	.75
255	Brockton shutter-worker	.65
256	Tucker awning blind hinges[1]	.87
257	Automatic blind awning fixtures[1]	.75
258	Byam's blind slat-adjuster	.25
259	Shutter-bars — bronzed-iron, 2-inch, per dozen	.84
260	Shutter-bars, bronze, 2-inch, per dozen	1.35
261	Morris' self-locking shutter-bar, bronzed-iron, 2-inch, per doz.	.60
	Morris' self-locking shutter-bar, bronze, 2-inch, per dozen	3.00
262	Sliding shutter-hook, bronze, each	.75

[1] For wooden house.

CHAPTER X.

Transom and Skylight Fittings.

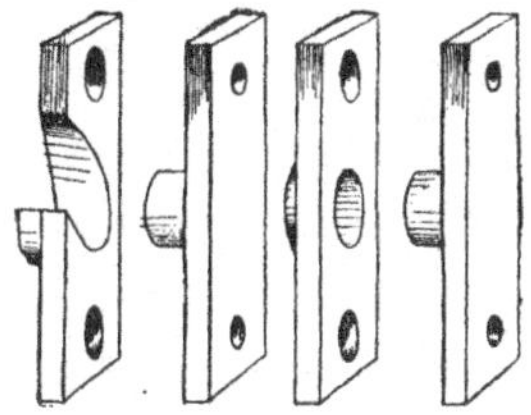

Fig. 263. Sash-centres or Transom-hinges.

TRANSOMS are hung by common butts at the top or bottom, or are pivoted in the centre horizontally. The ordinary hinges used for transoms are such as might be used for any purpose. These have been previously discussed. Sash centres or pivots are commonly mortised into the frame and into the sash. Figure 263 is the ordinary form. Figure 264 is another variety in which both pivots are exactly alike. This is secured in place by first fasten-

Chapter X.

Transom-centres.

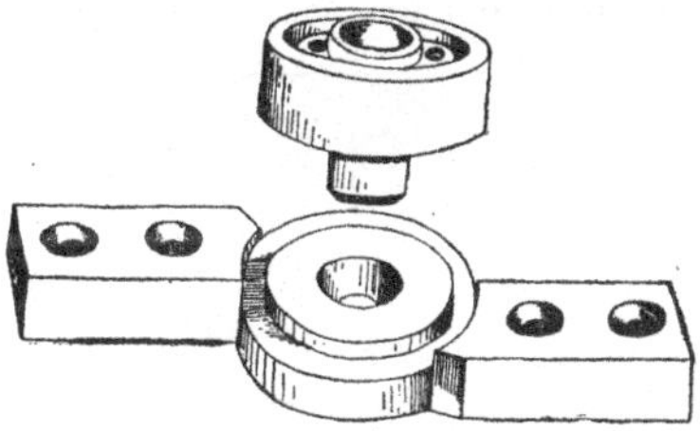

Fig. 264. Sash-pivot. A. G. Newman.

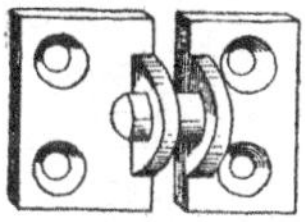

Fig. 265. Surface Sash-centre. P. & F. Corbin.

ing the round part of the pivot at entire end of the sash, and securing one socket-piece to the sash-frame. The other socket

Chapter X.

is then fitted to the opposite pivot, and the sash placed in position and turned at right angles, thus uncovering the second

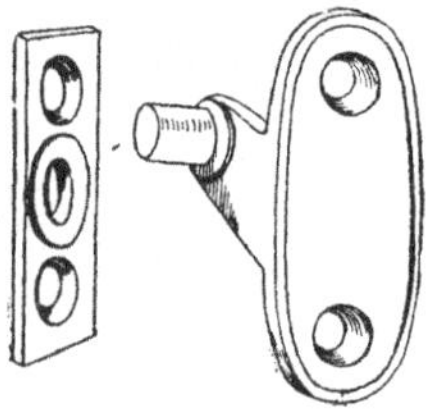

Fig. 266. Surface Sash-centre. J. F. Wollensak.

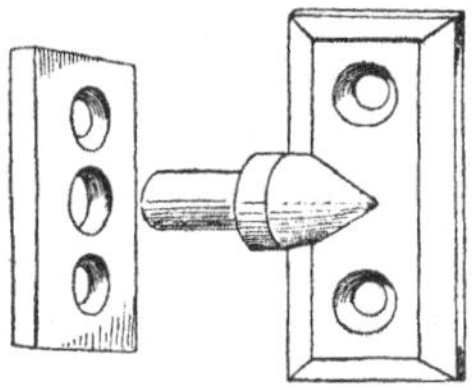

Fig. 267. Transom-pivot. Hopkins & Dickinson Mfg. Co.

socket, so that it can be screwed to the jamb. This form is claimed to be tighter and consequently more secure against draughts than the ordinary style.

Instead of either of the foregoing, it is sometimes desirable

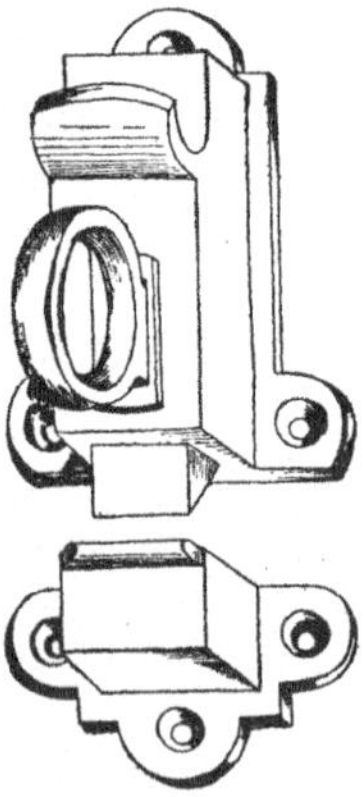

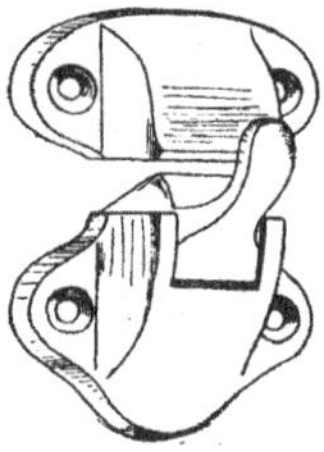

Fig. 269. Cupboard and Transom Catch. Ireland Mfg. Co.

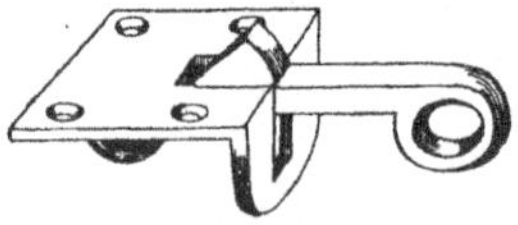

Fig. 268. Transom-catch. A. G. Newman.

Fig. 270. Transom-catch. J. B. Shannon & Sons.

to use pivots which do not turn on the line of the centre of the sash. Figure 265 illustrates a form which can be used in such a case, both pivot and socket being planted on the faces of the sash and the frame. Figure 266 and Figure 267 are other varieties sometimes met with. The different uses for which these various forms are applicable will readily be appreciated; the first being for a case in which the jambs and the sash are flush; the second, one in which the transom sets out from the

Chapter X.

Fig. 271. Transom-lift. J. F. Wollensak.

Fig. 272. American Transom-lift. American Mfg. Co.

jamb; and the third, one in which the jamb is too deep, or the transom set too far in to permit of the hinges being applied to the face of the jamb.

Transom-catch.

Transoms are usually provided with some form of spring catch to hold them closed. Figure 268 is a direct catch, the latch being secured to the transom. This is for use when the jamb and the sash are flush. Figure 269 is a transom-catch

Chapter X.

worked on a little different principle from the foregoing. The same form is also used for cupboards. This, as well as the first, is fastened onto the face of the transom. Figure 270 shows a transom-catch intended to be mortised into the edge of the transom, either at the top or the bottom.

Transom-lifts.

In the best work it is customary to provide some appliance for lifting the transom and holding it in position. With the ordinary catches previously described, a chain is attached at one side of the transom, permitting it to be opened down from

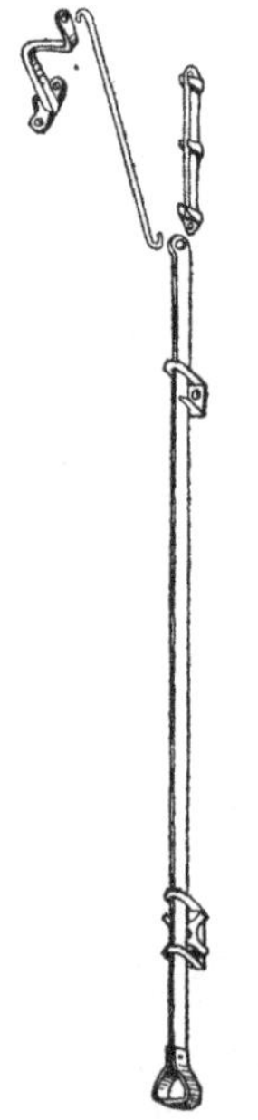

Fig. 273. Steller Transom-lifter. Russell & Erwin.

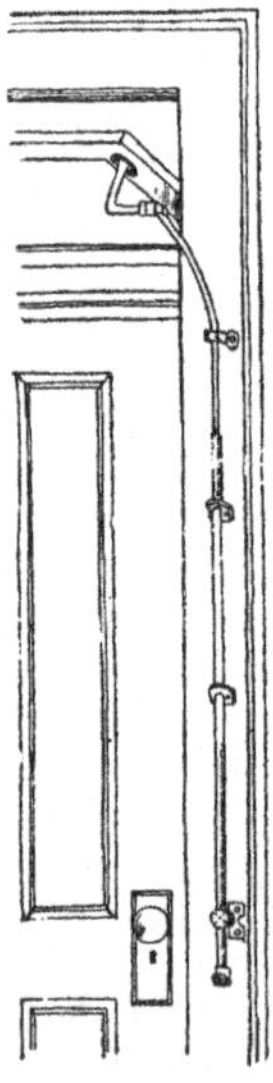

Fig. 274. Overell's Transom-lifter. P. & F. Corbin.

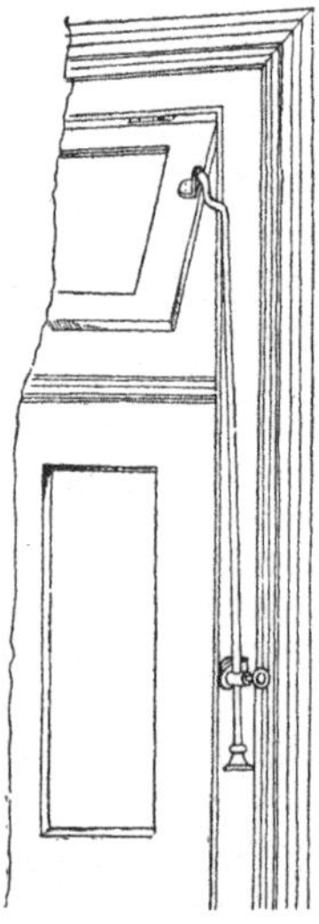

Fig. 275. Excelsior Transom-lifter. Russell & Erwin.

the top a certain distance only; but it is much more convenient to have some appliance that will permit the transom to be opened in either direction, and will hold it securely. The

Chapter X.

most popular, and one of the best known is the Wollensak transom-lifter, Figure 271. This consists of a straight rod with a hinged arm attached to it, the arm being secured to the edge of the transom, while the rod works up and down in a series of rings, being held at any given height by turning a button at the bottom binding on the rod. These are made for transoms either pivoted at the centre and swinging down, or pivoted and swinging up, or hinged at either top or bottom. Figure 272 shows another form, made by the American Manufacturing Company. The rod in this case is replaced by a flat bar, the attachment otherwise being essentially the same as in the previous example. The bar is notched at the bottom on the inner edge, and a catch on the lower guide-ring locks the bar

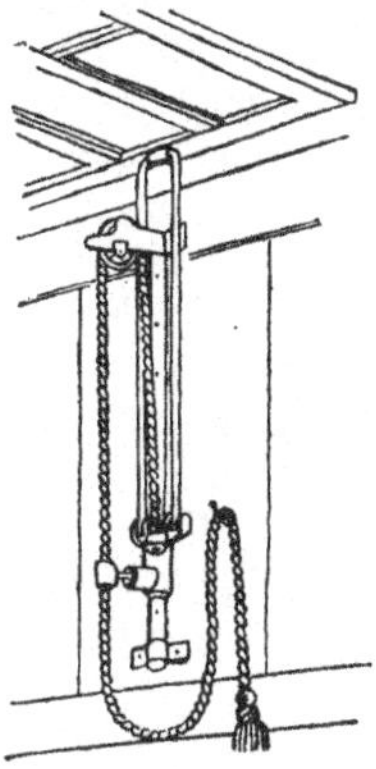

Fig. 276. Skylight Lift and Lock. J. F. Wollensak.

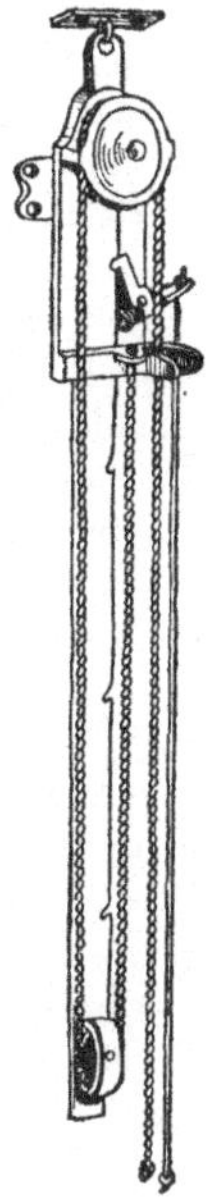

Fig. 277. Skylight-lift. S. L. Hill.

at any height. Figure 273 is another form manufactured by Russell & Erwin. In this case the bar is held in position by

Chapter X.

turning the button at the bottom. This transom is provided with a supplementary set of guides at the top, so that in shoving up the bar there will be no opportunity for the weight of the transom to deflect it sidewise. Figure 274 shows a form of transom-lifter manufactured by P. & F. Corbin, consisting of a straight rod, with a long, flexible steel attachment at the top. The rod is secured at any height by a turn-button in the same manner as in the first example, while the flexibility of the upper portion of the rod permits the transom to turn at any angle. There is yet another form, Figure 275. This consists of a single rod attached directly to the transom, and secured on the jamb only by a single turn-button, near the bottom. This turn-button is placed at an angle in such a manner as to allow considerable side-play on the rod, and so permit of the deflection necessary for opening the transom.

Skylight-lifters.

Closely allied to the transom-lifters are those which are used for skylights. Figure 276 shows a form manufactured by Wollensak. This consists of a double bar attached to a socket working on a slotted bar. The socket has attached to it a spring-catch which slips into the slots on the bar. The rope passes through the socket up over a pulley, and down through an eye in the end of the spring-catch. By pulling the bar out away from the socket, the spring-catch is released and the socket, and with it the skylight may be lifted or lowered, the spring-catch shutting back when the horizontal strain on the rope is relaxed. This is made in two sizes, with a length of eighteen inches each. Figure 277 shows another form of skylight-lifter in which a ratchet on the side of the upper framework fits into slots on the edge of the lifting-rod, the ratchet being worked by a separate cord. The ratchet is fitted with a spring to keep it in position.

The following table gives the retail prices of the goods described in this chapter.

TABLE OF TRANSOM-FITTINGS.

Fig.	Description.	Price.
263	Sash-centres, japanned, per dozen pairs	$ .62
	Sash-centres, brass, per pair	.62
264	Sash-pivots, 1¾-inch brass or bronze, per set	2.00
	Sash-pivots, bronzed-iron, per set	1.00
265	Surface sash-centres, P. & F. Corbin, brass, per set	4.00
266	Surface sash-centres, Wollensak, bronze No. 4, per set	1.00
	Surface sash-centres, Wollensak, bronzed-iron, per set	.17
267	Surface sash-centres, Hopkins & Dickinson, bronze, per set	.95
268	Transom-catch, per dozen	15.00
269	Transom and cupboard catch, bronze, per dozen	7.50
	Transom and cupboard catch, bronzed-iron, per dozen	.50
270	Transom-catch, bronze, per dozen	5.00
271	Wollensak's transom-lifter, bronzed	1.20
	Wollensak's transom-lifter, nickel-plated	2.50
272	American transom-lifter, coppered	1.10
	American transom-lifter, nickel-plated	3.15
273	Steller's transom-lifter, bronzed-iron	.53
	Steller's transom-lifter, bronze	2.50
274	Overell's transom-lifter, bronzed	.50
275	Excelsior transom-lifter, bronzed	.55
276	Wollensak's skylight-lifter, No. 12, each	2.00
277	Hill's skylight-lifter, each	1.50

Prices for transom-lifters are for a medium 4-foot rod and for a single fixture.

Chapter XI.

CHAPTER XI.

Locks.

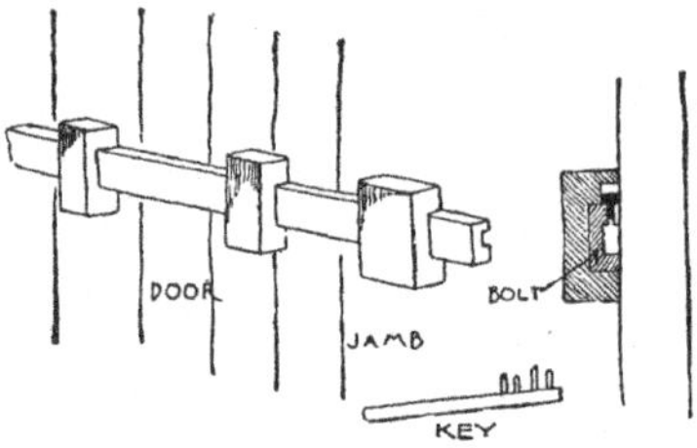

Fig. 277*a*. Egyptian Wooden Lock.

ANY one who should visit the mediæval museums of Europe, and should chance to see among the curiosities of iron-work some of the elaborately wrought and apparently intricate locks of the fourteenth, fifteenth and sixteenth centuries, would hardly think of comparing those unwieldly and cumbersome devices with the locks that are turned out in such quantities by our best modern manufactories. And yet, if the older contrivances are examined attentively it will be seen that the difference between the old and the new is one of finish and delicacy, rather than of idea or mechanism; and that, with the exception of a few noteworthy inventions for obtaining a greater security against picking by an ordinary thief, the locks of to-day are exactly the same, in principle and arrangement, as those which were made centuries ago. Indeed, it is rather strange that with all the inventions which have been made during the nineteenth century and especially within the present generation, and notwithstanding the inventive genius which

American industry has brought to bear upon the subject, the Yale system should be, after all, very nearly the only invention of practical utility which is a direct departure from the older methods of lock making. Probably a large proportion of the readers of this paper can distinctly remember the time when pin locks were almost unheard of. It might be said in explanation of the seeming fruitlessness of mechanical research upon this subject, that there was really very little that could be discovered or improved upon, as the real principle of a lock is too simple and too definite in its nature, not to have been thoroughly appreciated and exhausted long ago; but the same could have been said before Linus Yale set his Yankee wits to work upon the subject, and it would be impossible at present to foretell what discoveries may be made or what radical changes brought about in the appliances for locking our doors. Possibly our descendants may some day wonder at the locks of the nineteenth century, even as we wonder at the cumbersome pieces of mechanism and the ponderous keys of our great grandfathers. At any rate, it will not do to claim that our locks are perfect, or that the record of progress is entirely closed. A very few years ago the Yale lock was pronounced to be complete; but some very radical improvements have been made in it since then, and the opponents of the system claim it has yet many defects both in construction and idea. So it would not be strange if our best locks should one day become obsolete.

Finish and cost.

But if the progress which has been made in the essential, mechanical principles of lock manufacture is small, the improvements in finish and the reduction in the cost of the locks have been marvelous. Less than a century ago, locks were made entirely by hand, and very crude affairs they were, too, costing a great many times the price of a better article of to-day. At present, good, well-made, well-planned locks can be had at prices varying from twenty-five cents to five dollars, suited to all needs and all conditions; while the amount of real security afforded is of a much more tangible nature. And with the improvements in niceness and delicacy of arrange-

Chapter XI.

ment, it nas been possible to affect a change in the style and weight of the keys which the present generation can only faintly appreciate. The old-fashioned keys were heavy, cumbersome, and so large that no one ever thought of carrying them about the person. Now they are made so small that the keys for an entire house can be carried in one's vest pocket. Formerly the strength of a lock was judged by its weight, and it was considered essential to have heavy bolts or levers, and strong springs, requiring considerable force to operate; while now, all the parts are so well adjusted and so light, that a touch is sufficient to put the mechanism in operation.

Principles.

The fundamental principles forming the basis of all locking constructions, include a bolt which is moved by the direct action of the key, while secondary bolts or levers drop into such positions that the lock bolt cannot be forced back except by breaking some portion of the mechanism. The secondary bolt is usually termed a lever, and either acts by gravity or by the aid of a spring — usually by both. The key is so made as to first raise the levers, and then to shoot the bolt by a single turn of the hand. These principles have governed the manufacture of locks since the days of Adam, and apply equally to the ponderous locks of the Middle Ages and to the corrugated-key locks of the Yale & Towne Manufacturing Company. Complications have been added to the construction of locks in the shape of multiple levers, requiring nicely fitted keys, or fancy wards which would allow none but the right key to enter; and there have been special forms devised for bank uses, working by combinations of letters, by dials, or by clockwork; but in the locks used about an ordinary house, the principle is always the same — that of a key simultaneously lifting one or more levers and moving a bolt.

Egyptian Lock.

In order to clearly illustrate the antiquity of the principles upon which modern locks are constructed, it may be of interest in this connection to refer to a few of the older forms. A rude style of lock which has been used in Eastern countries for ages, no one can say how long, but certainly for over two thousand years, is approximately shown by Figure 277*a*. All the parts are of wood, including the key. The bolt is channelled on the inner edge, and slides through heavy wooden staples in which are arranged a number of pegs, of varying lengths, fitting into corresponding holes bored through the top of the bolt. The key consists of a flat piece of wood somewhat smaller than the channel which is cut in the bolt, and in use, is inserted lengthwise of the bolt. On the end of the key are pins spaced to correspond with the pegs in the staple. It is evident that while the pegs are caught in the bolt itself and in the staple, the bolt cannot be moved; but when the key is inserted, the pins will be directly beneath the holes in the upper part of the bolt, and by raising the key, the pins will lift the pegs just enough to clear the joint between the bolt and the staple, and the bolt can then be moved at will. In this lock, the action of the key is almost exactly the same as in the Yale lock; namely, to lift a series of pins of unequal lengths so as to bring the bottom of each on the same line, though the Yale key has other functions, as will be noted later.

Pompeian Lock.

Figure 278 shows a key which was dug up in Pompeii. It was evidently intended to operate a warded lock, a style which was in almost universal use up to thirty years ago. Figure 279 illustrates a fine old Elizabethan lock. This could be described as a fully-developed lever-lock, the springs on the levers being arranged in exactly the same manner as the locks which are sold over the counter to-day. Stripped of all the fancy cutting and misleading wards which have nothing to do with the efficiency of the lock, it will be seen that this is really a very simple contrivance, though quite complicated in appearance.

Elizabethan Lock.

Fig. 278. Key from Pompeii.

The number of antiquated examples might be multiplied indefinitely, but the foregoing will suffice for the purpose, as they may be taken as types of the three most markedly different arrangements for adding to the security of a lock; namely, with wards, with pins or with spring-levers.

Chapter XI.

The various parts of a lock will need some definition and explanation, in order to prevent any ambiguity in the terms.

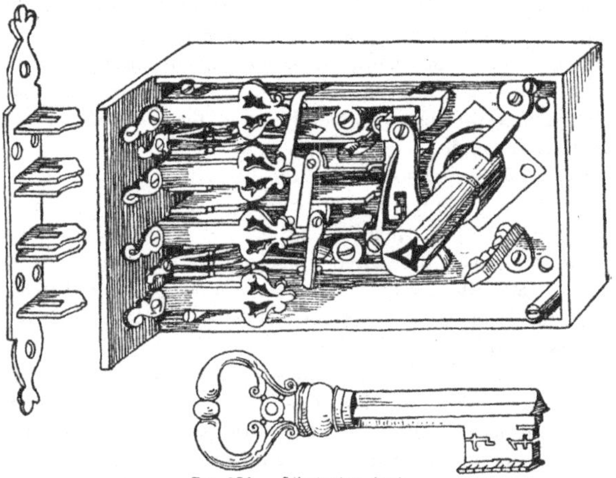

Fig. 279. Elizabethan Lock.

Parts of a key.

Figure 280 shows the general shape of the ordinary key, in which *A* is called the bow; *B*, the shank, and *C*, the bit. The difference between the keys of to-day and those of two or three generations ago has been already alluded to. Many of the hand-made locks are still provided with the old-fashioned, heavy brass keys, but the "Yale" locks have prejudiced people against anything but a flat key, and nearly all manufacturers use them in one form or another. A few lock-makers have keys which are arranged to fold up like a knife, to be used in connection with rim-locks, or with locks requiring a very long key, but generally the key is of steel, nickel-plated, with a flat shank and a thin bit. When the cuts on the bit are on the side or edge, as shown by the cut, they indicate

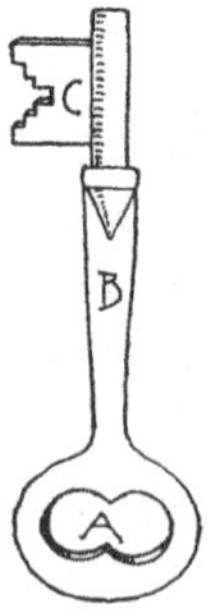

Fig. 280. Key.

a tumbler or lever-lock, while cuts on the top or bottom show that the lock is fitted with wards. Many of the old keys preserved in museums are made with very elaborate bits, cut in curious and intricate patterns. In some instances the cuts correspond to equally intricate wardings in the lock, but generally they are purely fanciful. When the shank of the key is tubular, it indicates a lock which can be operated from one side only, such as those used for drawers, etc. All keys for door-locks now have solid shanks.

The bolt which secures the lock, is generally made quite heavy where it projects beyond the face-plate, but is thinned down inside the lock so as to be as light as possible, and to give space for the levers. The talon, *A*, Figure 281, is the notch in the under side of the bolt in which the key works. The post, *B*, is the part which catches in the levers, preventing the bolt from being forced. Guide-posts on the case of the lock fit in the slots, *C*, one of the same posts often serving as a pivot for the levers.

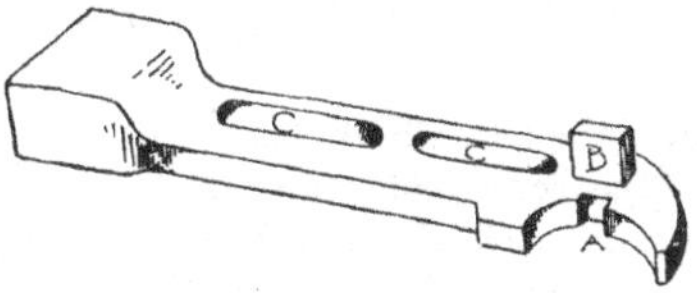

Fig. 281. Bolt.

Levers.

The most primitive form of lock would be one consisting simply of a bolt, which is shot back and forth by the key. But as any other key or even a wire would answer equally well, some obstacle must be interposed to prevent picking. This is done by combining with the bolt a series of levers or tumblers which permit only the proper key to be used. The two terms are used at present synonymously. Figure 282 illustrates a typical lever. There are from one

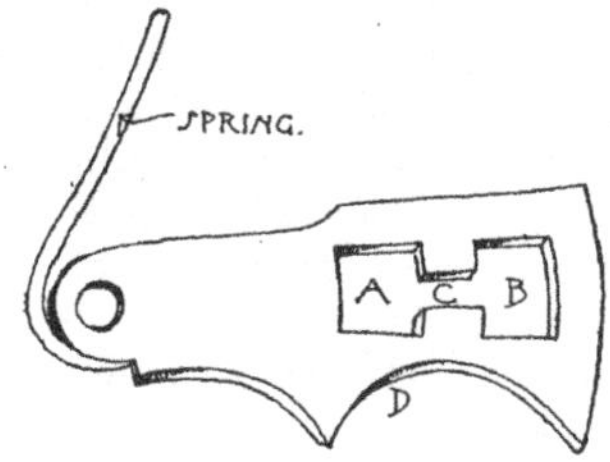

Fig. 282. Lever.

Chapter XI.

to five levers in an ordinary lock, and they are usually placed one over the other, pivoted over the guiding-post, and the bolt-post is so arranged as to fit through one of the cuts, *A*, when the bolt is thrown back, and through *B* when thrown out. The connecting gatings, *C*, are cut at different heights, so that the levers must be lifted unequally in order to permit the bolt to move. When the key is turned in the lock, the bits, which are cut to match the levers, bear against the bellies, *D*, lifting the levers simultaneously until the gatings are exactly on a line with each other. The key then catches in the talon of the bolt, the bolt-post passes through the gatings, and the levers drop as the key turns, catching behind the bolt-post and effectually preventing the bolt from being forced back. This is, generally speaking, the function of all lock-levers, though there are many variations from the form illustrated.

The levers, of course, slide one over the other, and in common locks they are laid closely together. In the best of hand-made work, however, and in a few of the machine-made locks, the levers are separated, either by side-wards cast onto the thickness of the lever, or by intermediate strips of brass which bear on each other and on the levers only at certain points, thus reducing greatly the friction between the parts.

Tumblers.

A somewhat different form has been much used in English locks, which is shown by Figure 283. In this case the levers are beneath the bolt. On each is a post which works in slots and through gatings cut through the bolt. Price, in his "*Treatise on Locks*,"[1] which is a very valuable and interesting work on the subject, as it was understood up to 1860, makes the distinction between levers and tumblers, applying the latter term to the device shown by Figure 283, and the former to that illustrated by Figure 282. His distinction seems to be a fair one, though seldom made in this country, where what he calls tumblers are little used.

[1] This work is entirely out of print, but can be found in most of the large public libraries. It is complete and thoroughly illustrated.

Chapter XI.

Changes.

A little reflection will cause one to comprehend the number of changes possible in a lever lock. The levers may be transposed, and within certain limits the heights of the gatings

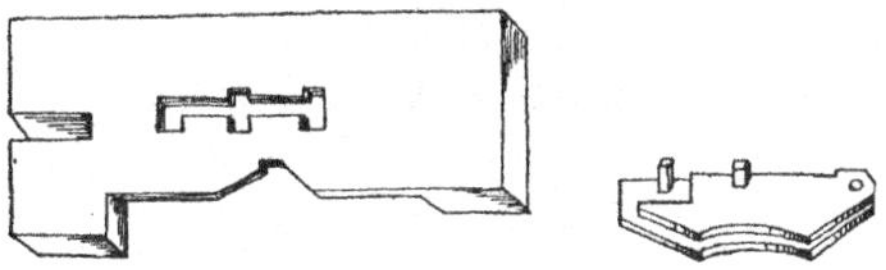

Fig. 283. English Tumbler.

may be varied, so that with six levers there can be as many as 7,776,000 changes, no two of which can be operated by the same key. Simple transposition, without any variation in the heights of the gatings, will give 720 changes.

Detector-springs.

A device has been used in some makes of locks, intended not only to increase the difficulty of picking but also to show if the lock has been tampered with. It consists of a spring so arranged that when one of the levers is lifted too high, as would naturally be done by any one attempting to pick the lock, it is caught and held in such a position that the bolt-post cannot possibly pass through the gatings. The spring is released by using the right key and turning the bolt out more, but no key can unlock the mechanism until the detector spring is released. This is a very ingenious arrangement, and at one time was considered absolutely burglar-proof, though it is now very seldom met with in the market.[1]

Wards.

The wards of a lock are fixed obstructions which are attached to the inside of the lock-case, so arranged that none but the proper key can pass and reach the levers. Formerly the confidence in warded locks was so great that levers and tumblers was used very little, but that feeling has entirely passed away. Modern locksmiths use wards very sparingly, and limit themselves to small shoulders or ridges, cast on the

[1] The detector-spring was an important feature of the celebrated "Chubbs" (English) locks.

Chapter XI.

inside of the upper and lower case-plates, which require corresponding cuts on the upper and lower edge of the key-bit. They do not add in the least to the burglar-proof qualities of a lock. At one time, however, locks were constructed with very elaborate wardings. Figure 284 illustrates the wards of a French lock about one hundred and fifty years old. The wards consist of two thin plates, one each side of the key-hole, with a series of ridges forming a semicircle on each, the ridges being star-shaped in section. The key-bit is cut out with a star pattern which has to exactly fit the wardings. This is one of the simpler forms which the ingenuity of French locksmiths at one time delighted in, and though seemingly proof against intrusion, can be opened with very little trouble, by a judicious use of a few stout wires.

Fig. 284. Wards of an old French Lock.

Springs.

There is a great difference in the quality and arrangement of springs used in connection with a lock. In regard to material, the best is, undoubtedly, phosphor-bronze; but springs of this material require to be so large in order to have the desired stiffness, that their use is not always practicable, especially as they can be used to advantage only in the shape of flat-bands. The springs which hold the levers in place against the bolt-post are usually made of round steel or brass wire, and are attached directly to the heel of the lever, as shown by Figure 282. A separate spring is necessary for each lever. It is sometimes desirable to attach the spring to a secondary lever acting directly on the top of the main lever, Figure 285, as in a case where the levers move up and down in the lock instead of being pivoted together. With such an arrangement the edge of the secondary lever should be grooved so as to fit over the top of the primary lever, thus obviating any difficulty of the levers slipping by each other, or of the wrong springs acting on the levers.

Chapter XI.

Latch.

The latch is a feature of the modern lock which our ancestors did not enjoy. Except in the case of store-doors, all door-locks are now made with some form of spring-latch. There are three distinct kinds of latches commonly used, the simple spring-latch, anti-friction latch and front-door latch. The cheapest form of ordinary spring-latch consists of a bevelled head, projecting from the face-plate of the lock, with

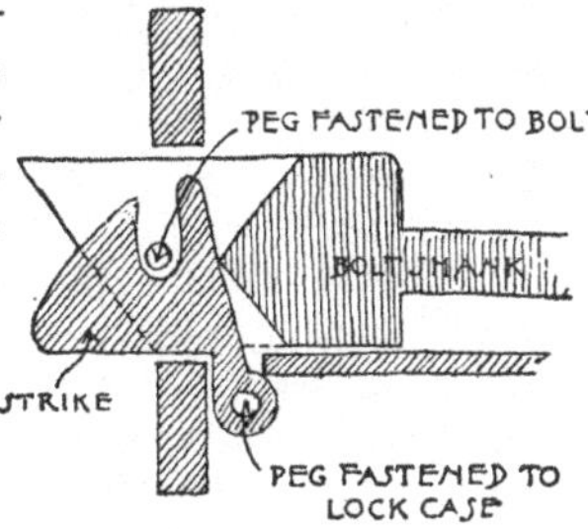

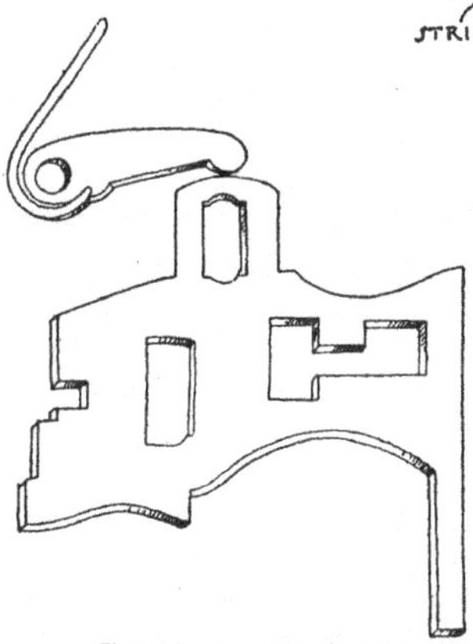

Fig. 285. Compound Lever.

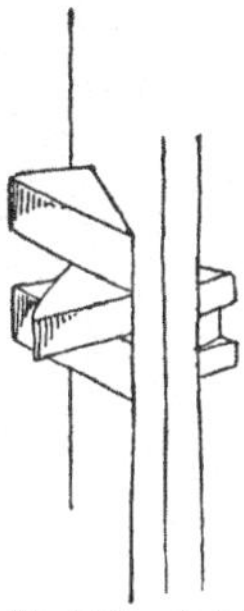

Fig. 286. Anti-friction Strike. E. Robinson.

a shank inside the lock, about which is coiled a strong spiral spring, keeping the latch pressed out. The inner end of the latch-shank is forked and hooks under each side of what is termed the follow, through which passes the spindle of the door-knob. Turning the knob either way draws back the latch. The objection to this arrangement is that while only a very slight spring is really necessary to keep the latch .in position, a pretty strong spring is required so that the knob shall

Chapter XI.

Easy Spring-latch.

not turn too easily; otherwise, every time the door-knob is touched the latch will be opened. Consequently in the better class of work a door-latch is usually fitted with two springs, one of which is operated when the latch is pushed back by the door being closed, while both springs are acted upon when the knob is turned. In this way the requisite resistance can be obtained for the knob, and, at the same time, the latch will close easily. A latch so arranged is termed an easy spring-latch. There are several methods of attaching the two springs. Ordinarily, spiral brass springs are employed. Hopkins & Dickinson and, we believe, a few others, are able to introduce into their locks springs made of phosphor-bronze, which, it is claimed, will keep its elasticity much longer than steel or brass. The different methods by which the springs are attached and the knob operated will be made clear when the various makes of locks are described, later on.

Anti-friction Strike.

The ordinary form of latch is made with a V-shaped bevel, the long side of the bevel striking against the jam-plate. Enoch Robinson, of Boston, was, it is believed, the first to patent an anti-friction strike, as it is called. Figure 286 illustrates the construction of his device, which is incorporated into all of the locks which he makes. It is simply an application of the principle of the old bell-lever crank. The action of the anti-friction strike is to raise the latch-bolt from the bed of the lock and carry it back without friction on the sides. Actual tests have been made proving that it requires less force, acting directly on the side of the anti-friction strike, to force the lever back, than is required to push back the latch by straight pressure against the apex of the bevel.

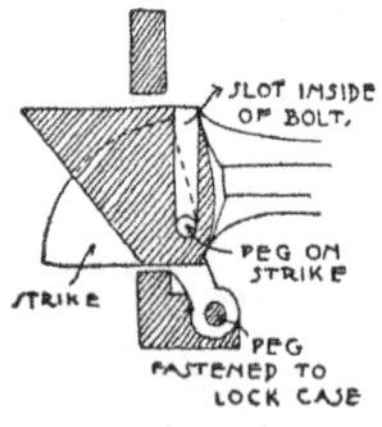

Fig. 287. Anti-friction Strike.

Figure 287 shows a form of anti-friction strike used by several other manufacturers. There is no difference in principle between this and the "Robinson" make, though the appearance is a little different, the "Robinson" strike being in the centre of the bolt, while the others are on one side, also in "Robinson's" strike the pin is on the latch and the slot in the Chapter XI.

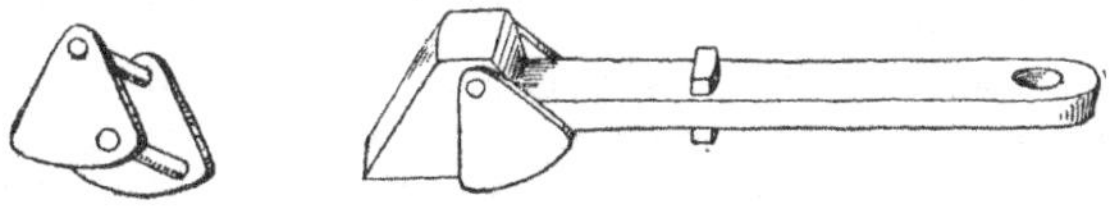

Fig. 288. Anti-friction Rocker Strike.

strike, while in the other anti-friction strike they are exactly the reverse. Figure 288 shows a form which is made by a few manufacturers, being listed in the catalogue of both J. B. Johnston and the Nashua Lock Company. It consists simply of a steel rocker attached by swivel pins to the bolt, the lower pin passing underneath the shank of the bolt. When the door is closed the latch, instead of moving straight back, swings on the lower edge of the rocker, being lifted from the lock-frame, and thus reducing the friction. The gain by this device is, of course, less than by the others previously described. Yet another form of so-called anti-friction strike is made. Figure 289 shows the pattern adopted by Hall, of Boston, for his spring-latches. It consists, essentially, of an adaptation of the well-known car-door latch, the latch-strike being hinged at the base and attached by a loose-pin to the latch-shank at the top, while the face of the latch-strike is slightly curved. This device makes really a very efficient anti-friction strike. The only objection to it is that the wide

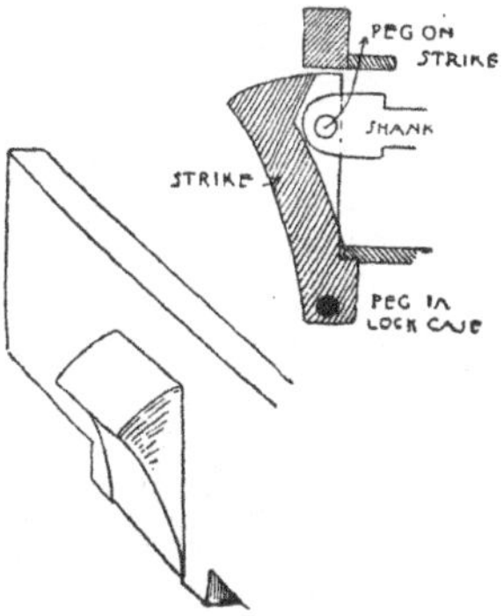

Fig. 289. Anti-friction Strike. Hall.

Chapter XI. plate necessitated by it cuts the door a great deal, and many persons do not like it on that account.

The custom in regard to latches varies in New York and Boston. In New York the outside knob is generally fixed firmly so as not to move at all, while in Boston the knobs are arranged with a swivel spindle permitting either to be turned without acting upon the other, and the mechanism inside of the lock is so devised that by pushing a button or a slide the outer knob can be held fast. In cheaper forms of front-door locks, the knob-spindle is made without a swivel, and security is obtained by a bolt on the inside.

Right and Left Hand Locks.

Locks are designated as being either right or left hand, though the distinction is one which is confined entirely to the latch. A left-hand lock belongs to a door fitted with left-hand hinges, as has been previously explained, the term right or left being decided by whether the door turns on the hinges

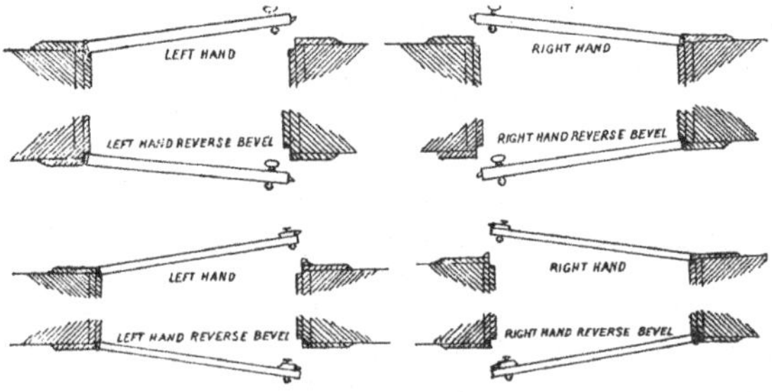

Fig. 290. Right and Left Hand Locks.

when opening either in the direction of the hands of a clock or the reverse. Locks are also designated as being either left or right hand reverse bevel, the reverse bevel applying to a door which swings out instead of swinging in. That is to say, in the case of a front door, for instance, if it swings out the night-latch would be on the outside, but the latch-bolt would be just the reverse in arrangement from what it would be, relatively, on an ordinary front door swinging in. Chapter XI.

Figure 290, will fix this distinction clearly in mind. The figure is taken from the catalogue of the Yale & Towne Manufacturing Company. It is believed that the distinction between right and left, and reverse bevels is seldom appreciated by architects.

Reversible Latches.

It is very often desirable to have a latch which can be reversed so that if any mistake is made in ordering, the lock will not be useless. Reversible latches are made in several ways, the latch-shank being generally of such shape as to permit its being turned over and worked in the opposite direction, without interfering with the action of the lock.

Wear on Locks.

Locks wear out not so much by actual failure or breaking of the parts, but by the lever and key wards being worn so that the key will not lift the levers and permit the bolt to pass. Key-wards are the slight projections which are cast on the inner face of the lock-plates to form an additional obstruction to the passage of strange keys. Of themselves they affect the value of a lock but little, as the key will operate as well without as with them, so that the only vital parts which wear out are the edges of the levers against which the key acts. The constant striking and turning, when a lock is used continually, will in time wear off the surface of the lever so that it will not rise quite sufficiently to allow the bolt-post to pass. The springs, also, sometimes become brittle, and the follows operating the latch will wear so as to work loose and rattle, but a little tinkering can remedy any of these difficulties. It costs but a trifle to have a new key made which will fit a partially worn-out set of levers. New springs are inserted at a trifling cost, and if the latch-spring is strengthened a trifle the rattling of the follows can be obviated; so there is, really, no reason why a fairly good lock should not last indefinitely. It is, also, a very simple thing to make a new combination of the levers when they cease to work smoothly, and renewed life can thus be imparted to an apparently worn-out set of works.

Chapter XI.

In judging of the intrinsic worth of a lock, therefore, the following conditions should be carefully observed.

Essential Conditions.

First: Good material for the use to which it is put.

Second: Careful adjustment, so that the parts will work easily and will stand any possible strain in use.

Third: The whole secret of the value of a lock is in the levers, which should be so made as to ensure a minimum of friction, of material not easily corroded nor easily worn away; and they should be adjusted to secure the greatest amount of security against picking, with springs not too easy, nor so hard as to bring undue wear on the levers.

Tests.

A very good test of the workmanship of a lock can easily be made by shooting out the bolt, removing the cap to the lock case, and then pressing in strongly on the bolt, at the same time lifting the levers, one by one. If the gatings are accurately fitted they should all bear equally against the bolt-post, so that the gating of no one lever would catch on the post as it is lifted by. Few of the ordinary locks will stand this test successfully.

Intricate combinations, made ostensibly to prevent the lock from being picked, add very little to its value for ordinary house work. It may be safely stated that any lock can be picked which is operated by a key, so that a good three-lever lock affords all the intricacy and gives one all the protection that could be desired. A lock has a personality of its own, and so much of its value depends on the maker that it is wise in purchasing to always get the best; keeping in view simplicity, and the points previously noted. A cheap, but well-made lock is better than an expensive one which is put together in a careless and indifferent manner.

It has not been the intention to consider in detail any articles of hardware which are not in actual daily use at the present time; but there are a few styles of locks which are entirely obsolete so far as the American trade is concerned, but which should be included in any study of the subject, if one wishes to thoroughly understand the principles of modern lock-making, and the processes of elimination and survival of the fittest which have brought the manufacture to its present state in this country. Chapter XI.

Bramah Lock.

Figures 291 and 292 illustrate the old English "Bramah" lock. This consists of a revolving cylinder in which is disposed radially a series of flat sliders working up and down through slots in a fixed horizontal plate. The sliders have notches on the outer edges, cut at different heights, so that the cylinder can revolve only when the notches on the sliders are on a line and level with the plate. The sliders are forced upward by a single central coiled spring. The key consists of a tube, on the sides of which are straight grooves corresponding to the desired depression of the slides, with a shoulder to turn the cylinder. The

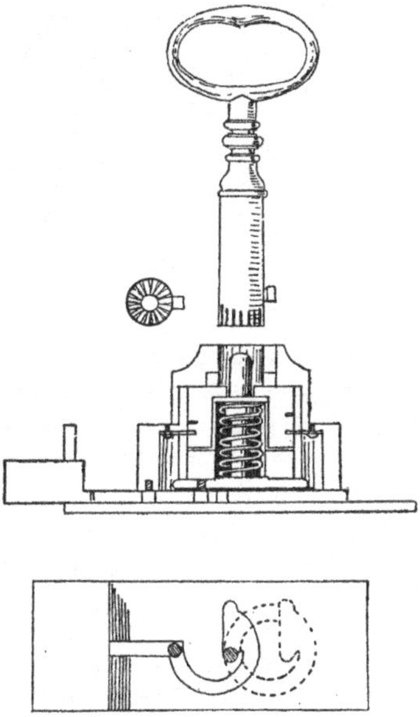

Fig. 291. The Bramah Lock.

locking-bolt is moved by an eccentric attached to the cylinder. The notches on the sliders are disposed as irregularly as possible, and false notches are added, with corresponding false widenings of slots in the plate. All of the sliders can be pushed in farther than is needed to bring the notches on a line with the plate, so that the lock is picked with great difficulty.

Cotterill's Lock.

"Cotterill's" lock, Figures 293, 294 and 295, is another example of English ingenuity. The portion which is acted upon by the key consists of a rotating flat disk or cylinder containing ten or more slides moving in radial grooves and pressed

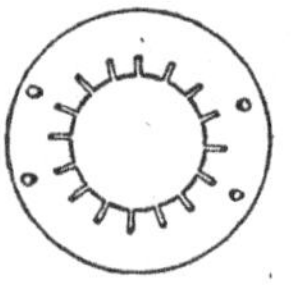

Fig. 292. The Locking-plate of the Bramah Lock.

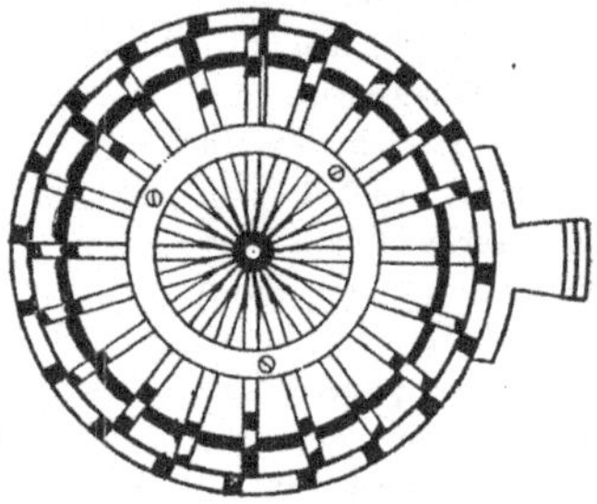

Fig. 293. Plan of Cotterill's Lock.

towards the centre by springs. A fixed ring or plate is fitted to a circular groove on the face of the disk, and has slots corresponding in position to the radial slides. There are also slots cut on the edges of the slides, so that when the key is in place the slots on the slides coincide with the circular

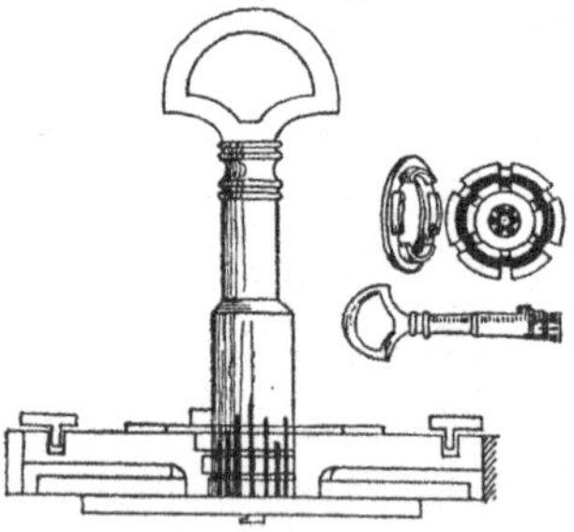

Fig. 294. Section of Cotterill's Lock.

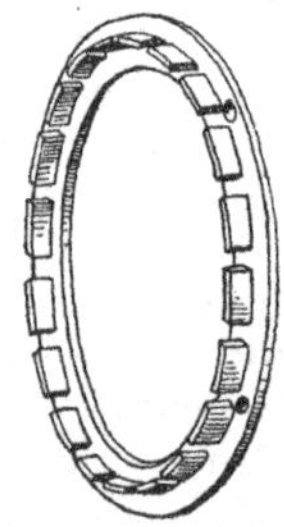

Fig. 295. Locking-plate. Cotterill's Lock.

groove on the disk, permitting the whole to be revolved. When the key is withdrawn the slides are forced in different degrees towards the centre, so that the solid portions intercept the groove in the disk, in which position it is held fast by the

Chapter XI.

fixed ring. It is believed that this lock never has been picked. A lock which in its time was a strong competitor with the "Bramah" and "Cotterill's" locks, and was equally impregnable, is "Day & Newell's" Perautopic bank-lock, an American invention which was in great demand at one time, but has long since ceased to be manufactured. It has the curious property that the key, which is made with movable bits, can be changed at will, so that the lock can be opened only by the key which was last used to shoot the bolt. The lock has never been picked. Figure 296, which is taken from *Price*, is too complicated to fully illustrate the workings. Figure 296*b*, while not exactly like the lock, embodies the same

Day & Newell's Lock.

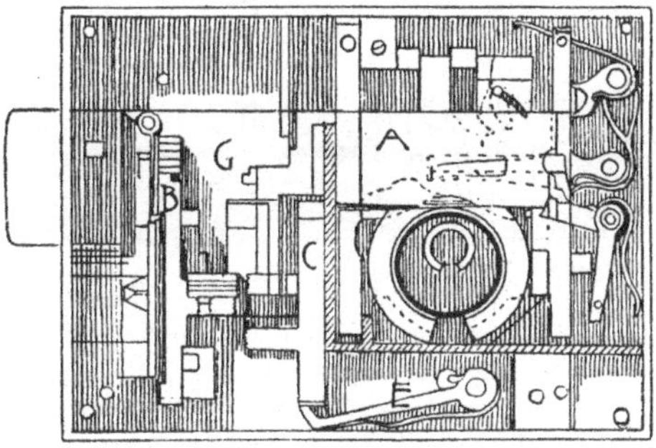

Fig. 296. Perautopic Lock, open.

arrangement and will serve to make the construction understood. The letters refer to both figures. There are three distinct sets of levers, *A*, *B* and *C*, each admitting of a sliding or lifting motion up and down, the levers *A* having springs which keep them pressed down, *D*, and the levers *C* being constantly forced up by a spring of lesser strength *E*, so that the levers *C* will always move up and down exactly as *A* are raised or lowered, the tops of *C* bearing against the bottom of extensions to *A*. The levers *B* have no springs, and slide up and down between studs attached to a wing of the bolt-tail, so that

Chapter XI.

when the bolt is shot, the levers *B* move with it. *F* is a dog or lever, which is hinged to a stud on the bolt at the top, and hinged with a bent elbow attached to the lock-case at the bottom. On this dog, *F*, is a tooth, and on the edge of each of the tumblers *B* are notches corresponding in mutual distance with the difference in lengths of the movable bits of the key. Furthermore, the levers *A* are each made with an arm

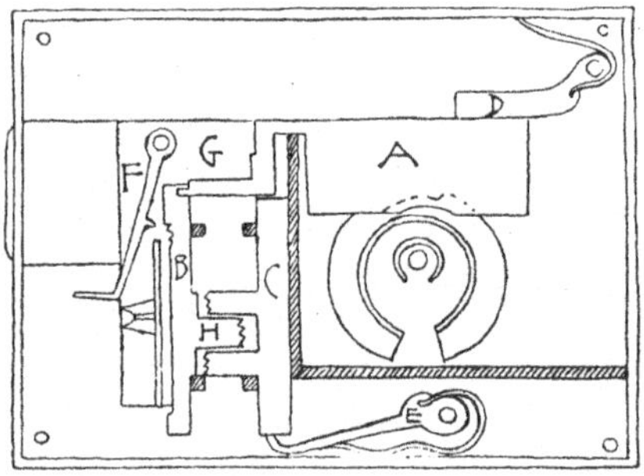

Fig. 296*b*. Perautopic Lock, shut.

G which fits into a corresponding notch in the levers *B*, and the levers *B* have each an arm *H* which exactly fits between two arms on each of the levers *C*. Figure 296 shows the lock with the bolt thrown, and Figure 296*b*, shows it drawn back. When the key is turned in the lock, the bits, no matter in what order they may be arranged, lift the levers *A*. These, by means of the arms *G* and *H*, lift the other sets of levers in exactly the same proportion. The key then forces out the bolt, and the levers *B* are withdrawn from the arms *G* and *H*, but before the arms *H* are entirely free from the arms on the levers *C*, the notches on *B* are caught on the tooth of the dog *F*, the levers *B* being then held at exactly the relative heights to which they were raised by the action of the key on levers *A*. The key, continuing to turn, then allows levers *A* and *C* to drop to their original position, and the bolt is then locked. It is evident that only the proper key will answer to unlock the combination, as unless the levers *A* and *C* are raised in exactly the proportion they were when the bolt was shot, the arms *H* cannot enter between the arms on levers *C*, and the bolt cannot be moved. There are several other features of the lock, such as detector plates, wards, etc., which need not be noticed here. A circular curtain protects the keyhole, and a solid partition entirely prevents access to the levers, while if any attempt is made to discover the combination by applying pressure to the bolt and tentatively rising the levers *A*, the arms on the levers *B* and *C* which have notches on the ends will catch on each other and be immovable as long as the pressure remains on the bolt. With an eight-lever lock and eight-bitted key, over 5,000 different combinations can be made.

Parnell's Defiance Lock.

A very ingenious idea which seems not to have survived the the test of years was embodied in another English device—"Parnell's" Defiance lock. The peculiarity here is in the key, which is made with expanding bits. When out of the lock it has the appearance of a key-blank. Eccentrics in the lock force out the proper bits to act on the levers, and the keyhole is guarded in such a manner that a key which could enter and was without expanding bits, would simply turn without affecting the lock; whereas a key with fixed bits which would be right to move the levers could not enter the keyhole.

As previously stated, none of the foregoing are now used in this country, but from them several of our best locks have been derived. Prior to 1851 all of the best locks used here were of English make, but American locks came to the front about that time, and to-day an English lock would be looked upon as a curiosity in our hardware trade.

Varieties.

Dead-locks.

Mortise Locks.

Turning then to our own current manufactures, there are several varieties of locks which are commonly found in the market. The "dead-lock" consists simply of a bolt thrown by the action of the key on the levers, but does not include any knob or latch. A "mortise lock" is one which is mortised into the frame of the door, and always includes, as commonly

Chapter XI.

understood, both bolt and latch. A mortise lock is generally operated from either side. A "rim-lock" is one that is planted on the face of the door. It is generally made with a nicer-looking case than the mortise locks, and requires longer keys and a little different adjustment of the knob-spindles. A dead-bolt may be either mortise or rim, but, generally speaking, rim-locks are understood to have both latch and bolt. A "rebated lock" is one which is mortised into the door-frame like an ordinary mortise lock, but the face-plate is rebated so as to fit the rebates of the door to which it is attached. This form of lock is used only for front double-doors. In the East it is customary not to rebate the front doors, but, we believe, generally speaking, in the West such locks are necessary. Special locks are usually made for front and vestibule doors. The lock for the front door includes a dead-bolt and a latch operated by a knob from within, and worked by a key from without. The vestibule lock consists simply of a latch worked by a knob from the inside and a key outside, the same night-key answering for the latches of both front and vestibule doors. Hotel locks are understood to be those which are so arranged that they can be opened from either the inside or the outside, but when locked from the inside cannot be unlocked from the outside. There are many varieties of hotel locks. Generally they are made in sets of fifty, one hundred, two hundred, or more, as desired, and are master-keyed, that is to say, the tumblers are so arranged that one key will unlock the whole series, though the individual keys of the different locks will not unlock each other. Again, they are sometimes made so that the lock can be locked from the inside with one key, and an exactly similar one can unlock it from the outside, but the master-key cannot unlock it after the bolt has been thrown from the inside, and after the bolt has been thrown twice from the inside nothing can open it from the outside. Such locks are intended to be used where two persons room together, but do not come in at the same hour, each wishing to be secure against intrusion, and yet leave the lock so it can be opened by his comrade.

Rim-locks.

Rebated Locks.

Front-door Locks.

Hotel Lock.

Hand and machine made Locks.

Locks are made both by hand and by machinery. Boston, at present, seems to lead the country in lines of hand-made locks. Indeed, it is doubtful if in any other city such an industry could so long survive the extended application of machinery to labor which has so strongly marked this century. But in Boston the old ideas are slow to go, and the people are loath to give up a thing once tried and proved, merely because there is something else in the market, even though the something else may be cheaper. There is no question but that a hand-made lock, if the manufacturer is thoroughly conscientious, is better than one made by machinery, especially as the hand-made lock manufacturers, thus far, never have catered to a cheap trade, and have always kept their goods up to the very highest mark. In the hand-made locks the levers are carefully adjusted, nearly all the interior fittings are made of brass, and, while in some respects hand goods may be inferior in fineness of polish and smoothness of exterior appearance, no one ever denies their excellence. But, on the other hand, the cost of hand-made goods is so much higher than of those made by machinery that the former are gradually being driven out of the market, especially since some of the best of the machine-lock manufacturers have succeeded in turning out such admirable goods. To the uninitiated the best of the machine-made locks are quite as good as any that are turned out by hand, while the progress of machinery has been so great that it is possible to obtain almost any desired accuracy of adjustment. Of course, the best of locks, even those which are nominally machine-made are fitted by hand. Only in the cheapest forms are locks left as they come from the machine.

Prices.

In regard to price, machine-made locks may be divided generally into six classes. This division, of course, is not absolute. Locks are made in all grades, and are of all prices. Some very good locks are made in cheap form, and some very poorly designed locks are listed at a high price; but for general comparison this division will be satisfactory: —

First, the cheapest form of lock made, with iron face and

Chapter XI.

bolts, steel spring, and a single lever; P. & F. Corbin have a lock of this description which sells in the market for a $1.50 a dozen.

Second, a lock with brass face and bolts, all the rest of the construction iron, one lever; average price $4.00 to $4.50 a dozen.

Third, brass face and bolts, all the rest iron, with two levers; $7.00, or with three levers $8.00 per dozen.

Fourth, anti-friction latch, brass face and bolts, three levers, $17.00 per dozen.

Fifth, front-door lock and latch, $1.50 to $4.50 each.

Sixth, hotel locks, $2.50 to $5.00 each.

Hand-made locks may be divided according to cost into five classes: —

First, single lever with brass face and bolts, $1.50 each.

Second, three levers, brass face and bolts, $2.50 each.

Third, anti-friction strike, three levers, brass face and bolts, $3.00 each.

Fourth, anti-friction strike, all brass-work, $5.00 each.

Fifth, front-door locks from $8.00 up.

The foregoing classification of machine and hand made locks according to price does not imply two classes in regard to either efficiency in working or nicety of plan. The machine and hand made locks are designed on exactly the same principles, and the differences are but slight. Still the hand-made locks are, throughout, better than a relatively corresponding grade of machine-made locks.

ORDINARY MODERN DOOR-LOCKS.

In considering the locks at present in the market, it is manifestly impossible to even mention all of the styles and varieties, nor has it been found practicable to gather reliable data concerning all of the different makes. It is believed, however, that those illustrated will serve as fair criterions of what the market is producing. The descriptions will be limited chiefly to such as are used about an ordinary building. Time-locks, bank-locks, safe-locks, prison-locks, etc., are too complicated to come within the scope of this treatise, and are, besides, quite outside the line of what could fairly be termed builders' hardware. Chapter XI.

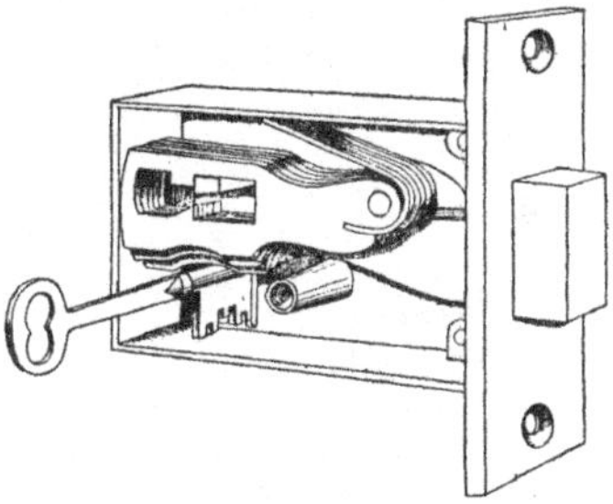

Fig. 297. Mortise Dead-lock. Russell & Erw.n.

An analysis of the various styles of locks can be best followed by taking the different examples according to the use to which each is put. They may, then, be classed as:

First, dead-locks.

Second, ordinary lock and latch combined.

Third, front-door locks.

Fourth, vestibule-locks.

Fifth, hotel-locks.

Any of these, except the first, may have anti-friction strikes, and may be mortise, rim, or rebate, and all can be master-keyed. Consequently in these five categories can be included all ordinary house-locks.

DEAD-LOCKS.

Dead-locks.

Figure 297 is a type of the most simple form of dead-lock, manufactured by Russell & Erwin, having five plain, pivoted levers, permitting of 120 changes in the lock by transposition of the levers. The same style of lock is made with as few as one lever. A. G. Newman manufactures a very good store-door lock, Figure 298, in which the levers slide up and down but are not pivoted together. Figure 299 illustrates the "Standard" store-door lock, manufactured by the Yale & Towne Mfg. Co., a very strong, well-made, and almost unpickable lock. The bolt-tail is the full thickness of the bolt but is made with a shell so that the tumblers work within the bolt, as it were, and the key, instead of acting against the under side

Chapter XI.

of the four levers, works through the centres; and, instead of acting directly upon the bolt, simply rotates an irregularly-shaped cam. The side figure showing the bolt and the cam alone, will illustrate how this lock works. The levers in this example are of steel, as in all the "Standard" locks.

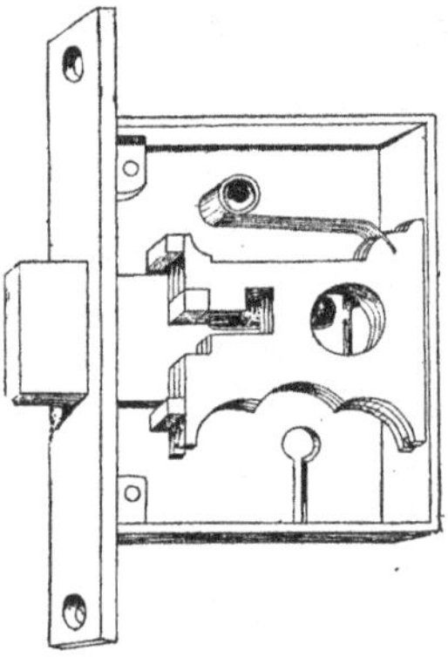

Fig. 298. Mortise Dead-lock. A. G. Newman.

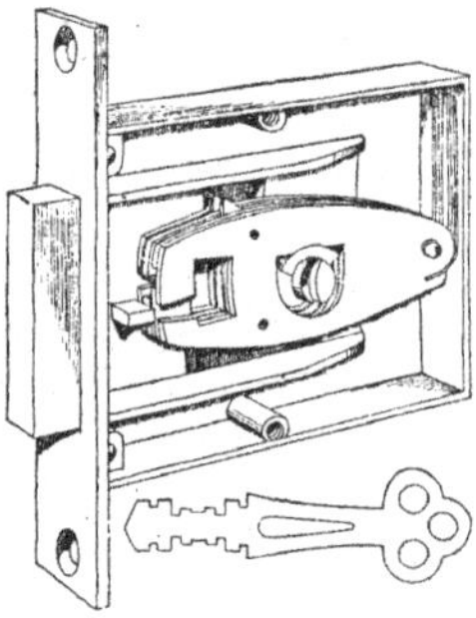

Fig. 299. Standard Store-door Lock. Yale & Towne Mfg. Co.

Store-lock with notched gatings.

Neither of the foregoing offers any special protection against picking, except such as results from careful fitting, or, in the Standard lock, from the difficulty of reaching the levers through a small key-hole. Figure 300 shows a "Robinson" store-lock, in which the inside of the bolt-post is cut with a square notch. If an attempt is made to pick the lock by exerting a pressure on the bolt while the levers are raised tentatively in succession, the notch in the post will catch in corresponding notches on the edges of the lever gatings, holding the levers so they cannot be moved in either direction. Two of the levers only are so notched, the uppermost lever having plain gatings to prevent the posts from catching when the proper key is used. This is a hand-made lock, with all the works made of brass except the bolt-post.

Chapter XI.

Store-lock with sliding post.

Figure 301 shows another "Robinson" lock in which the post and gatings are notched in the same manner as the preceding example, but in which additional security is obtained

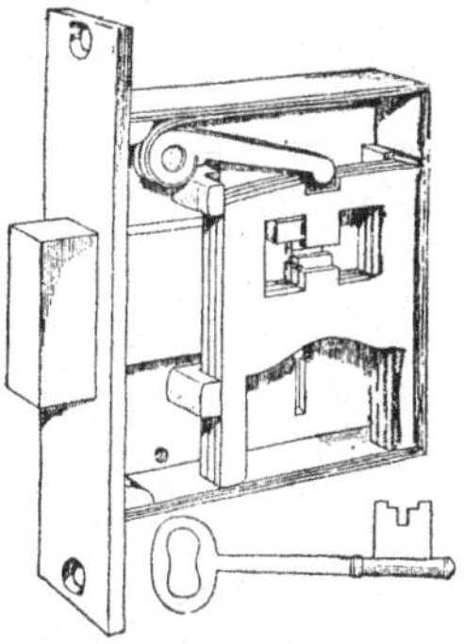

Fig. 300.

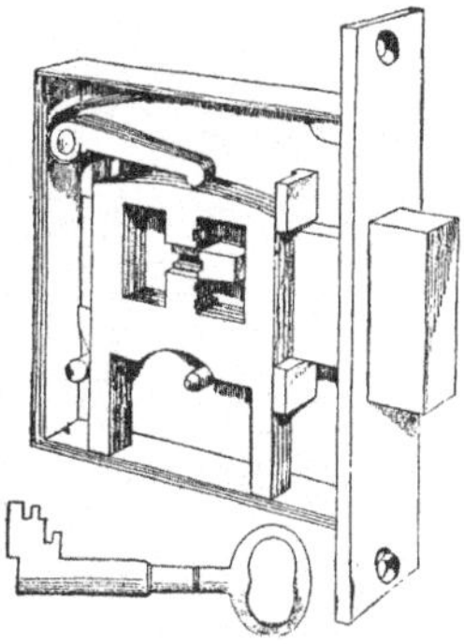

Fig. 301.

Store Locks. E. Robinson.

by attaching the post to a thin plate, sliding up and down in the bolt-tail, but held down by a spring lever such as those which work against the main levers. The post and the gatings are so arranged that if the levers could be so lifted as to bring the gatings exactly in a line, the bolt could not be moved, as the post would be too low down to pass. The post, as well as the levers, has to be raised, and on account of the notches, which prevent any tentative picking, this can be done only by the proper key. The works of this lock are all of brass, except the sliding parts of the bolt and the bolt-post which are of steel. The key is tubular, and the lock can be opened from one side only. It is an old style, and is little used at present.

Dietz Lock.

A lock which is asserted to be absolutely proof against picking, is the "Dietz" lock, Figure 302. In this the locking-levers are not touched at all by the key, being separated from the key-hole by a curtain or partition on the bolt-tail, so that no wire or picking instrument can reach the levers through the

Chapter XI.

key-hole. There are two sets of levers, exactly corresponding in thickness and bearing against each other only at the shoulders, as shown by the figure. The key-bits first lift the primary-levers, which are fitted with the stronger springs. The springs of the secondary or locking-levers then force the latter down in proportion as the primary-levers are raised. The secondary-levers are so arranged that the gatings are above the line of the lock-post, rather than below it as in ordinary locks, and it is evident that by raising the primary-levers to the proper heights the gatings of the secondary-levers can be brought exactly in line to permit the bolt-post to pass.

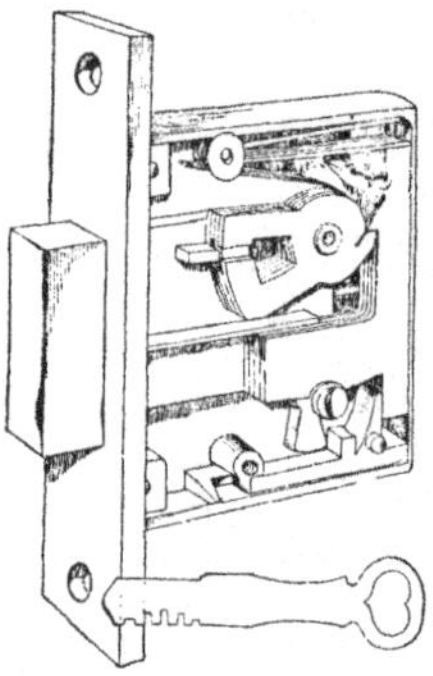

Fig. 302.
Store Lock. A. E. Dietz.

But to prevent picking by the tentative process, one of the secondary-levers is made with plain gatings but the others are finely notched to correspond with notches on the post, so that if any attempt is made to force the bolt, the levers become fixed. The bolt is moved by a key-cam similar to that shown by Figure 299. The small slide at the bottom of the lock is simply to prevent the cam from turning too far. The "Dietz" lock is machine-made, but is first class in every respect, with all-brass inside works. The agents maintain that this lock never has been picked. The description may seem complicated, but the lock is very simple in action, and it is one of the most satisfactory of its kind in the market.

ORDINARY LOCK AND LATCH.

One of the cheapest locks in the market, and one which, considering the price, is a very fair article, is manufactured by

P. & F. Corbin.

P. & F. Corbin, Figure 303. Everything about this lock is of cast-iron except the springs. The single lever, under the bolt-tail, shown by Figure 304, has a small shoulder instead of

Chapter XI.

P. & F. Corbin.

of gatings, and the latch has only one steel spring. It is a lock that offers no real security, but it is worth all it costs, $1.50 per dozen. It works easily, and is so simple in construction that it seems capable of withstanding considerable wear, perhaps more than a better article. Figure 305 is a more expensive, one-lever lock by the same manufacturers, having double springs for the latch. The form of follow, *A*, and the arrangement of springs in this ex-

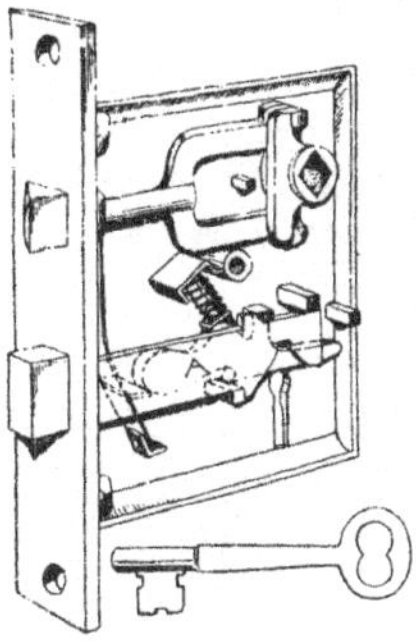

Fig. 303. Lock. P & F. Corbin.

Fig. 304. Lever of same.

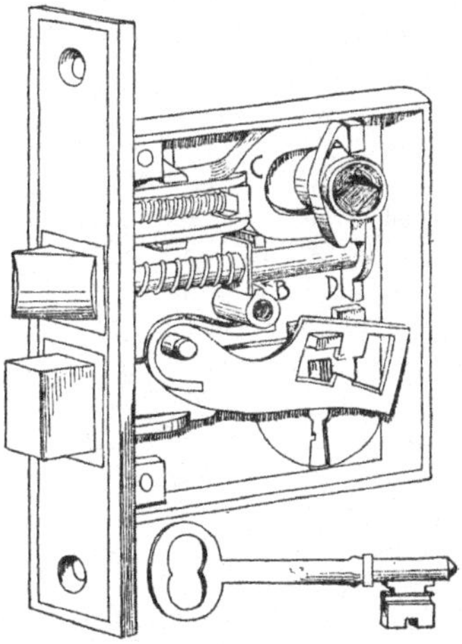

Fig. 305. Lock. P. & F. Corbin.

ample is that which has been found to give the best results, generally speaking, and which has been adapted to a great many varieties of locks. When the latch is forced back, upon closing the door, the lower spring alone is compressed, reacting against the plate and posts at *B*, but when the door-knob is

Chapter XI.

turned in either direction the follow forces back one of the arms of C, compressing the upper spring, while a shoulder on the lower part of C catches on D, which is attached to the latch-bolt, thus bringing both springs into play. This would be termed an easy spring latch, in that the knob can be turned with equal ease in either direction.

Nimick & Brittan.

Figure 306 illustrates a lock manufactured by Nimick & Brittan, in which the lever and bolt are essentially the same as in the preceding example, but which has a follow arranged upon a different principle, lugs being cast on the top and bottom so as to bear against the irregular spring-lever A, and the latch-bolt being pinned to an extension of the lever. The follow and lever shown in Figure 307, a lock by

. B. Shannon & Sons.

J. B. Shannon & Sons, is of much the same description. In both of these, the knob can be turned more easily to the left than to the right by reason of the unequal leverage against the piece A, though the difference in resistance is partially compensated for by making the shoulders on the follow of unequal lengths. The lock shown by the last figure has three levers, and is catalogued as being handmade. In Figure 306 the latch is reversible so that the lock can answer for either a right or a left hand door.

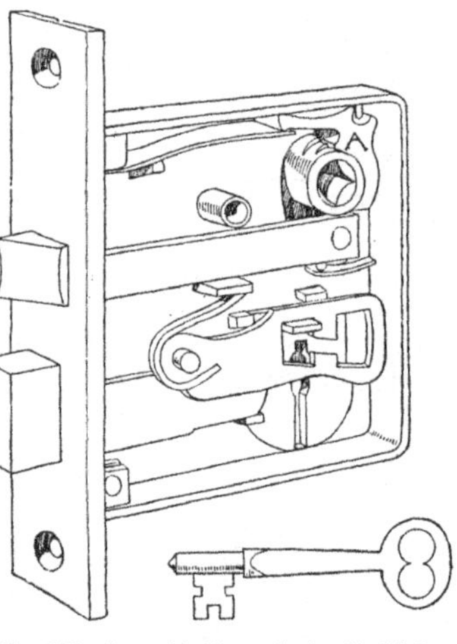

Fig. 306. Reversible Mortise Lock. Nimick & Brittan Mfg. Co.

The "Niles" locks, of which Figure 308 is a type, are all made to be operated by knobs having a follow cast solid onto the spindle. The action of the knob will be referred to later on. The figure shows only the follow, which is inserted from the back. The "Niles" locks have the name of wearing very well. The levers are of steel and are pretty well fitted for a machine-made lock, and the springs are also of steel, the bolt being the only portion of the mechanism for which brass is employed. As in some of the previous examples, the knob turns more easily towards the left than the right. If instead of the irregular, hinged lever, *B*, a form were adopted similar to that shown in Figure 305, the "Niles" locks would leave little to be desired, and would compare favorably with anything else in the market.

Chicago Hardware Co.

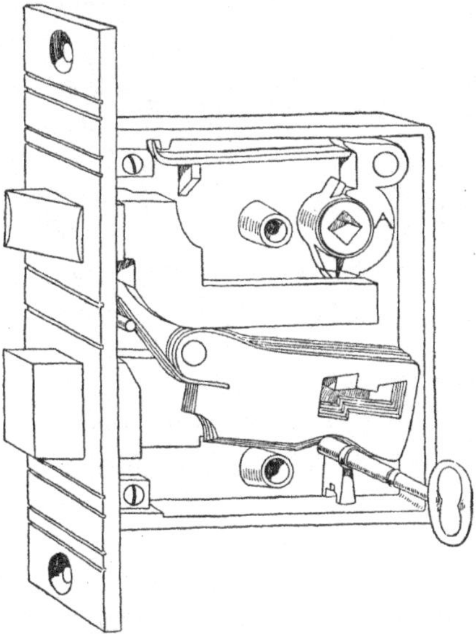

Fig. 307 Mortise Knob-Lock. J. B. Shannon & Sons.

An examination of the figures will show that, except in the very cheapest example, the face-plate of the lock is screwed to the lock-case in such a manner that it can be moved slightly and set at whatever bevel may be desired in order to fit the door. Figure 309 shows a lock of the Ireland Manufacturing Company in which all the parts can be reversed. The latch is simply drawn out and turned over. The bolt-tail is in two

Ireland Mfg. Co.

sections and the outer part can be unscrewed and reversed to

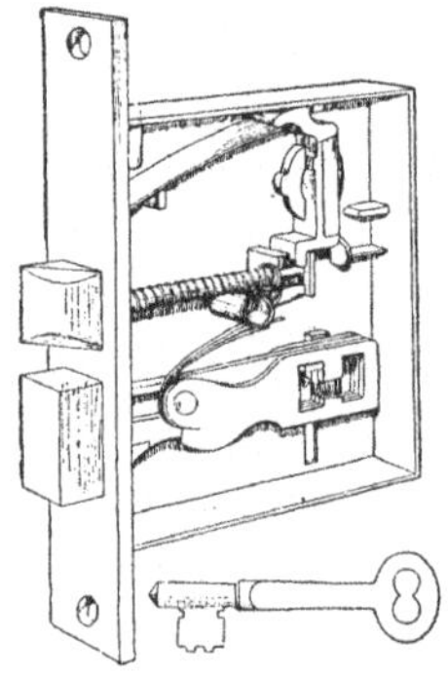

Fig. 308. Niles Lock. Chicago Hardware Co.

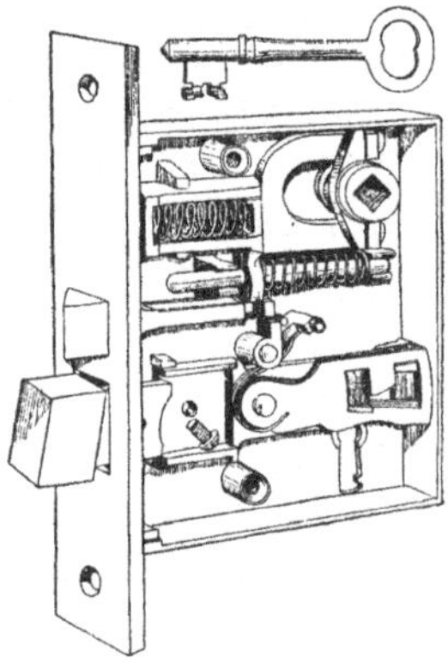

Fig. 309. Reversible Lock. Ireland Mfg. Co.

match the change in bevel. Otherwise this lock is of the

Fig. 310. Reversible Lock. Ireland Mfg. Co.

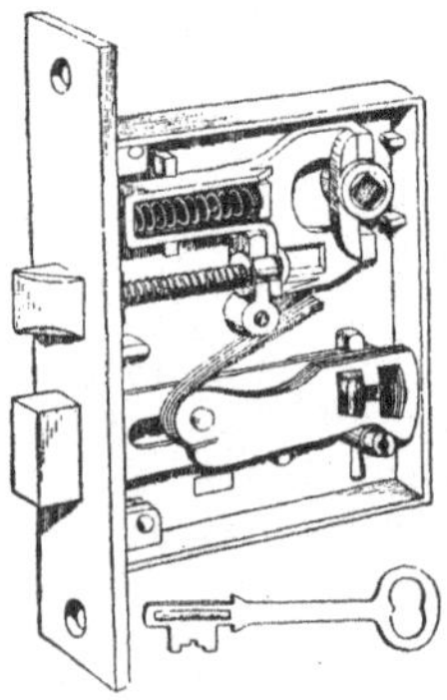

Fig. 311. Three-Lever Lock. Hopkins & Dickinson Mfg. Co.

ordinary type. Figure 310 shows another lock manufactured

by the same company, in which the hand can be changed by Chapter XI.

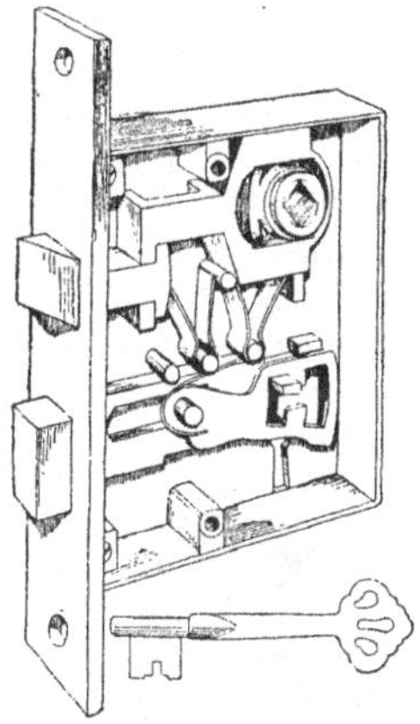

Fig. 312. Reversible Lock. Hopkins & Dickinson Mfg. Co.

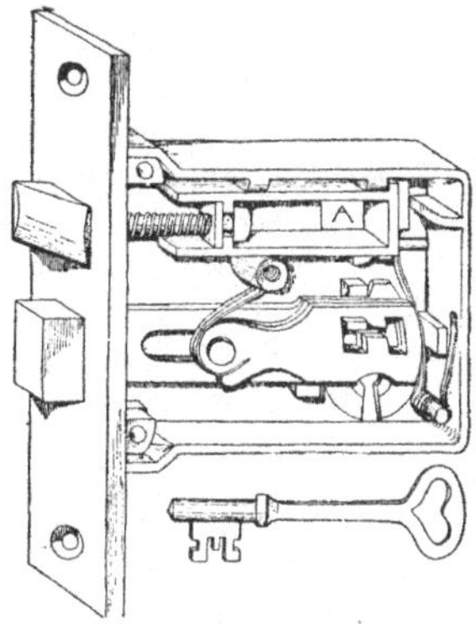

Fig. 313 Gilbert Lock. Gilbert Lock Co.

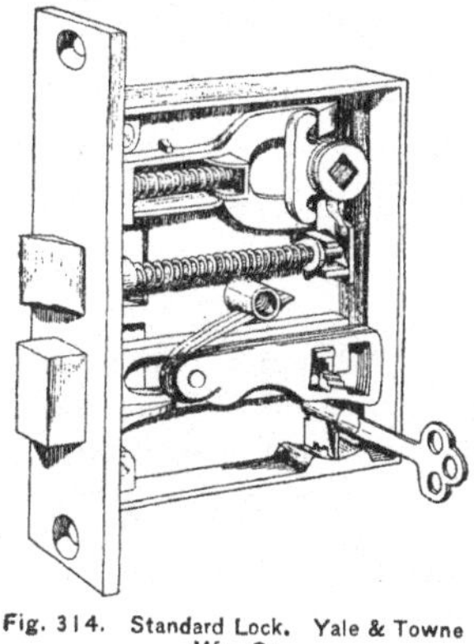

Fig. 314. Standard Lock. Yale & Towne Mfg Co.

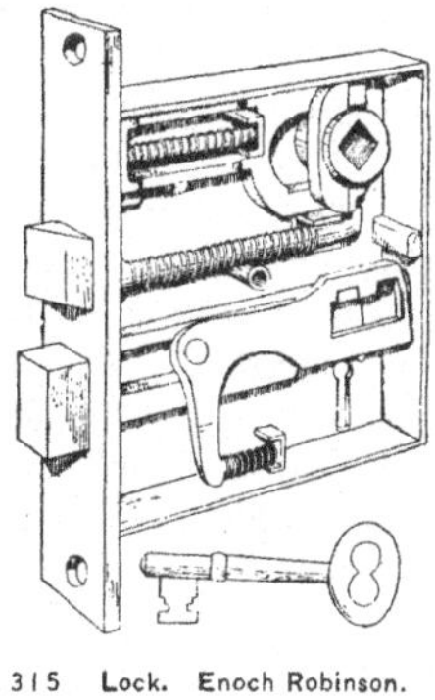

Fig. 315 Lock. Enoch Robinson.

turning the latch over.

Figure 311 illustrates a very satisfactory three-lever lock

Chapter XI.

Hopkins & Dickinson Mfg. Co.

made by the Hopkins & Dickinson Manufacturing Company. The key-hole in this example is protected by a small rotating curtain similar to those described in connection with the store-door locks, intended to aid in securing the levers from being tampered with. Figure 312 is another lock by the same company, in which the latch-springs are of phosphor-bronze, and quite ingeniously, though very simply arranged so as to give an easy spring-latch. The latch is reversible. The lock is shown with a single-lever, but is also made with three, if desired. Both of these locks are excellently finished.

Gilbert Lock Co.

Figure 313 shows a lock in which the latch is operated by a peculiar form of knob having no spindle or follow, but working against the latch mechanism with a lever at *A*. It has the same disadvantage as the "Niles" locks, that the ordinary form of knob and spindle cannot be used with it. Aside from the latch, this lock presents nothing out of the usual line.

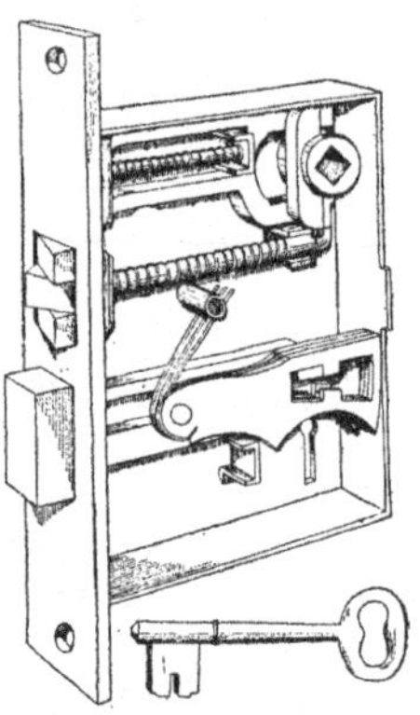
Fig. 316. Lock. Enoch Robinson.

Yale & Towne Mfg. Co.

Figure 314 is a type of a make of locks which for simplicity of design, carefulness of execution and for good lasting qualities is hardly excelled by anything in the market, except the best hand-made work. The Yale "Standard" locks, as they are termed, to distinguish them from the ordinary Yale pin locks, are made with steel levers, and brass springs, bolts and follows. They are so perfectly simple as to require no description. The best forms of springs, levers, and follows are used in these locks, so that they seldom fail to give satisfaction.

E. Robinson.

Excepting Figure 307, all of the foregoing locks are machine-made, the levers being hand-fitted only in the best grades. Figure 315 shows one of "Robinson's" cheapest hand-made locks costing $1.25 each, fitted with a single iron lever, bronze or brass being used only for the follow and the bolts. Figure 316 is a better example of Robinson's work, costing $3.50 per lock. In this the levers, as well as the bolts and the follow are of bronze, and the latch is fitted with an anti-friction strike. The interior of a machine-made lock usually is finer looking than that of one made by hand, as in the latter all the care is conentrated on the adjustment of the mechanism. There is no denying the excellence of the "Robinson" locks, at least it would be difficult to persuade many Boston builders that they are not the best to be had, and although the locks are much more expensive than the best of the Yale "Standards" or the Hopkins & Dickinson locks, they are used a great deal on all kinds of work. It is a satisfaction to know that there is one corner of this country where careful, conscientious work can command its own price, in the face of the competition which exists in the hardware trade.

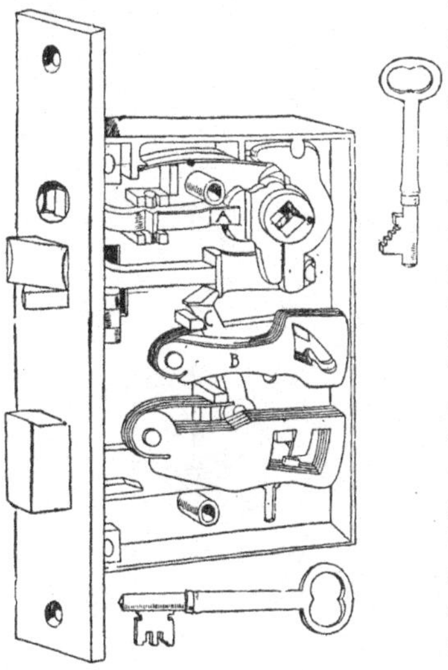

Fig. 317. Front-Door Lock. P. & F. Corbin.

FRONT-DOOR LOCKS.

The greatest amount of care and ingenuity has been expended upon the locks which are used for the front-doors of dwelling-houses, and the largest degree of complication is

Chapter XI. usually found in these goods. They afford, generally speaking, a greater security against picking than do the locks which are employed for inside-doors. The conditions of an outside-door lock are that it shall have two sets of mechanisms operated by keys, to move either bolt or latch at will, and shall have the

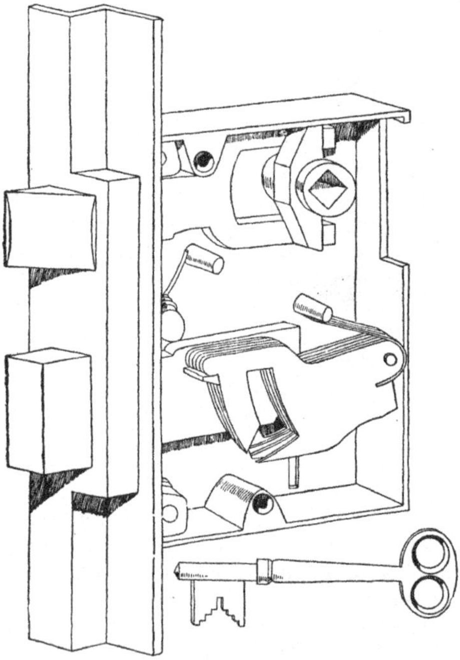

Fig. 318. Mortise Knob-Lock. P. & F. Corbin.

knob-spindle so arranged that the latch can be moved by turning either knob, and that the outside knob can be made immovable, while the inner one is free to move. Front-doors are usually two inches or more thick, and the lock can consequently be made quite thick, so as to permit of multiplication of the levers, and a stronger mechanism than for inside-doors. A front-door lock should always have an anti-friction strike. Figure 317 shows a form of front-door lock manufactured by P. & F. Corbin. This is fitted with an anti-friction strike, has four levers for both the lock and the night-latch. The follow is in two pieces. When the small catch on the face-plate over the latch is pressed to one side, the lever, *A*, is moved so as to fit in a slot on the side of the outside follow, as shown by the figure, thus holding the follow, and with it the outside-knob and spindle, so they cannot be moved. The night-key operates by first lifting the levers *B*, and by moving the lever, *C*, which carries back with it the latch-bolt. Figure 318 is a form of rebated-door lock by the same manufacturers. It is inserted here merely to show the manner in which mortise-locks are fitted to a rebated-door.

Chapter XI.

P. & F. Corbin.

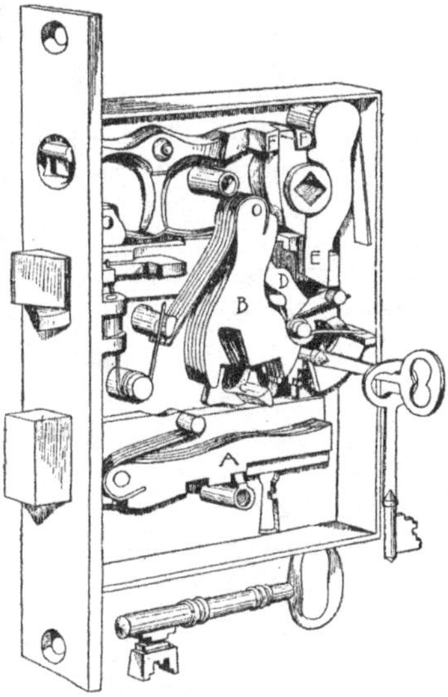

Fig. 319. Front-Door Lock. Russell & Erwin.

Figure 319 illustrates a front-door lock manufactured by Russell & Erwin. The levers on the locking-bolt, *A*, are attached to the bolt, and move with it, not being particularly proof against picking, however. In operating the night-latch, the levers *B* are pushed to one side until the gatings are on a line to permit the post to pass, the post forming part of a bent lever, the end of which shows at *D*, which portion acts

Russell & Erwin.

Chapter XI.

directly against *E*, and so draws back the latch. In order to secure the outside knob, the catch on the face-plate is pushed up, throwing the slots on the lever *F*, *F* over a shoulder on the outside-follow. Figure 320 is another front-door lock by the same manufacturers.

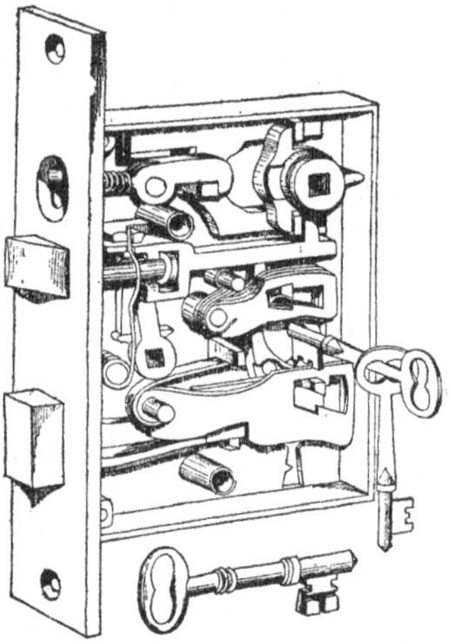

Fig 320. Front-Door Lock. Russell & Erwin.

Figure 321 is a very excellent lock manufactured by J. B. Shannon & Sons, so arranged that the knob comes between the night-latch and the lock-bolt. It will be noticed that the levers and the posts are notched in the same manner as was explained for some of the dead-locks. The latch is moved by means of a lever, *A*, underneath the upper set of levers, *A* being attached to the latch-bolt. This is a very secure lock.

J. B. Shannon & Sons.

Fig. 321. Front-Door Lock. J. B. Shannon & Sons.

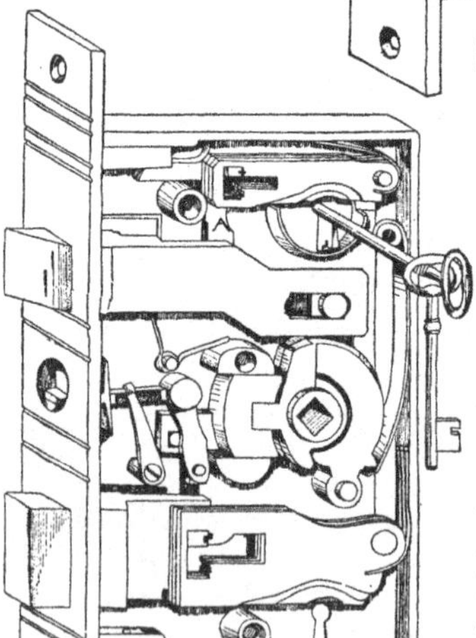

Chapter XI.

Figure 322 shows a variety of the "Niles" front-door lock, which is quite simple in its arrangement. The latch is worked by the lever *A*.

Niles.

A very simple but efficient lock is shown by Figure 323. The latch-key works through a curtain, *A*, raising the levers until the post, *B*, and with it the plate and the latch can be

A. G. Newman.

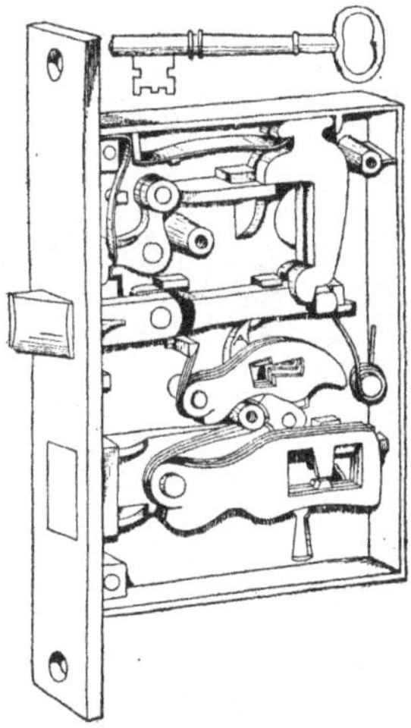

Fig. 322. Niles's Front-Door Lock. Chicago Hardware Co.

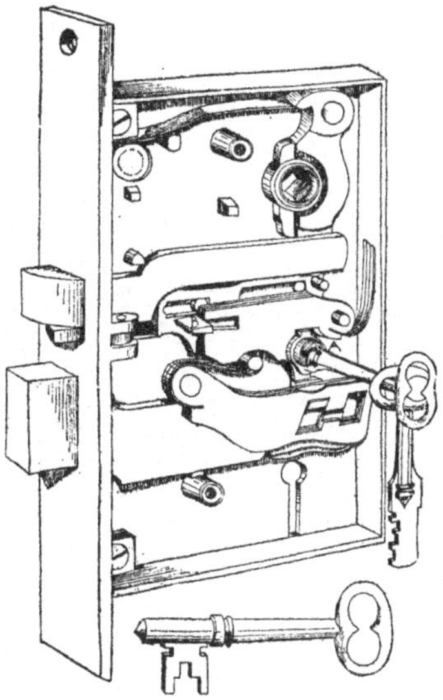

Fig. 323. Front-Door Lock. A G. Newman.

drawn back. This lock is made in the "New York" style, with a single follow, intended to receive the spindle of the inside-knob.

Hopkins & Dickinson Mfg. Co.

Figures 324 and 325 illustrate two styles of front-door locks by the Hopkins & Dickinson Manufacturing Company. The

Chapter XI.

former is rather a light lock, the latter especially strong and heavy, and fitted with five levers to both latch and lock.

Yale Standard.

Figure 326 shows one of the best of the front-door locks, the "Standard," by the Yale & Towne Manufacturing Company. There are three steel levers for both the latch and the lock. The night-key pushes the levers *B* to one side and

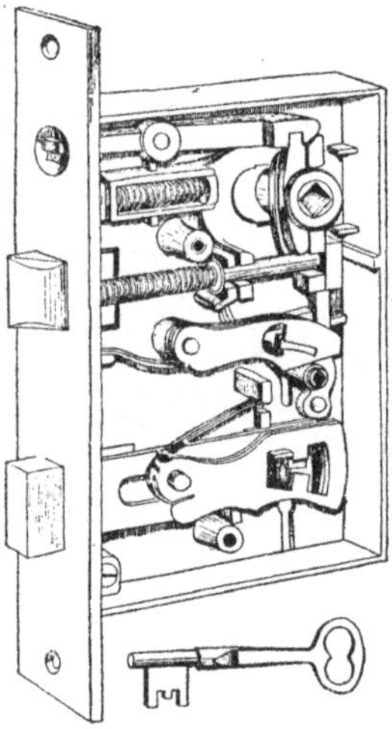

Fig. 324. Front-Door Lock. Hopkins & Dickinson Mfg. Co.

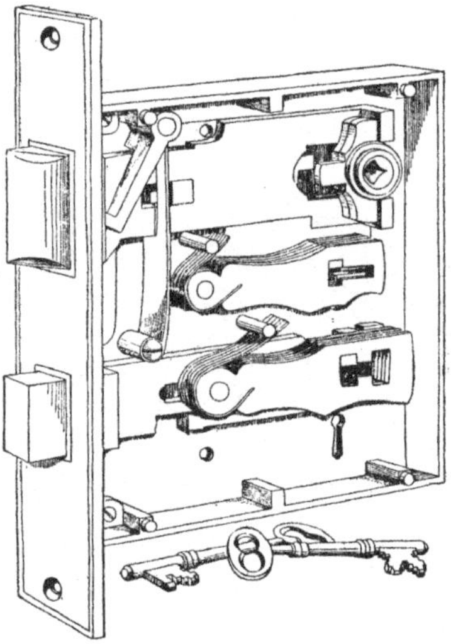

Fig. 325. Front-Door Lock. Hopkins & Dickinson Mfg. Co.

moves the bent piece *A*, which forces back the latch-bolt. The tongue, *C*, which locks the outside-knob, is pushed in or out by the buttons on the face-plate. It is not intended to use this lock with a swivel-spindle, but when the knob is locked by the tongue *C*, a spindle and cam at *D* serve to throw back the latch from the inside of the door. The arrangement of the levers *B* is defective in this lock, in that they will not work should the springs give out. Levers which act by gravity, as well as with springs, would seem to be more suitable. Chapter XI.

The lock represented by Figure 327, is one of "Robinson's" best make, being sold, with the corresponding vestibule lock, at $14 per set. It is a hand-made lock, all the mechanism being of brass. In the examples previously considered, there have been two sets of levers to each lock. In this case, however, there is Robinson.

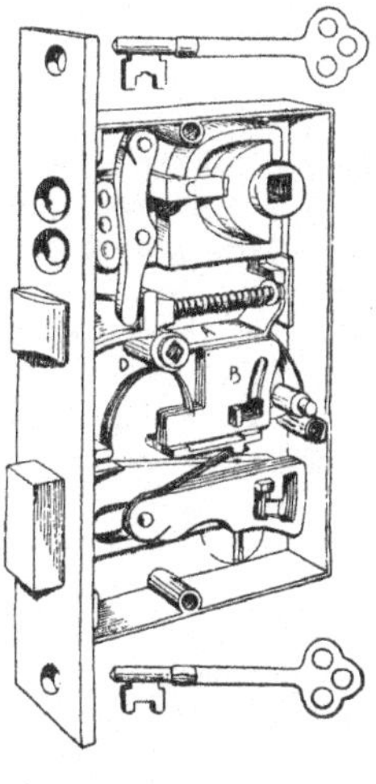

Fig. 326 Yale Standard Front-Door Lock. Yale & Towne Mfg. Co.

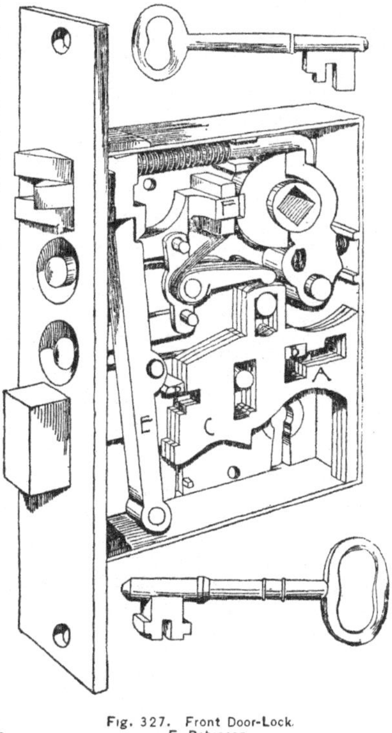

Fig. 327. Front Door-Lock. E. Robinson.

but one, the holes for the night-latch and the dead-lock key being side by side. The shape of the levers will explain the arrangements, two sets of gatings and rackings being cut

Chapter XI.

on each. The dead-lock key acts against the edges at *A*. *B* is the post on the bolt-tail, which passes through the gatings in the ordinary manner. The night-key acts against the edges *C*. The post *D* is attached to a sliding-plate, working between the levers and the dead-bolt tail. The lever *E* is pivoted to this plate and also to the lock-case. When the levers are raised so as to allow the post *D* to enter the ratchings, the plate and the lever *E* are drawn back together at the same time as the latch. The follow is made double, to permit of swivel-spindles, and the outside is locked by the arm *F*. The latch has a very easy spring, the follows being stiffened by a spring.

Hall.

Hall manufactures a front-door lock almost exactly like Figure 327, but with his peculiar anti-friction strike.

VESTIBULE-LATCHES.

Vestibule Latches.

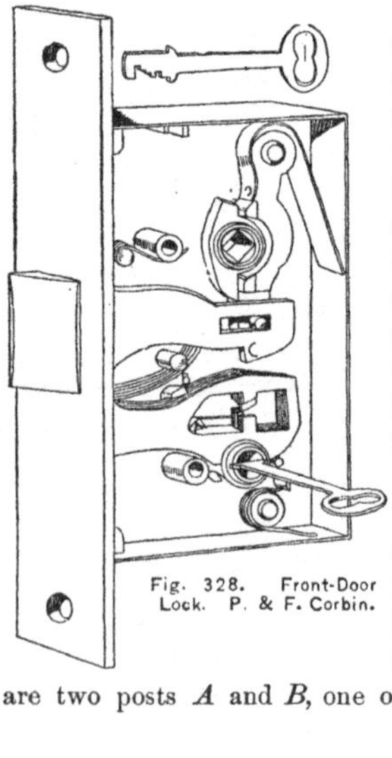

Fig. 328. Front-Door Lock. P. & F. Corbin.

These are always sold in sets, with a front-door lock, and the levers are so arranged that the same latch-key will open both, the vestibule-lock having no dead-bolt. But, more generally speaking, a vestibule-latch may be considered as any spring-lock having no dead-bolt. When used for a vestibule-door the latch should have swivel-spindles and levers to lock the outside-knob.

Figure 328 is a pattern which P. & F. Corbin list as a front-door lock, but which seems to be more properly a vestibule-latch. The key lifts the levers and moves a plate on which are two posts *A* and *B*, one of which must pass the gatings before the other can reach the shoulder on the latch-bolt *C*, and force it back. Chapter XI.

Figure 329 is the vestibule-latch sold with the front-door lock represented by Figure 327.

Figure 330 is a Standard knob-latch manufactured by the Yale & Towne Company, which is not, properly speaking, a vestibule-latch, but which is worthy of consideration in this connection. It is provided with triple-springs, thus permitting a very easy action on the part

Yale & Towne Mfg. Co.

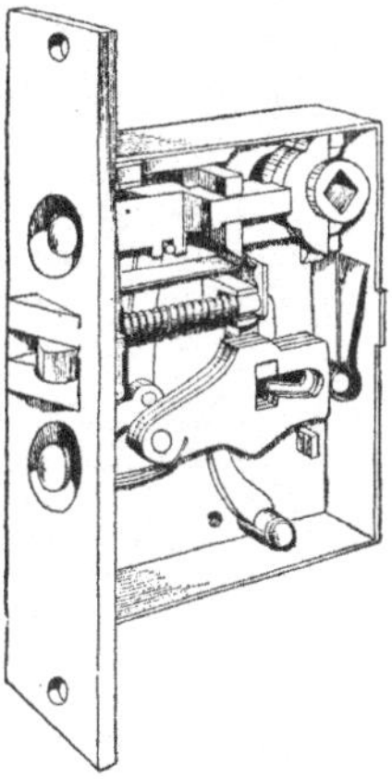

Fig. 329. Vestibule-Latch. Enoch Robinson.

Fig. 330. Standard Knob-Latch. Yale & Towne Mfg Co.

of the striker while giving all necessary strength to resist the turn of the knob. This can be adjusted to either right or left hand doors.

HOTEL—LOCKS.

Hotel-locks are usually made to order, and master-keyed in sets. In a large hotel all the locks on a floor can be opened with one key. In smaller buildings all the room-locks are master-keyed in a single series. The protection afforded by locks which are master-keyed is, of course, less than it would otherwise be, as a master-keyed lock can very easily be picked if the principle of master-keying is understood, and in most

Hotel-Locks.

Chapter XI.

cases master-keying benefits no one but the hotel-keeper. Except with the "Yale" and the "Hopkins & Dickinson" cylinder-locks, there has not yet been devised a really satisfactory system of master-keying. The two exceptions will be described in a subsequent chapter.

P. & F. Corbin.

The simplest and also the cheapest method of master-keying is illustrated by one of "Corbin's" locks, Figure 331. The

Fig. 331. Master-keyed Lock. P. & F. Corbin.

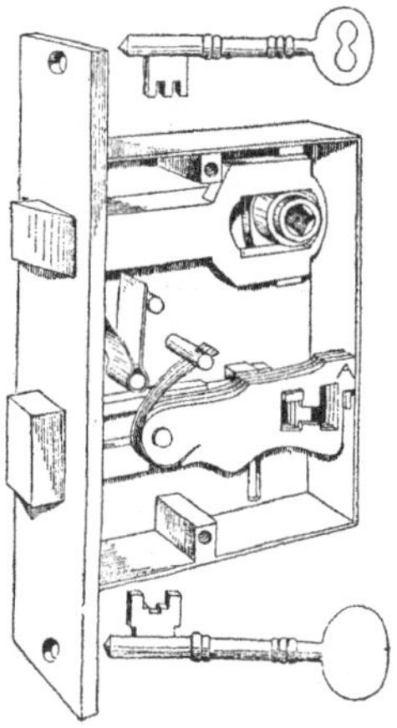

Fig. 332. Master-keyed Lock. Hopkins & Dickinson Mfg Co.

gating on the one lever is made so wide as to admit of fifty different positions, in any one of which the bolt-post could pass. The room-key raises the lever so as just to clear the top of the gating, and the master-key allows the post to clear the bottom of the gatings. A bent wire would serve quite as well for opening the lock as either of the keys. Fortunately for occupants where such locks are used, it is customary to fit hotel-locks with a small bolt, worked from within. Figure 332 is much better. The levers are exactly like those of any ordinary lock, except that there is a shoulder *A* at the back of

Hopkins & Dickinson Mfg. Co.

each. Beneath the bolt-tail is a fourth lever, with an arm on it, rising so as to catch under the shoulders *A*. This lever is protected by a ward about the key-hole. The room-key lifts the levers and shoots the bolt without disturbing the fourth lever. The master-key lifts the fourth lever without touching

Chapter XL.

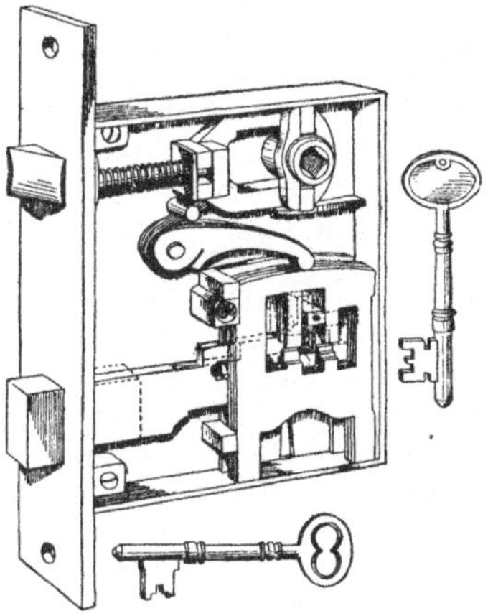

Fig. 333. Master-keyed Lock. Hopkins & Dickinson Mfg. Co.

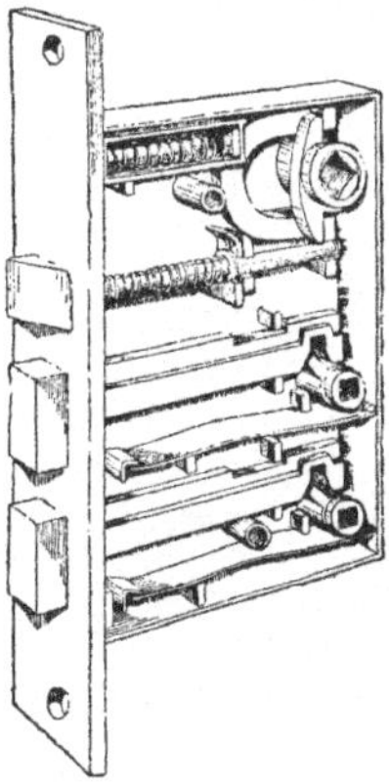

Fig. 334 Hotel-Lock Hopkins & Dickinson Mfg. Co

the others, the shoulders being so sized that the master-key lever will bring the gatings on the locking-levers into line.

Figure 333 shows another form of master-keyed lock by Hopkins & Dickinson. In this instance the regular key and the master-key work from either side of the lock in the same key-hole on the same tumblers and bolts. Still, each has a different set of tumbler-rackings and a different post in the bolt. When the master-key is used the bolt-post for the regular key is thrown down by a patent device, and another post brought up in the second rackings of the tumblers. When the master-key is removed the lock is set in use for the regular key. It is

Chapter XI.

claimed that 1,200 of these locks can be made, all different, each lock with a key of it own which will fit no other, and with master-key to pass all. This is a rather expensive lock, however, and on that account is not used a great deal. The idea is an exceedingly ingenious one.

Figure 334 shows a Hopkins & Dickinson lock, or rather bolt, used for hotel and office doors between connecting rooms. This is intended to be used when it is desired to have the door definitely locked from either side, so that it cannot be unlocked from the other side, and, accordingly, the handles which operate the bolts are placed on opposite sides of the doors. The same company also manufactures a hotel-lock which is so arranged that the locking-bolt can be operated from the inside by a turn-button, instead of a key. When the door is locked from the outside it can at any time be opened from within by turning the button, so that it is impossible for an occupant to be locked in the room.

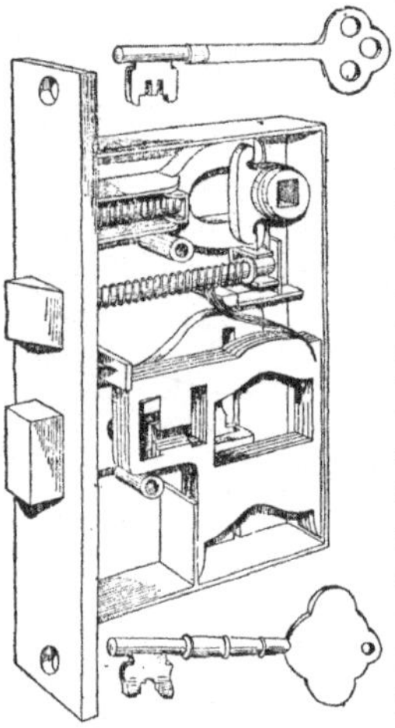

Fig. 335 Standard Hotel-Lock.
Yale & Towne Mfg. Co.

Figure 335 shows the construction of a Yale "Standard" hotel-lock. In this case the master-keying is provided for by a second set of rackings cut in the levers, so that almost any number of variations can be had in a given series of locks, the variation being entirely in the lower set of rackings. The room-key lifts the levers exactly the same distance as the master-key, but as the proportion between the lengths of the bits, and the height of the lever bellies above the lower key-hole is different in each lock, it is easily understood why no two locks can be opened by the same room-key.

Yale & Towne Mfg. Co.

Chapter XI.

CYLINDER LOCKS.

The broad and general principle which distinguishes the ordinary lever-lock from the style of lock manufactured under the Yale patents, is that in the latter the mechanism upon which the key directly operates is entirely distinct from the lock itself, being enclosed in a cylinder or escutcheon. The function of the key consists simply in so arranging certain movable pins, slides or other obstructions, that the mechanism is free to rotate, and by its movement, to operate on the locking-bolt. This variety of lock is by no means without a prototype, as we have already seen in the case of the "Egyptian," the "Bramah" and the "Cotterill" locks; but in its application it has been simplified and reduced to a marketable form chiefly in this country, and can be fairly claimed as a product of American ingenuity.

Yale Patent.

Linus Yale invented the lock which bears his name, about thirty years ago. His original patents covered substantially only the use of a flat key to operate a locking mechanism, a series of vertical pins of unequal lengths being lifted by means of certain nicks or irregularities on the upper edge of the key, so that the ends of the pins were brought on a line. Within recent years an important change has been made in the construction of the Yale escutcheon. The slot through which the key reaches the pins is now cut in sharp corrugations, the key being corrugated longitudinally so as to exactly fit the slot. By this simple device, the "Yale" locks have been rendered practically proof against any but the most expert lock-pickers. The external appearance of the "Yale" lock is presumably familiar to every one, but the internal construction will require some explanation.

Yale Mechanism.

Figure 336 shows a cross and a longitudinal section through a typical Yale escutcheon, together with the exposed face of the same. It will readily be seen that the action of the mechanism is very simple. There are two barrels or cylinders, one rotating within the other, but eccentric with it. When

Chapter XI.

the key is withdrawn the lower cylinder is held from rotating by means of five sets of round pins which are fitted in vertical grooves extended partially through the two cylinders, and pressed constantly downward by five spiral springs. In each groove are two pins of unequal lengths, one over the other. When the proper key is inserted all the pins are raised simultaneously, but to varying heights, so that the joints between the

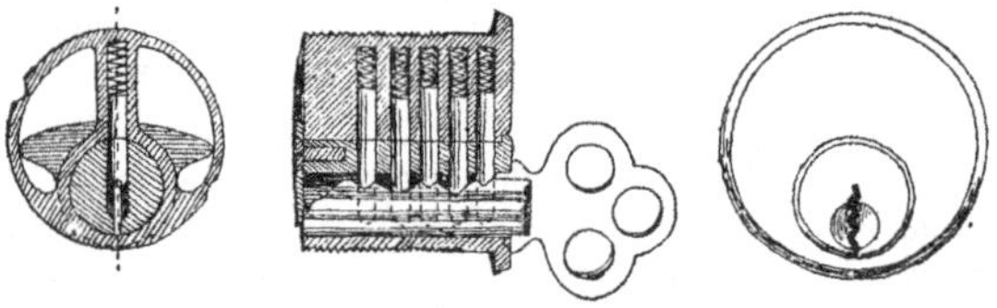

Fig. 336. Mechanism of the Yale Lock. Yale & Towne Mfg. Co.

upper and the lower pins are brought exactly on a line with each other. It is evident that as the inner cylinder, categorically designated as the plug, is exactly fitted to the bore in the shell, an almost imperceptible variation in the height to which any one of the pins is raised, will prevent the plug from turning; whence it follows that an immense number of locks can be made with such mechanism without duplication. From this results the unrivalled capacity of the "Yale" lock for permutations, with its proportionate safety against any accidental interchange of keys.

It will be seen that in this lock the key acts only as an adjuster of the pins. Motion is communicated to the locking-bolt of the lock simply by means of a hub on the back of the rotating plug, or, in the case of a rim-lock, by a flat key extending from the plug through the door. Some of the opponents of this system consider that in it, too much is demanded of the key, but when the locks are otherwise as nicely arranged and evenly balanced as the "Yale & Towne" goods are usually found to be, the amount of twisting strain required to move the bolt is really not a great deal. In no well-made lock should there be any great strain on the key,

Chapter XI.

much less in such a device as this, wherein there are no strong lever-springs to work against.

It will easily be appreciated that this device has almost revolutionized the lock-trade in this country. Not only has it opened the way for many valuable inventions of a similar nature, but it has stimulated the perfecting of the ordinary lever-locks, and was instrumental in the abandoning of the old style of heavy door-keys, so that one's pockets are no longer burdened with such keys as were thought indispensable forty years ago.

Advantages of Yale Locks.

The advantages claimed for the Yale lock are as follows:

First, a key of the smallest size and most convenient form.

Second, immense capacity for changes or permutations, so that more thousands of changes are possible than an equal number of dozens with the old systems.[2]

Third, great safety against picking.

Fourth, uniformity of size of the key for locks of all kinds and for all purposes.

Fifth, protection against accidental interchange of keys by reason of the great capacity of the lock for permutations.

In regard to the third point claimed, it must be remembered, however, that with all its security the Yale lock does not offer an exception to the general rule that any lock can be picked which is operated by a key. Still, very few persons have the nicety of touch necessary to raise the pins by means of fine instruments inserted through the key-hole, and bring them exactly to the position necessary for moving the plug. There are experts who claim to be able to open any "Yale" lock which has been made, but for all practical purposes a lock of this sort affords absolute security, as the time required to pick it renders it very unlikely that any thief would be so indiscreet as even to make the attempt.

It will be understood that the zig-zag corrugations extend

[2] Assuming that a variation of one-fiftieth of an inch in the length of a pin is sufficient to lock the plug, 267,331,200 locks can be made on this system, no two of which can be operated by the same key.

Chapter XI.

entirely through the length of the plug. In a measure, this feature prevents any duplicate key from being manufactured by persons not authorized to do so, as it requires very heavy and specially made machinery to produce one of these keys, and unless the corrugations exactly correspond with the lock, the key cannot enter. The plugs are cut by a peculiar form of band-saw specially designed by the manufacturers; and altogether it seems as if every precaution had been thought of which could render the lock more inviolable.

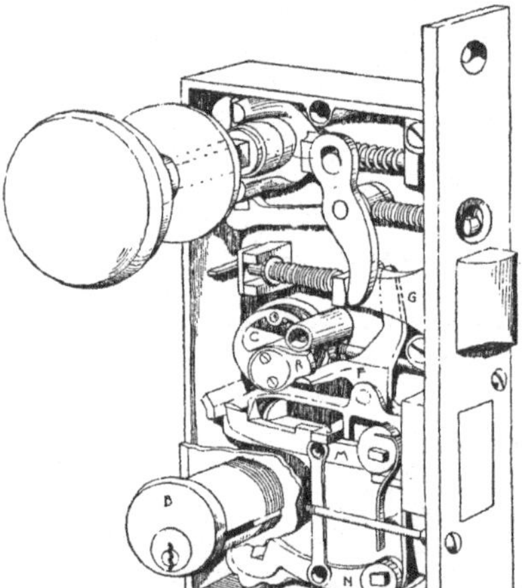

Fig. 337. Yale Front-Door Lock. Yale & Towne Mfg. Co.

Like a great many other successful inventions, the Yale locks are remarkable for their simplicity. The whole of the mechanism being practically combined in the escutcheon, there is no necessity for any complicated system of levers or springs in the lock proper, and there remains very little to get out of order. The older plugs, made with a straight slot, would allow a certain amount of vertical play to the key, so that it would rock in the cut and would not always exactly lift the pins; besides which the slot permitted the lock to be picked with comparative ease. This is entirely obviated by the corrugated

Chapter XI.

slot, as already explained. It will be noticed also that the lock is not in any way dependent upon the springs, as the pins would act by gravity, even should the springs give out entirely.

Yale Master-Keyed Lock.

It would seem almost an impossibility to master-key a series of Yale locks, and yet it is accomplished in two different ways. The first is to fit each lock with a separate master-escutcheon, practically making a double lock, though both sets of escutcheons act on the same locking-bolt. By this method a million locks could be master-keyed in a single series, if desired. The second way is to use three pins in each slot instead of two, the lengths of the pins being so adjusted that, throughout the series, the upper joints can be brought on a line by the master-key, while the lower jointings are all different, and fitted to the individual room-keys. This method necessitates a larger and more cumbersome plug and cylinder, and is seldom used.

Yale Front-Door Lock.

Yale locks are manufactured in all styles and for all purposes, but the escutcheon is always arranged in exactly the same manner, whether intended to operate a night-latch or a desk-lock. The variations consist mainly of differences in the form of the latch or of the lock. A single example will be sufficient to illustrate the whole. Figure 337 represents one of the most perfected forms of Yale front-door lock. *C* and *B* are the two escutcheons, each with a cam, *R*, attached to the back of the plug. *M* and *N* are two levers hinged to the bolt-tail. *L*, *F* is a bent lever, hinged to a flange of the bolt-tail, and catching under a hub on the bolt of the latch. The dead-bolt can be operated from either side, the cams first depressing the levers so as to pass the post, *S*, and then shooting out the bolt in the same manner as with an ordinary key. When the dead-bolt is unlocked the end of the lever *F* takes the position shown by the figure. If the cam *R* is then turned to the left, it so acts on the lever as to cause it to draw back the latch, *G*. Consequently a single key serves both to unlock the dead-bolt and to draw back the latch.

The "Yale" lock has, of course, won for itself a host of

Chapter XI. imitators in the hardware trade. The closest approach to the "Yale" system is embodied in an escutcheon lock manufactured by P. & F. Corbin. Figure 338 illustrates this. The internal arrangement is exactly the same as in the "Yale" lock, so far as relates to the pins, etc., but the plugs are cut with square-edged, instead of zig-zag slots. These slots, also, are not carried entirely through the plug, but extend only through a thin face-plate, behind which is a wide slot exactly like that of the original "Yale" locks. This seems like an imitation of, but in nowise an improvement on the original, and is considered by the Yale & Towne Manufacturing Company as an infringement on their patents.

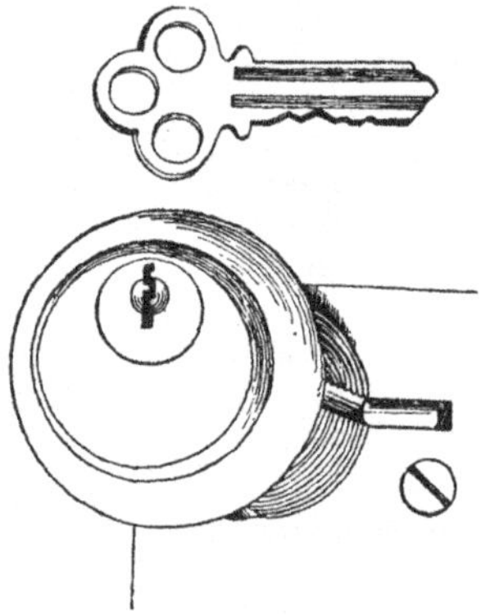

Fig. 338. The Harvard Lock. P. & F. Corbin.

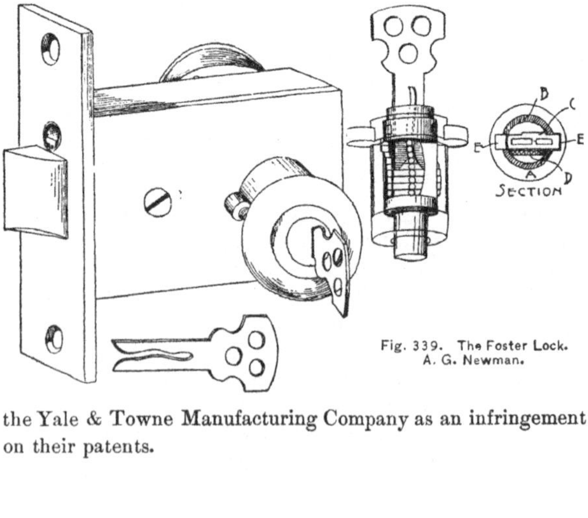

Fig. 339. The Foster Lock. A. G. Newman.

Foster Lock.

Figure 339 illustrates the "Foster" lock, manufactured by A. G. Newman, a very ingeniously devised lock, which is harder to pick than the "Yale," and, as put on the market, shows the greatest of care in workmanship and finish. The cross-section of the escutcheon shows the internal construction. The outer shell, *A*, is fixed to the lock-case. The plug, *B*, is hollow, and fitted with ten slides *C*, which work through cuts in the side of the plug and catch in slots, *E*, *E*, cut in the shell, so that the plug cannot rotate until the slides are withdrawn. Half of the slides protrude from the plug towards the right and half towards the left; each slide being fitted with a small brass spring, *D*. The key is cut with an irregular cleft, and the slides are cut out, with a cross-piece near the centre. The cross-pieces, and the sinuosities of the cleft in the key are so mutally spaced that when the key is inserted all of the slides are drawn in and the ends no longer protrude but are flush with the surface of the plug, which is then free to rotate. It is believed that this lock is unique of its kind, and, though in outward appearance much like a Yale lock, it is decidedly original in every other respect.

Hopkins & Dickinson Cylinder Lock.

A form of cylinder-lock has recently been put on the market by the Hopkins & Dickinson Manufacturing Company, which partakes somewhat of the nature of the old "Bramah" lock, previously described. Figure 340 illustrates the external appearance as well as the internal construction of the escutcheon or cylinder, whose functions are the same as in the Yale lock. The shell, *A*, is secured to the lock-case so as to be immovable. The plug *B*, rotates inside of this, being held in place by screws, *C*, turned through the outer shell. Inside of the plug are five

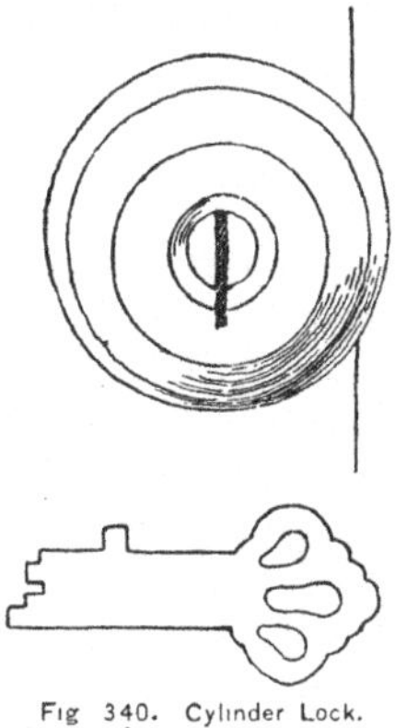

Fig. 340. Cylinder Lock. Hopkins & Dickinson Mfg. Co.

Chapter XI.

slides, *D*, working in a closely fitted groove, with a separate spring to each slide. The springs are on opposite sides, in separate slots, so that there is no chance for the slides to rock. The key is flat, with five notches on the end corresponding to the five slides. It is inserted through a straight slot in a capping-piece, *E*, and bears against the bottom of slots in the centre of the slides. At the back of the plug is a flat piece of metal, known as a fence, *F*, working up and down

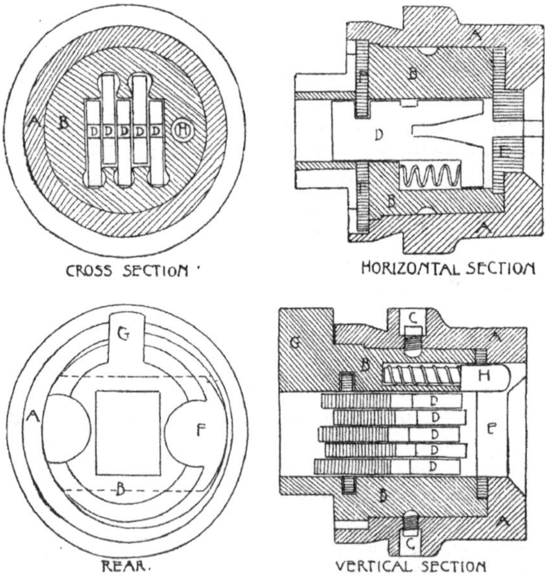

Fig. 340. Cylinder Lock. Hopkins & Dickinson Mfg. Co.

in grooves, with a hole through the centre sufficiently large to allow the ends of the slides to protrude by it. The top of each slide has one notch in it the same width as the thickness of the fence, at varying distances from the key-hole, besides one or more false notches of lesser depth. The plug is extended with an arm, *G*, by which the lock-bolt is operated. Chapter XI.

The mechanism operates as follows: The fence is in the plane of an eccentric groove or ward cut on the back of the shell, as shown by the figure. This eccentric groove is so located with reference to the centre of rotation of the cylinder that when the plug is turned, the longer arm of the fence is forced to one side, the amount of eccentricity being sufficient to firmly wedge and hold the plug, in case the fence should not be free to move laterally. When the key is inserted, a shoulder on it first presses back a pin, *H*, which works in a slot so as to hold the plug and the shell together and prevent accidental rotation. The cuts on the end of the key then force back the slides in such ratio that all the deep notches are brought exactly on a line with the plane of the fence. The key is then turned, rotating the plug, bringing the fence to bear against the walls of the eccentric groove, and forcing it down into the notches of the slides, these notches being of sufficient depth to allow the fence to entirely follow in the eccentric groove. The arm, *G*, can thus operate on the locking-lever.

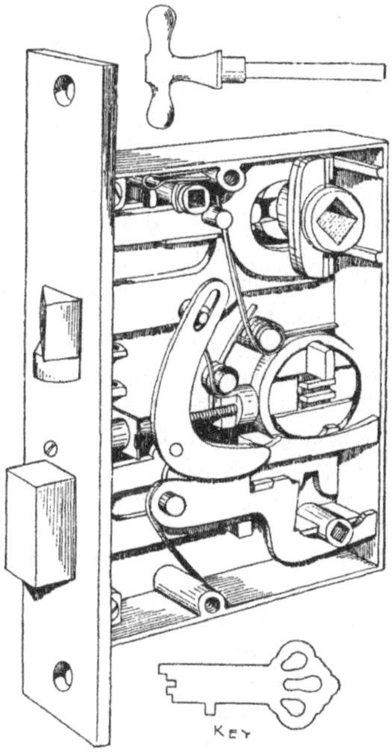

Fig. 341. Cylinder Front-Door Lock. Hopkins & Dickinson Mfg. Co.

Chapter XI.

The shallow notches on the slides are intended as a safeguard against picking. By turning the plug with a knife blade, the fence can be brought to bear against the slides. Slight inequalities in the width of the slides cannot be avoided, and the widest slide will bind most firmly against the fence, so that by depressing the slides successively with a fine pick one might in time be able to catch all the notches over the fence, and so undo the lock, were it not for the false notches which are so confusing that it is extremely difficult, and for most persons, impossible to pick the lock.

Hopkins & Dickinson Front-Door Cylinder Lock.

Figure 341 illustrates an adaptation of this escutcheon to a front-door lock. The works are ingeniously arranged so that the key will operate both the dead-bolt and the latch, while at the same time the dead-bolt can be shot back by a turn-button and spindle from the inside of the door. The illustration is too clear to require any detailed description. This kind of escutcheon or cylinder can, of course, be applied to any form of lock, though thus far it has been used by the manufacturers only in connection with front-door and office-door locks.

Many improvements have been made in the mechanism of this lock during the past six months, and the most thorough study and care have been given to perfect it in every way. The first samples put on the market were deficient in many respects, but the lock as now offered to the trade is about as perfect in every way as anything of the kind which has thus far come before the public. It has excelled everything except the Yale locks, and indeed there is little that can be said of the "Yale" which does not apply with equal force to the Hopkins & Dickinson cylinder-lock. It is well made, compact, not liable to get out of order, easily repaired and practically burglar-proof.

Winn Lock.

The patents to a very interesting cylinder-lock are controlled by the Yale & Towne Manufacturing Company. The "Winn" lock, Figure 342, is so peculiar in its workings that even after taking it apart it is hard to follow the movements it makes in unlocking. The outer cylinder is secured to the lock-

Chapter XI.

case and to the door, so as to be immovable. Inside of it rotates the plug, a section of which is cut away to allow for a slide-holder, *A*, which is free to move in and out. Inserted in the face of the holder is a pin, *B*, projecting sufficiently to catch in a groove which is cut out from the inner surface of the outer cylinder-barrel, the groove following a waved line, so that when the plug is rotated, the slide-holder is first drawn

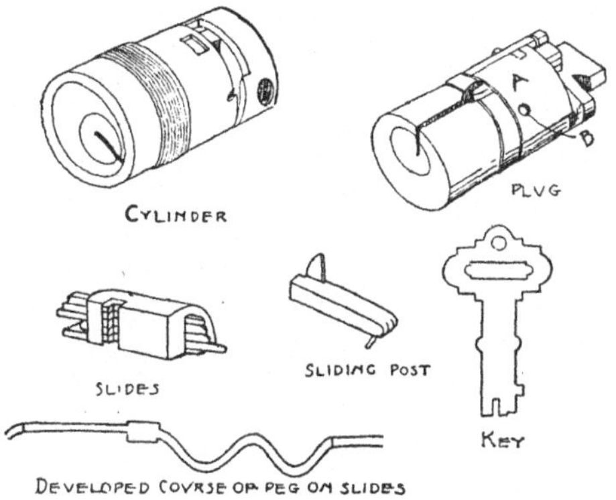

Fig 342. The Winn Cylinder Lock. Yale & Towne Mfg Co.

Winn Lock.

away from the key-hole, then back, then away again. The slides are flat pieces of steel, one-twelfth inch wide at the ends nearest the key-hole and one-sixth inch at the other, and are each notched on one edge, at varying distances from the end. There is also a sliding-post which passes through the plug behind the slides, which is a little longer than the diameter of the plug, so that one end of the post must project through a short slot in the outer cylinder-barrel. The key, when inserted in the plug, sets the slides by means of the nicks on the end, bringing the slots exactly on a line. The plug being then rotated, the peg, *C*, carries the slides and the slide-holder away from contact with the key, the notches remaining set on a line. After performing a quarter revolution with the plug, the projecting end of the sliding-post encounters an obstacle tending to force it out on the opposite side of the plug, and the notches on the slides being on a line, a fence on the sliding-post slips into the notches, and the plug can continue to rotate. Before a complete revolution is effected, the slides encounter a fixed obstacle which forces them back to their original position, the alignment of the notches being destroyed. The connection between the plug and the bolt of the lock is the same as in all the cylinder-locks.

A little reflection will convince one how futile would be any attempts at picking this lock. The key simply sets the slides and acts as a lever to rotate the plug. The slides are all pointed on the ends towards the key, and a very slight experience is sufficient to show that the lock cannot be picked at all. Indeed, this is the worst thing about it from a commercial point of view, as few people care to have a door-lock so impregnable that the door has to be broken in every time the key is lost.

There are several other styles of cylinder-locks, in which the key operates on levers instead of pins; also several varieties which have much the same appearance as the Yale locks. None of these, however, present any striking peculiarities, and being used more for cabinet work than for doors, they hardly came within the scope of this discussion.

COMBINATION DIAL-LOCKS.

Dial-locks are used almost exclusively for safe and vault work, and so cannot be included under the general topic of Builders' Hardware. But, representing, as they do, the highest degree of perfection in the line of locks, a brief statement of the principles upon which they are constructed and worked, may not be out of place.

The external appearance of a dial-lock is familiar to every

Chapter XI.

one, consisting of a rotating disk, graduated around the circumference either with letters or with numbers. To operate the lock, the knob attached to the dial-disk is turned a certain number of times to one side, then to the other, etc., stopping each time on a certain number or letter, until the combination is set, when a single turn of the knob draws back the bolt. The internal arrangement consists of a series of flat, circular disks or tumblers, which rotate freely on the spindle of the dial-knob. In the edge of each tumbler is a notch, and the inner-most tumbler is made with a dog which catches the tooth of a lever attached to the bolt. This inner tumbler is made fast to the spindle. On each face of each of the tumblers is a small peg, all the pegs being placed at the same distance from the centre of rotation ; so that when the spindle is turned, the peg on the first tumbler strikes against the peg on the second tumbler, causing the latter to rotate, and in turn to start the third, and so on, so that with a four-tumbler lock, turning the spindle four times to the left moves the fourth tumbler to any desired number; turning next three times to the right adjusts the third tumbler, but does not disturb the adjustment of the fourth; then turning twice to the right adjusts the second, but does not disturb the other tumblers. When the slots in all the tumblers are brought to a line, a bar drops into them, permitting the bolt-lever to catch in the teeth of the first or locking-tumbler, when a single revolution will draw back the bolt.

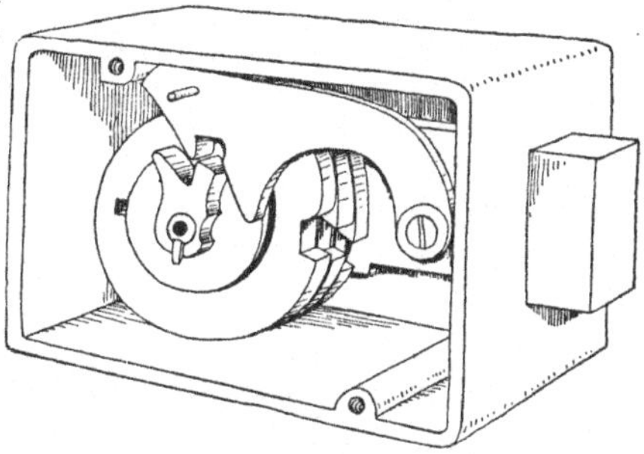

Fig. 343. Dial-Lock. Damon Safe and Lock Works.

Chapter XI.

Damon Dial-Lock.

A single lock will illustrate the subject sufficiently for our purpose. Figure 343 shows the works of one form of safe-lock, used by the Damon Safe and Lock Works; and though this is a cheap lock, it embodies all the essential principles of every combination lock. This lock is susceptible of 755,000 different combinations, but some bank-locks afford as many as 134,000,000 changes.

There is absolutely no way to pick such a lock as this, except by "ringing the changes," that is to say, by making successively all the possible combinations, until the right one is found.

Combination locks cost from five dollars for the cheapest kind, to several hundred dollars for the most perfect styles of time locks.

MISCELLANEOUS LOCKS.

In addition to the regular lines of lever and cylinder locks, there are several forms which may be considered in this connection.

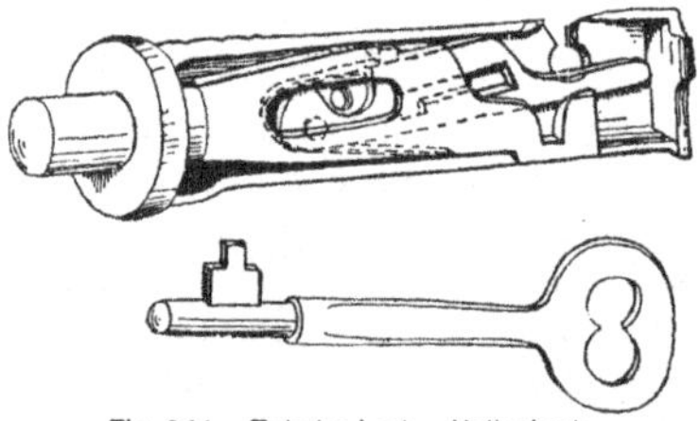

Fig. 344. Tubular Lock. Hollenbeck.

Tubular Locks.— Some cheap styles of lock are manufactured of such form that all the mortising can be done with an augur, being essentially the same in principle as the mortise door-bolts described in a previous chapter and illustrated by Figure 56. Figure 344 shows the construction of the "Hollenbeck Tubular lock." It is too simple and cheap to afford any very great degree of security as compared with an ordinary three-lever lock, but for some cases it would answer very well, as it saves seventy-five per cent of the labor ordinarily necessary to fit a common lock to a door. It is held firmly in place by the lugs at top and bottom, so it cannot work loose. Hollenbeck also manufactures a tubular latch on essentially the same principle. Several other firms have tubular locks listed in their catalogues, but they are too much alike and too simple to require further illustration.

Electric Locks.

Electric Locks. — It is often desirable to have a lock which can be operated by any one at a distance from the door. In apartment-houses, clubs, etc., it is well to fit the front-door with a lock so connected with an electric battery that when a knob is pressed in an upper story a catch in the lock is drawn by the action of an electro-magnet, permitting the lock or latch to be moved. Any form of lever-lock might be adapted to this purpose, but there are a few forms of specially designed electric locks which are more commonly used. Properly speaking these are all electric-latches, as none of them have a locking bolt. Figure 345 illustrates "Thaxter's" electric lock. The pressure of a button closes the circuit through the electro-magnets, *A*. These act on the bent lever so as to release the arm, *B*, from its catch on *F*. The spring at *C* draws back *F* and *D* from the follow, *E*. The outside knob can then be turned and the door opened. When the latch is drawn back by closing the door, it carries with it the arm *F*, which resets itself so that the bolt *D* catches in the follow and locks the door. The latch is also fitted with a set of levers, so it can be operated by a key, independently of the knob.

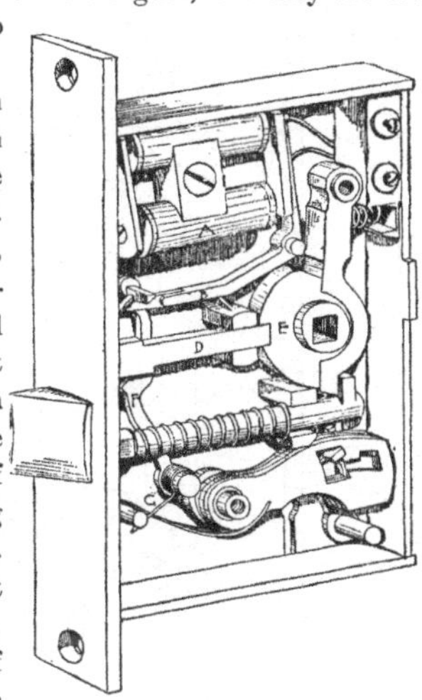

Fig. 345. Electric Lock. Thaxter.

"Fuller's" electric lock, Figure 346, is a trifle simpler. The

Chapter XI.

magnets draw the armature *A* away from the cam, *B*, permitting the knob to be turned. When the door is closed the latch lifts the bent arm, *C*, and forces back the armature under *B*.

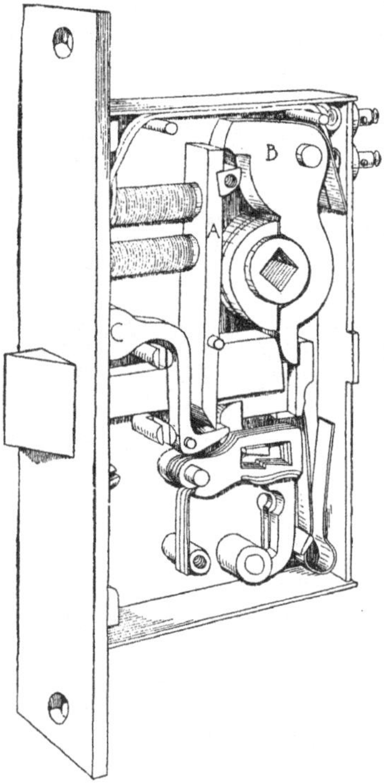

Fig 346 Electric Lock. Fuller & Holzer.

The "Thaxter" and the "Fuller" locks are the ones most commonly employed in and around Boston, though there are several other makes in the market, most of which are, however, asserted to be infringements of the patents.

Sliding-Door Locks.

Sliding-door Locks. — Figures 347 and 348 illustrate two types of sliding-door latch and lock. The locking mechanism used for this purpose is usually quite cheap in its construction, as a finely fitted lock is seldom required for sliding-doors. Indeed in many cases no lock at all is necessary. The bolt is curved and hooks down into the face-plate on the opposite door or on the jamb. The door-pull is either in the form of a hinged-lever, as in Figure 347, or a straight pull reinforced by a concealed spring, as in Figure 348. Both pulls can be pushed in flush with the face-plate. In some localities it is thought desirable to use

Chapter XI.

knobs on the sliding-doors, one set of knobs working the hook-latch, while the other knobs are simply dummies. In this case the key is used to lock the latch-bolt.

Drawer and Wardrobe Locks.

Drawer and Wardrobe Locks. — These are more properly associated with cabinet-work than with builders' hardware, and will not be considered at any length. Drawer-locks are made

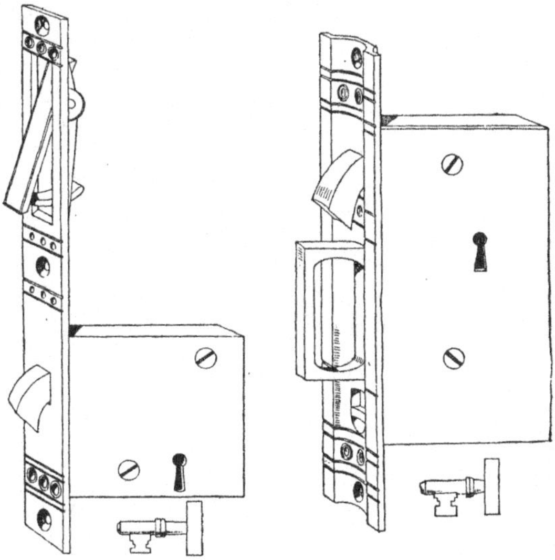

Fig. 347. Sliding-door Locks. J. B. Shannon & Sons. Fig. 348.

in a great variety of sizes, from one to three and a half inches deep, and in all grades, from a simple bolt worked by the key, without levers of any sort, affording no real protection against intrusion, to the locks which are operated by Yale cylinders, having all the latest improvements of the Yale system, and being practically unpickable. Figures 349 and 350 illustrate two good types from the great variety of locks used for wardrobes and small closet doors. The first shoots a bolt up and

Chapter XI.

down and is a fair two-lever lock. The second shoots a double bolt horizontally. Both are gained into the inner face of the door.

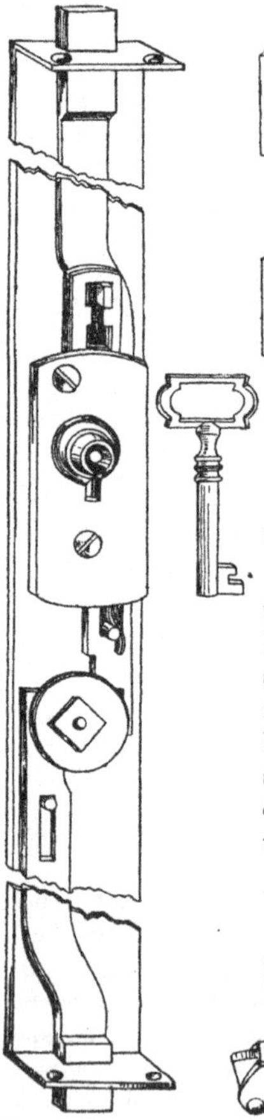

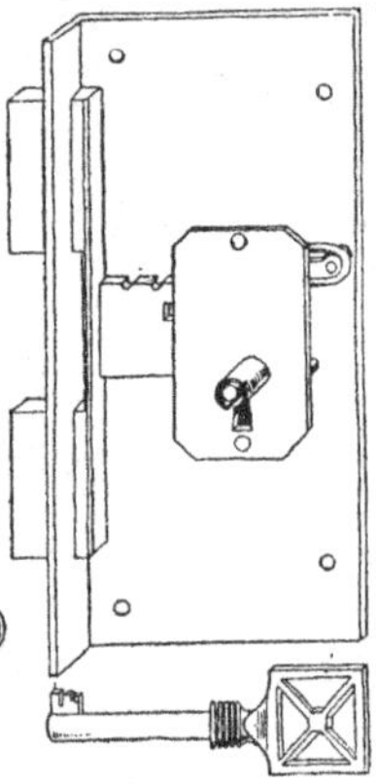

Fig. 350. Wardrobe Lock. J. B. Shannon & Sons.

Corbin Post-Office Lock.

The Corbin Cabinet Lock Company has recently put on the market a very ingenious change lock, intended specially for post-office boxes. It is somewhat upon the principle of the Day & Newell "Perautopic" lock previously described. Figure 351 shows the lock with the face-plate removed. Each lock can be locked by any one of a series of keys which can be extended in number almost indefinitely, all the keys being different in the arrangement and spacings of the notchings. But the bolt can be unlocked only by the key which was last used in locking it, so that the

Fig. 349. Wardrobe Lock. A. G. Newman.

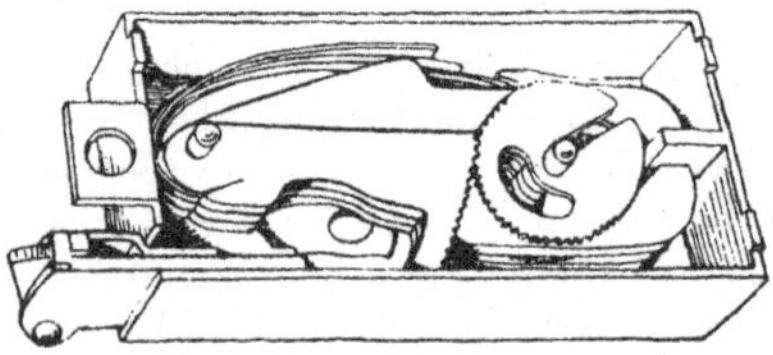

Fig. 351. Post-office Lock. Corbin Cabinet Lock Co.

Chapter XI.

key can be changed as often as desired. In case the key is lost, an arrangement at the back of the lock permits the post-master to open the box and throw back the bolt, when a new key can be used, without in any way changing the lock, and the key which was lost would not then work the lock at all. Furthermore, the bolt is so arranged that it will turn back only sufficiently to permit the box to be opened, but not enough to allow the key to be withdrawn, unless the bolt is forced back by external pressure. The working is as follows: The upper levers are pivoted so as to permit of a rotary as well as a longitudinal motion. The second set of levers moves only

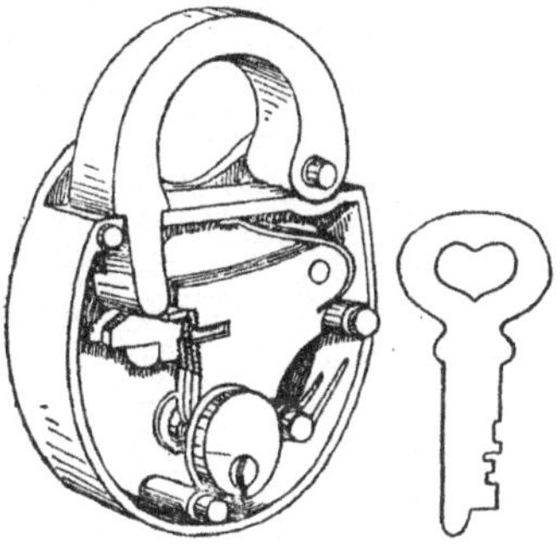

Fig. 352. Padlock A. E. Dietz.

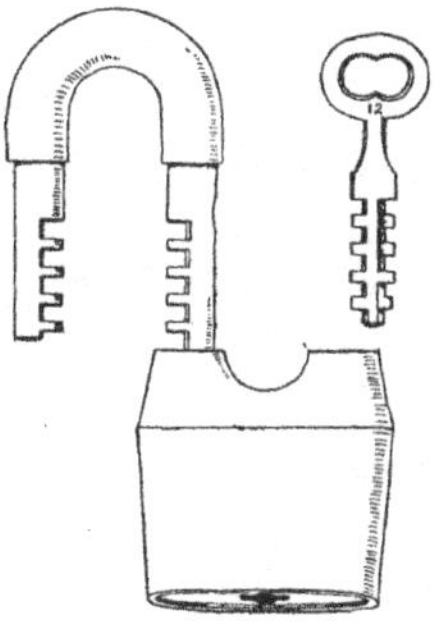

Fig 353. Scandinavian Padlock. Russell & Erwin.

laterally. The opposite edges of both sets of levers are notched, the width of the notches corresponding to the difference between the notches on the keys. Suppose the bolt to be unlocked: when the key is turned, the lower levers are first pushed to one side varying distances, corresponding to the notches of the key, and the upper levers are then drawn down and away from the post.

As the key continues to revolve the levers interlock and the lower ones are forced sidewise by the springs, carrying with them the pivoted upper levers, which rotate so that the slot in

Chapter XI.

each lever no longer comes opposite the post. At the same time, the bolt is shot out. It is evident that the action would be the same, no matter what key were used, only the sets of levers would not interlock in exactly the same relation. It is also evident that the only key which will rotate the upper levers so as to bring each slot opposite the post and permit the key, in turning, to draw back the bolt, is the key which last made the combination between the two sets of levers.

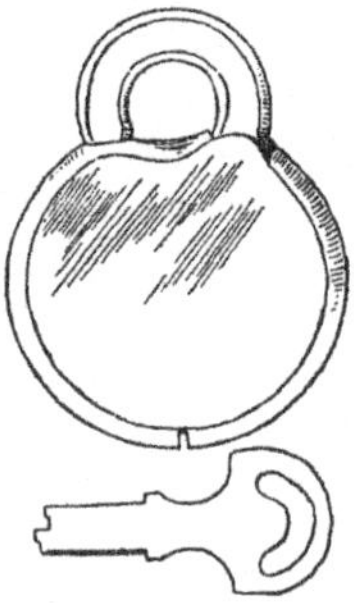

Fig. 354. Giant Padlock. Smith & Egge Mfg Co.

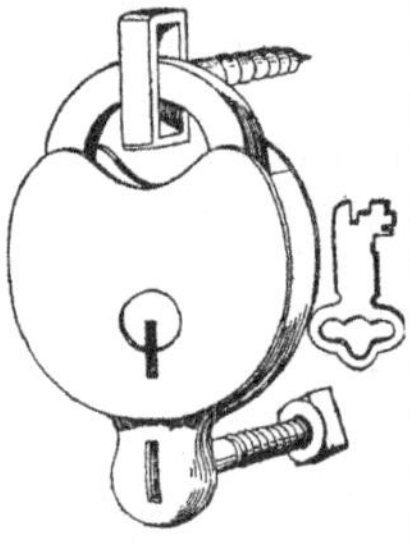

Fig. 355. Hasp Padlock. Stoddard Lock & Mfg. Co.

This lock hardly comes within the scope of builders' hardware, but it is too ingenious to pass unnoticed. The Corbin Cabinet Lock Company also makes a change lock for drawers, operating on much the same principle.

Padlocks. — The subject of padlocks is one which might be illustrated indefinitely, as there are quite as many different varieties as have been noted in regard to lever-locks, though with a few exceptions all padlocks are on essentially the same principle, consisting simply of spring-levers and a shooting-bolt, operated by a key in the same manner as an ordinary door-lock. Padlocks are now used but little about a house, as mortise or rim locks are usually more convenient, and at the same cost, are more secure. Only a few of the market forms will therefore be considered.

Figure 352 illustrates the internal arrangement of a very secure padlock manufactured by A. E. Dietz, the key, notched levers, etc., being somewhat similar to those in the Dietz store-lock previously illustrated. Figure 353 is a form made

Chapter XI.

by nearly all the leading lock-manufacturers. The key is inserted at the bottom of the padlock and rotates a set of levers which catch in the slots on both of the arms of the hasp. One arm is swivelled into the padlock case. Figures 354 and 355 are two other well-known padlocks, the former being used a great deal for government work and the latter having the hasp, staple and lock in one piece. The more common makes of padlocks are too well-known to require illustration.

LATCHES.

The ordinary door-latches have already been described in connection with the locks, but there remains quite a variety of latches which are made without any locking appliances, being intended simply to hold the door in position. Figure 356 shows the commonest form of latch used for elevator-doors, consisting simply of a bent lever, the lower arm of which is counterbalanced so that the lock will drop by gravity and

Elevator-Door Latch.

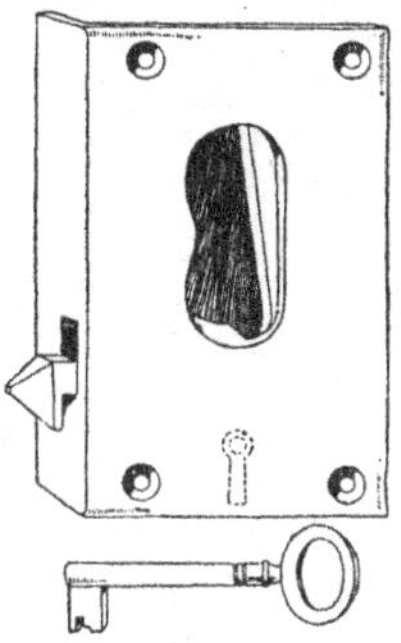

Fig 356. Elevator-door Latch. J. B. Johnston.

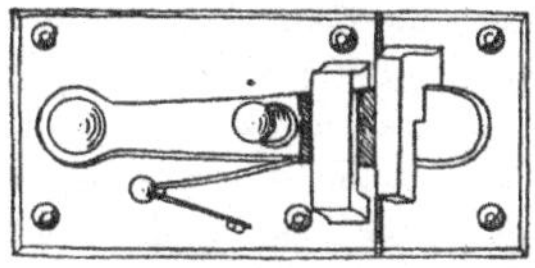

Fig. 357. Rim Sliding-door Latch. J. B. Shannon & Sons.

remain closed until drawn back by pressure on the upper arm. Figure 357 is a very simple rim sliding-door latch; and Figure 358 is a very good rim door-catch which is self-acting, the hook being released by raising the lever *A*, either directly or by aid of the spindle, *B*, from the outside of the door; while it is locked from within or without, the slide *C* being moved so that *A* cannot be raised. Figure 359 represents one of a great variety of styles of thumb-latch, a very simple, old-fashioned form which is very suitable for some cases. Figures 360 and

Sliding-Door Latch.

Thumb-Latch.

Chapter XI.

Screen-Door Catch.

361 are cheaper forms of thumb-latches, intended to be used only on screen-doors. Each of these styles has a lever of some sort, *A*, which serves to lock the latch. All of these patterns act by gravity. Figure 362 shows a spring-catch which is released by lifting or pulling out the handle on one side or by

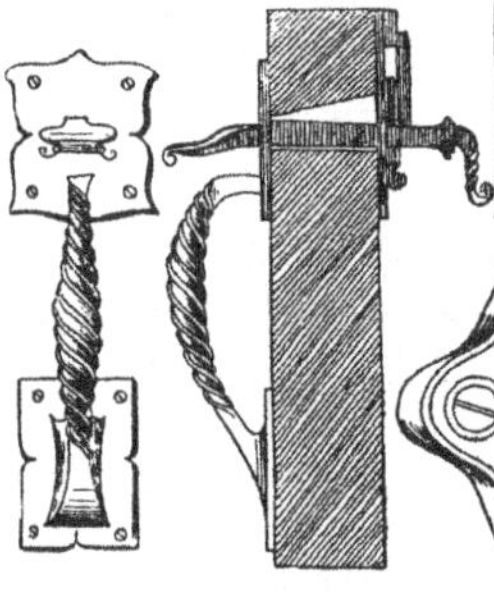

Fig. 359. Brass Thumb Latch. J. B. Shannon & Sons.

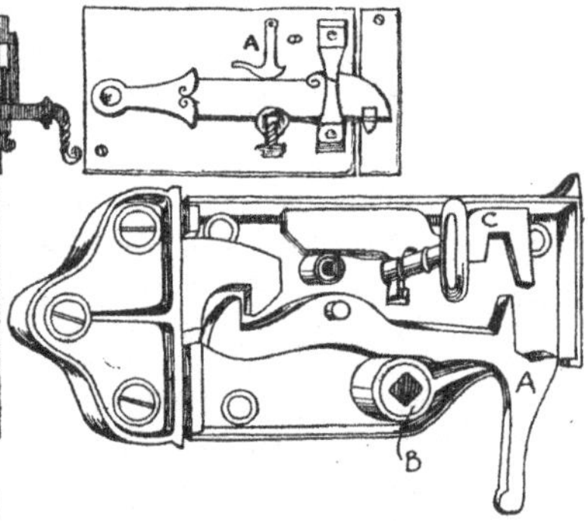

Fig. 358. Mack & Redway's Barn-door Lock. Nimick & Brittan.

depressing the thumb-latch on the other, the latch being locked by the swing-lever *A*.

For French windows and cupboard-doors or for light screen-doors, one of the styles represented by Figures 363, 364 and 365 are employed. Figure 364 can be locked, and it and Figure 365 work with a spring.

PRICES OF LOCKS.

It has not been deemed advisable to publish in this connection any summary of the market prices of the locks which have been illustrated and described, as, without such an acquaintance with the subject as can come only by examination and comparison of the actual samples, any prices which might be given would be misleading, and would often be unfair criteria of comparison. The real value of a lock depends so largely upon the care with which the levers are fitted, and the care taken with such details differs so much with the various manufacturers that the price ought to be the last thing to be considered in selecting the locks for a house. A good lock by a Chapter XI.

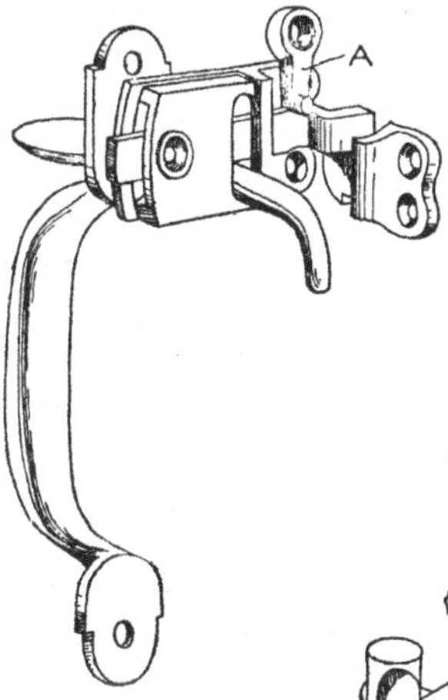

Fig. 360 Roggin's Latch. P. & F. Corbin.

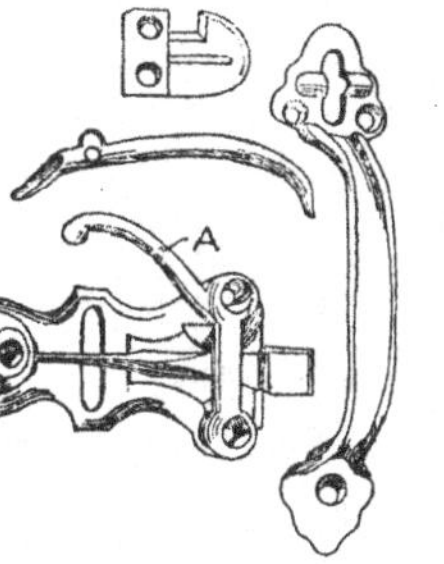

Fig. 361. Crown Screen-door Latch. Van Wagoner & Williams Co.

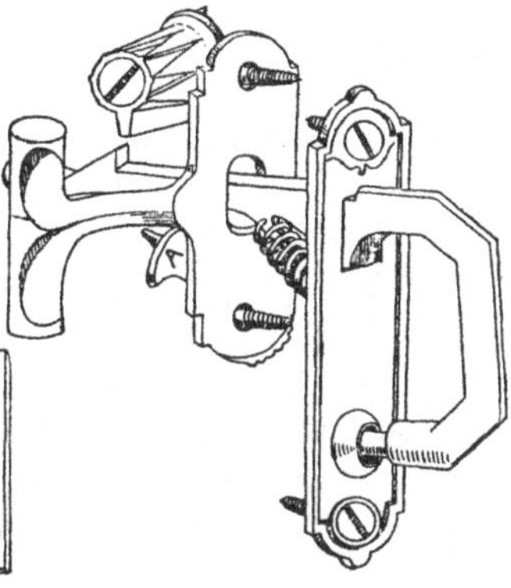

Fig. 362. Screen-door Catch. P. & F. Corbin.

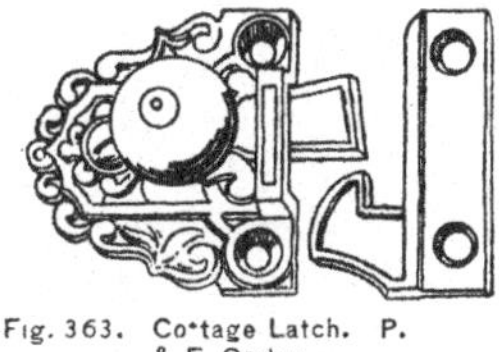

Fig. 363. Cottage Latch. P. & F. Corbin.

thoroughly reliable firm can always be matched by a lock sold for considerably less money, which has the outward appearance of being exactly as reliable, and yet which is totally inferior. Surely the difference between good and bad workmanship

Chapter XI.

could not be fairly illustrated by even the best of drawings, and it would never be wise to select merely from a trade catalogue. The only approximation which can be presented here is that previously given in the classification of locks by prices. It is of course very general, and consequently somewhat vague, and liable to exceptions; but it was prepared in conjunction with one of the largest hardware dealers in the

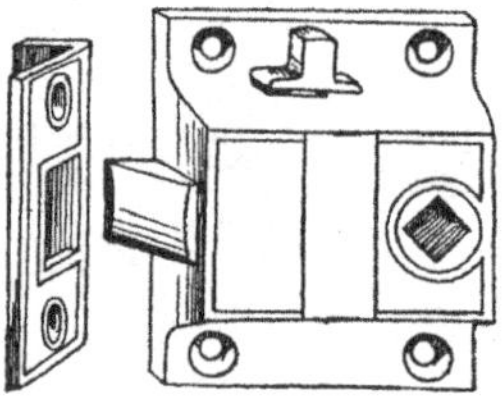

Fig. 364. Screen-door Catch. Reading Hardware Co.

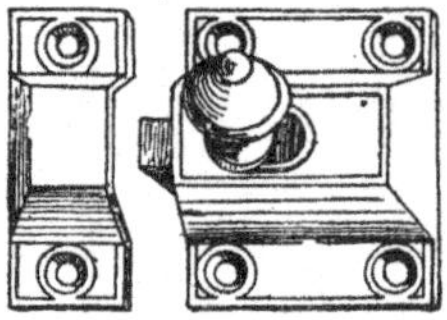

Fig. 365. French Window Catch. Reading Hardware Co.

country, and is sufficiently exact to serve as a guide to the general prices which should be paid, bearing always in mind that the wisest plan is to select only from the work of the best manufacturers and then only by samples.

The seventy-five or more locks which have been illustrated and described must be considered as types rather than as an exhaustive selection. A simple, three-lever lock is common property and several manufacturers whose names have not been mentioned in this connection turn out locks which are quite as good or better than those which have been selected for illustration. The difference would be entirely in the fitting or the finish, neither of which can be shown by the illustrations. All that can be hoped for is that this chapter may serve as a summary to guide in the general selection of the goods.

CHAPTER XII.

Door-Knobs.

Ordinary Spindle Attachment.

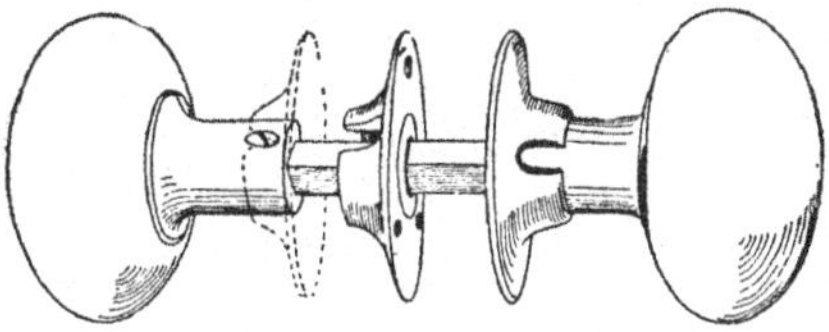

Fig. 366. Knob with Spindle-screw partly covered.

THE ordinary appliance for operating a door-latch consists of a knob on each side of the door, made of porcelain, wood, composition, or metal in various forms, but generally in the shape of a flattened sphere. The knobs are attached to metal shanks serving to hold them away from the door, and to prevent their pushing in, and the two knobs are connected through the lock by a square spindle. The spindle is firmly attached to the shank of one knob, and on the other side of the door it fits loosely in the shank, considerable length being allowed for the adjustment to various thicknesses of doors, the inner knob being finally secured in place by a screw on one side of the shank which passes entirely through the spindle, and sometimes is also made of sufficient length to

Chapter XII.

Roses.

Washers.

turn into the opposite side of the shank. The hole in the door through which the spindle passes is covered by a metal disk technically designated as a rose. The rose is secured to the door by screws, and as the shank of the knobs is made to fit closely against the rose, if well put on there will be no strain on the lock when the knob is pulled from the opposite side, all strain being gathered on the rose itself. Knobs are usually provided with a number of small washers, so that the adjustment between the bearing-surfaces of the roses and the ends of the shanks can be made exact, and thus any rattling be obviated. In many instances the shanks are secured to the spindle with screws on each side of the door, so that the knob can be taken off from either side of the door. For front-door and vestibule work the outer knob should always be securely attached to the spindle, so that no screw is necessary, as otherwise, if the shank is held by a screw it can be removed from the outside of the door, the spindle pushed in and the inner latch follow turned back. For interior work, however, it makes little difference whether screws are used on one or both sides, though many consider the use of screws as altogether objectionable, owing to their liability to work loose; and, aside from any questions of design, the ingenuity of hardware manufacturers has been chiefly expended upon securing a better connection between the knob and the spindle. Still, few of the patented forms of attachment have been very generally received, and the old style of screw attachment seems to meet with the most favor, if we may judge by usage. It is not the question of cost alone which has decided this in the minds of many builders and architects, but rather a belief that a tangible fastening like a screw, which is easily placed and easily removed, is, after all, more satisfactory than any concealed device.

The objections to the old style of fastening are, however, easily appreciated. One trouble is that the spindle will work and wear away so as to be loose in the follow, and rattle every time the knob is touched. This is particularly noticeable in very old work, in which the parts are sometimes so worn as to admit of as much as half an inch play at the end of the knob. In new work, the spindle, the follow and the roses can be fitted so that any rattling is impossible, though with the old styles of fastenings this is accomplished only by the best manufacturers.

With the old style, the screws are apt to work loose, as applied by ordinary mechanics. In cheap work they nearly always do so; still, if proper care is taken and the screws turned up with a drop of thick shellac in the threads there will be little trouble, and none that cannot easily be remedied with a screwdriver.

There are other objections of less moment, such as the fact that considerable time is occupied in fitting the washers necessary to a proper adjustment of the spindle and shank; and the proper attachment of the screws takes time also. It is further found that when the spindle and shank wear away there is apt to be a strain brought upon the lock-plate through the door, thereby endangering the proper action of the levers. We have said, however, that these objections are by no means vital, and are such as might be due to careless or indifferent workmanship. One of the best evidences that the old style is the most satisfactory, is that every manufacturer has it on his catalogue-list. Anything else is really an exception, and we know of only one instance in which a manufacturer has undertaken to push exclusively a single form of knob attachment differing from the common style. It must not be thought, however, that no clever or good devices have been thought out. It is hard to simplify simplicity, and the screw connection, all things considered, gives eminent satisfaction.

Knob and Spindle, Screws partly covered.

The first variation from the old style has been to enlarge the rose, extending it out over the shank so as to partially or completely cover the screw-hole, a slot being left at each side through which the screw can be applied, the rose subsequently being turned and secured against the door so as to completely cover the screw. Figure 366 shows such a form. This device renders it absolutely impossible for the screw to become

Chapter XII.

detached, though it does not prevent it from being a little loose, and so permitting the knob to rattle; and as the difficulties of getting at the screw are increased by this method, the probabilities are that most people would let the knob rattle instead of taking the trouble to tighten the screw. Still, this is an improvement, and when well applied is very satisfactory. The next step has been to cover the screw entirely. Figure 367 shows one mode in which this has been accomplished. The rose is made in two portions, one consisting of a flat piece resting against the door, and serving as a bearing-plate for the shank, while the other portion of the rose which would show in the finished work consists of a thin shell curved out so as to entirely cover the screw. The screws which hold the rose to the door pass through both the outer shell and the inner plate.

Spindle-Screws covered. Russell & Erwin Mfg. Co.

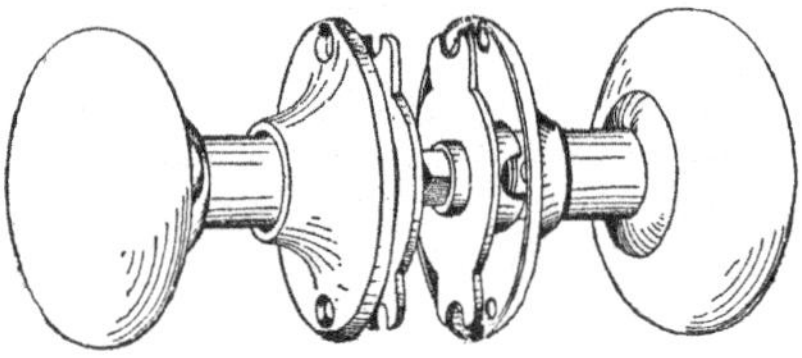

Fig. 367. Knob with Spindle-screws covered. Russell & Erwin Mfg. Co.

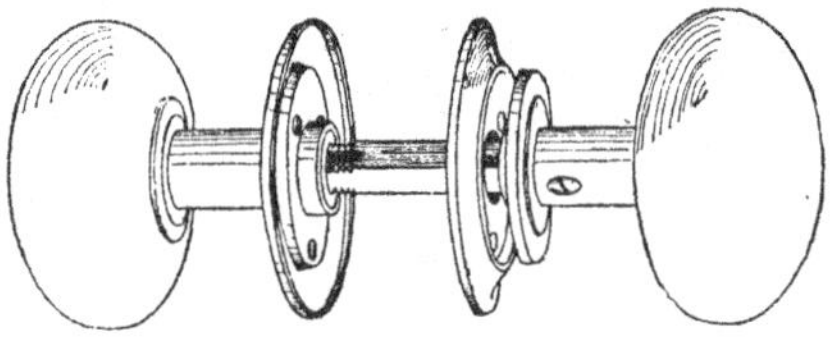

Fig. 368. Knob with threaded Spindle. Stoddard Lock Mfg. Co.

Threaded Spindle. Stoddard Lock Mfg. Co.

Figure 368 shows another form in which one screw is done away with. The spindle is cut with screw-threads. The rose is made in two portions, one being screwed to the door, and the other acting as a binding-screw or washer, screwing onto the threads of the spindle at the same time that the shank of the knob screws behind it, the two locking, and preventing the knob from being unturned except by forcible means. As the spindle is held in the latch, the knob can, of course, be turned

457

474

459

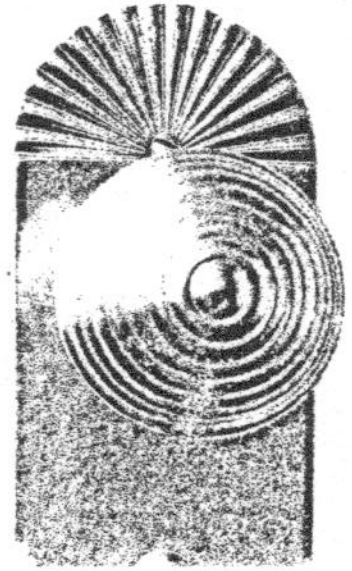

467

466

469

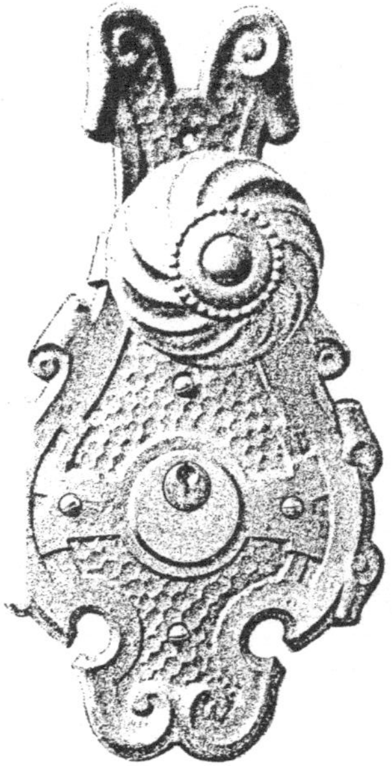
458

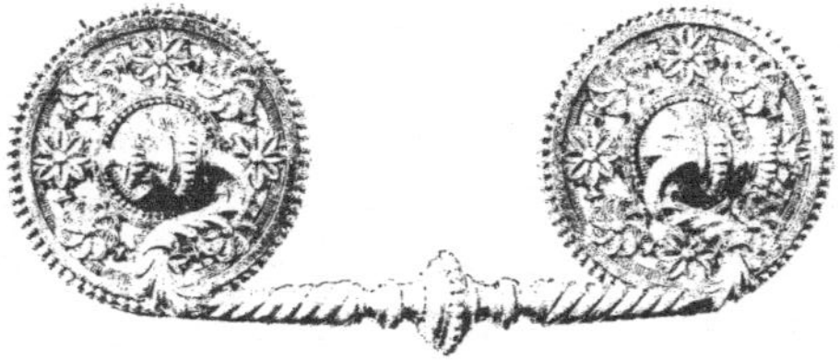
468

460

··ARTISTIC·HARDWARE··

Chapter XII.

but half way in either direction. Another form of knob substitutes a continuous ratchet on one face of the spindle for the screw-holes of the common form. These knobs are made by the Boston Knob Company, and outwardly appear like an ordinary knob. The advantage is that the knob can be adjusted at any point without the aid of washers, the screw catching onto the ratchet in any position of the shank.

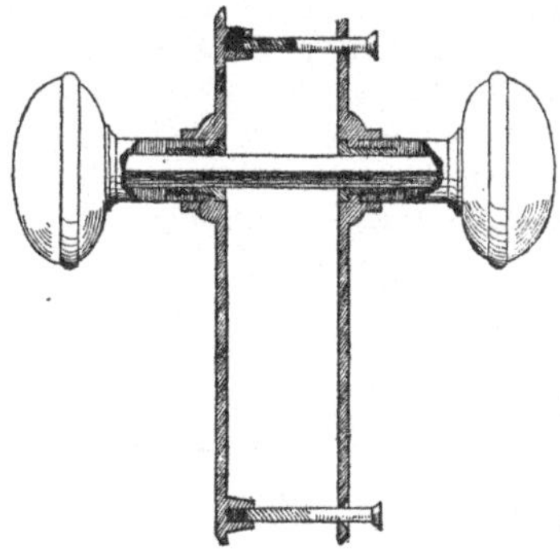

Fig. 369. Screwless Door-knob and Escutcheon combined. P. & F. Corbin.

Screwless Knob. Russell & Erwin.

Figure 369 shows a form which does away with the screws entirely. The key escutcheon and the rose are combined in a single plate on each side of the door. Inside of the rose is a hub which is cut with a screw-thread. The spindle passes through this and into the shank of the knob, which is cut with a thread corresponding with the thread on the hub. In applying this fixture the knobs are simply screwed on until they bear slightly on the edges of the rose. The escutcheon-plates are then screwed together through the door as shown on

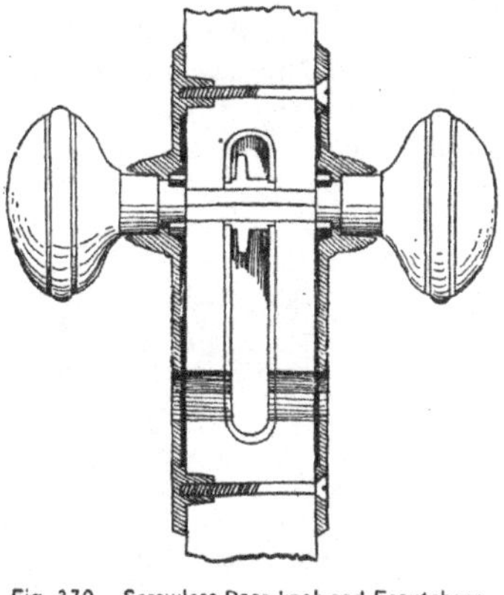

Fig. 370. Screwless Door-knob and Escutcheon combined. Russell & Erwin.

Chapter XII.

the drawing. As the spindle passes through the latch it will readily be seen that the knob cannot be unscrewed except by removing the escutcheon-plates, and as these plates bear on each side of the door above and below the lock, it is almost impossible to bring any strain on the lock-plate itself.

Figure 370 is a somewhat similar form as regards the escutcheon-plates. The knob, however, is attached by means of

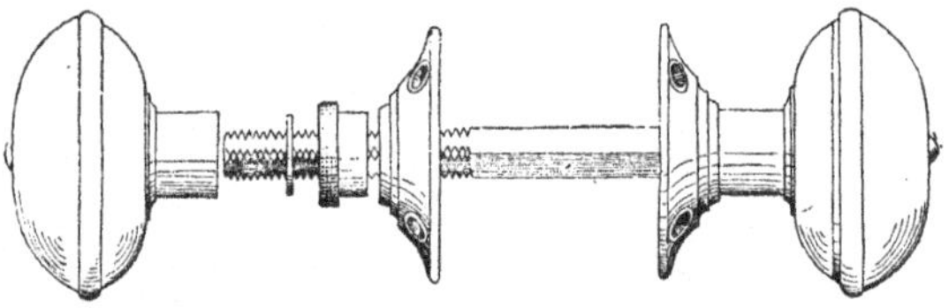

Fig. 371. Mathes's Adjustable Screwless Door-knob. Nimick & Brittan.

lugs on the shank, which in one position of the knob will slip into the hole in the rose; but when half turned will catch on the inner side of the plate, thus rendering it impossible for the knob to be removed except by unscrewing the face-plates

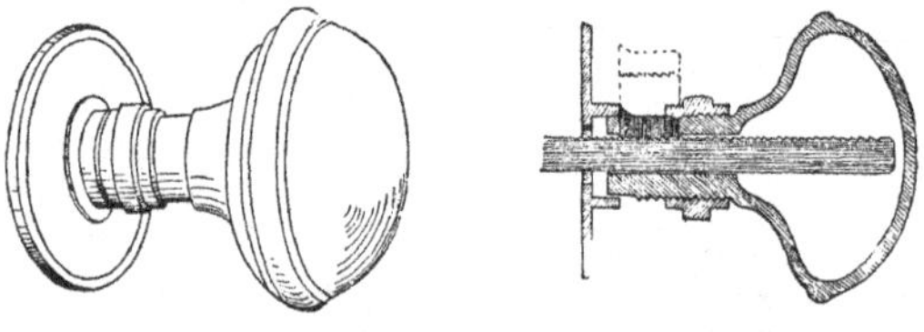

Fig. 372. Morris Patent Door-knob. Ireland Mfg. Co.

from the door. Figure 371 is a device practically the same as that shown by Figure 368. Figure 372 is still another variety of the same general style of attachment, using a steel binding-screw to hold the knob-shank in position.

Screwless Knob. Yale & Towne Mfg. Co.

The Yale & Towne Manufacturing Companv has recently put on the market a form of screwless knob-shank shown by

Chapter XII.

Figure 373. In this case the spindle is turned round at each end and threaded. The knob is provided with a swivel-nut, *D*, which fits the thread of the spindle. In applying, the nut is turned up until it bears slightly against the face of the rose, and is then left in that position, a washer being interposed between the rose and the nut. The nut takes the place of the ordinary shank, and as this portion of the knob is seldom touched, there is little liability of the nut working loose, especially as it can be turned up pretty tight, and is made so as not to work too easily.

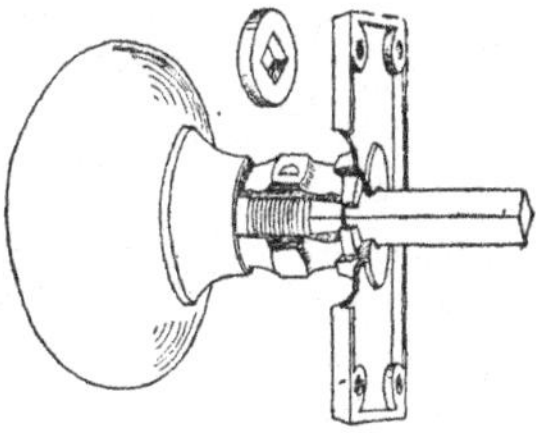

Fig. 373. Screwless Knob-shank. Yale & Towne Mfg. Co.

Hopkins & Dickinson.

A very simple and effective form of screwless knob-fastening has been recently brought out by the Hopkins & Dickinson Manufacturing Company. In this device, the spindle, which is cut with a screw-thread, is rigidly attached to the shank and the knob on one side. The opposite shank has a swivel connection with the knob, and is threaded inside to screw over the spindle. In application, the loose shank is screwed onto the spindle until two dogs or teeth on the inner end of the shank are engaged in slots in the rose. The rose, which projects somewhat from the face of the door and has a milled-edge, then serves as a set-screw, drawing the two shanks together and binding against a washer on the door, so that while the knobs can be rotated freely, the rose washer and consequently the shanks, will not work loose.

Screwless Knob. Bardsley.

Figure 374 represents still another variety of screwless knob-fastening. The nut, *C*, forces the washer, *B*, against a shoulder inside of the shank, *A*, binding the latter firmly to the rose and to the door. The knob is then slipped over the spindle, and the shank, *A*, screwed over the shank, *D*, until the knob is drawn up tightly. The only chance of the fastening working loose is by accidental turning of the shank, *A*, which is not likely to occur.

The Yale & Towne Manufacturing Company has a device illustrated by Fig. 375 which is on a very different principle from any of the foregoing, as it does not depend upon screws of any kind.

Screwless Spindle.

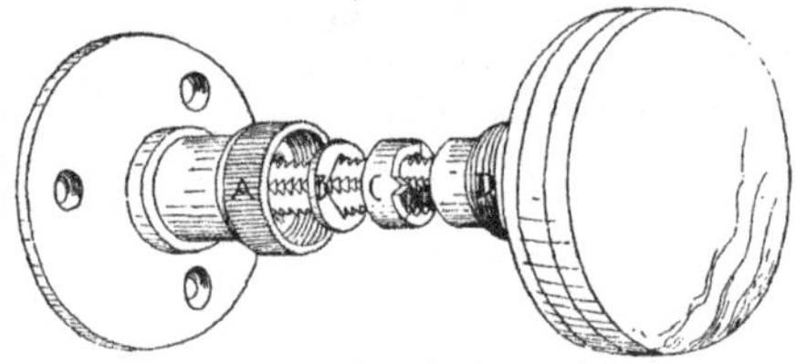

Fig. 374. Screwless Fastening Knob. J. Bardsley.

In this case the knob-shank is cut out with an eccentric socket or bore. The ends of the spindle are turned down to exactly the same contour as the bore of the shank; so that while the knobs on either side of the door can easily be slipped over the ends of the spindle, they can be fastened by simply rotating them in opposite directions, when the fine pitch of grade of the eccentrics causes a great pressure to be exerted, which results in binding the knobs rigidly to the spindle. This is the simplest form of knob attachment in the market, and if properly applied, will always remain in order, though great care must be taken that the knobs are turned up firmly.

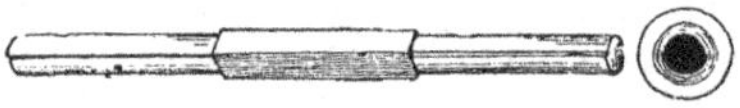
Fig. 375. Screwless Spindle and Socket. Yale & Towne Mfg. Co.

Swivel Spindle.

Fig. 376. Swivel Spindle.

For front-door locks and latches it is necessary to have some form of spindle in which the two extremities may be worked independently, so that the outer knob may be locked while the inner one is free to rotate. The commonest form is to connect the two halves of the spindle by a swivel joint, Figure 376. Corbin has in the market a spindle in which the two halves screw together, thus permitting of very careful adjustment to the thickness of

Chapter XII.

the door. The pitch of the screw-threads is so slight that the quarter turn necessary to open the latch does not throw out the knob from the door.

Knob Fasteners.

There are various methods of attaching the head of the knob itself to the shank. When porcelain or mineral composition is used, the shank is leaded into the knob. Hemacite, zylonite, etc., are cemented or screwed to the shank, as are the cheaper forms of wooden knobs. Metal knobs are blind riveted, cast solid to the shank, or shrunk on. Glass knobs are commonly leaded, but in some cheaper forms are cemented or even puttied.

There are, however, some devices which are intended to attach the knob more firmly to the shank. Figure 377 is one which is used in connection with wooden knobs. The shank is cut with a screw-thread which turns into a corresponding

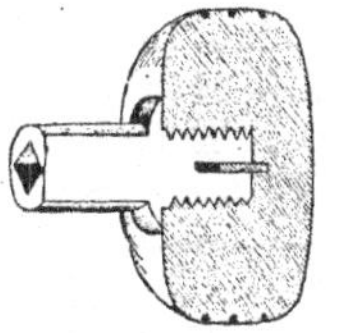

Fig. 377. Knob Fastener. J. Bardsley.

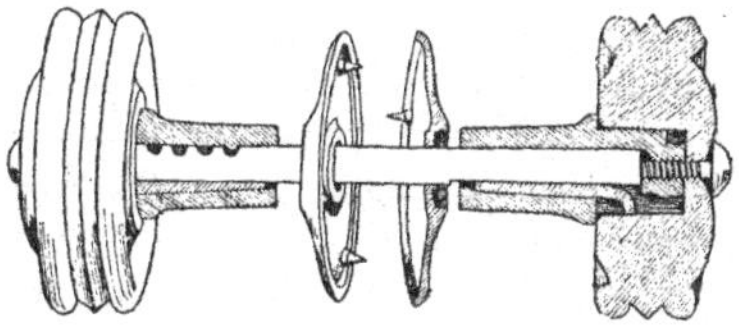

Fig. 378. Phipps's Patent Door-knob. Milford Door-knob Co.

thread cut into the knob. Before the shank is screwed in, a metal key extending through the shank is placed in the slot, and after the knob is firmly screwed on the key is forced into the wood by means of a punch placed in the opening of the shank, the key thus effectually locking the shank into the knob.

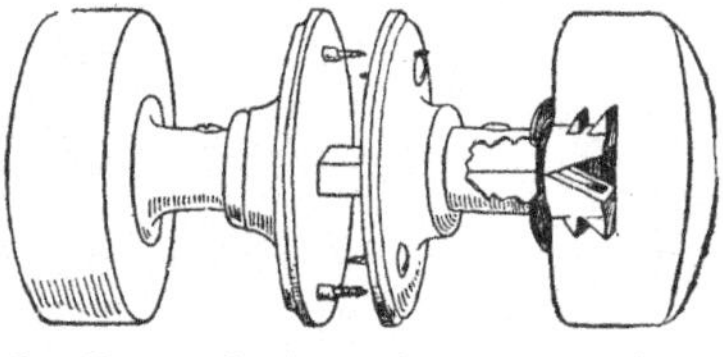

Fig. 379. Wooden Door-knob Attachment. J. B. Johnston.

Figure 378 shows a form of attachment for either wood or metal. In this case the knob is held by a screw passing from the knob through the upper portion of the shank and into the

Chapter XII.

head of the spindle. The spindle can be adjusted for any thickness of door by means of a small wedge which can be driven in before the knob is attached, in such a manner as to hold the shank at any given position. Figure 379 is a form of attachment designed for wooden knobs. The shank is split lengthwise and the ends of the two pieces cut away from each

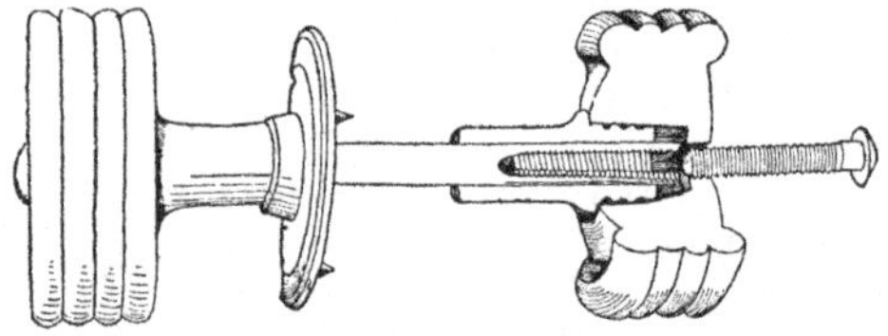

Fig. 380. Hollenbeck's Expanding Spindle Door-knob.

other on a bevel, with wedges or lugs on the outside, the bevels being so cut that when the surfaces are brought together the lugs can enter a hole in the wooden knob the same diameter as the main portion of shank. By then bringing the opposite ends of the shank together the lugs are forced sidewise into the wood so strongly that they cannot be drawn out except by breaking the parts. A light thimble fits over the shank and into the rose, securing the whole.

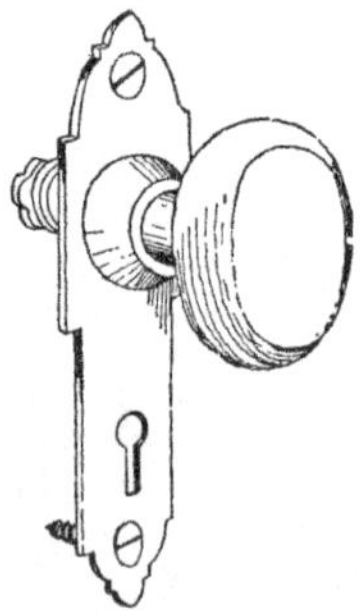

Fig 381. Niles Patent Knob. Chicago Hardware Co.

Expanding Spindles.

Figure 380 is a form of knob secured by a screw entering the head of the spindle, at the same time wedging it to any adjustment by reason of the screw being slightly larger than the hole in the spindle.

Niles Knobs.

All the foregoing knobs are constructed with spindle extended through the door and continuous from knob to knob Some locks are so devised that the spindle is done away with, each knob acting independently of the other by means of shoulders or extensions on the shank. Figure 381 illustrates the form of knob which is used with all of the "Niles" locks. The end of each shank is provided with

Chapter XII.

a shoulder of about the same shape as the ordinary lock follow, acting directly against the latch-lever. The shanks rotate freely in the escutcheons. To apply the knob, the shank is passed through the escutcheon plate and the shoulder or follow inserted in the lock, the latch-lever being pressed back with a flat blade or a screw-driver until the follow can be snapped into position, which is easily accomplished by inserting the shank at an angle. The knob is then brought around square with the face of the door and the escutcheon plate screwed in position, holding the knob so it can be removed only by moving the plate. The chief advantage of this arrangement is that there can be no rattling in the lock. The latch operates the moment the knob is turned, be it ever so little; nor are there any screws to work loose.

Gilbert Knobs.

The Gilbert Lock Company manufactures a knob especially designed for their locks,[1] the construction of which is illustrated by Figure 382. In this, as in the preceding example, there is no spindle. The knob-shank is secured to the escutcheon, which is boxed out sufficiently to allow play for a lugged plate, turning with the shank and acting against a lever. The latter is hinged at the top and fitted with an arm at the bottom which works in a slot through the lock, drawing back the latch by a direct, lateral action. The escutcheon is secured by long screws above and below the lock. There are some excellent points about this device. There is no spindle to work loose and rattle, no screws in the shank to drop out, and no adjustment of washers or screws, as the knob has a perfect adjustment to any thickness

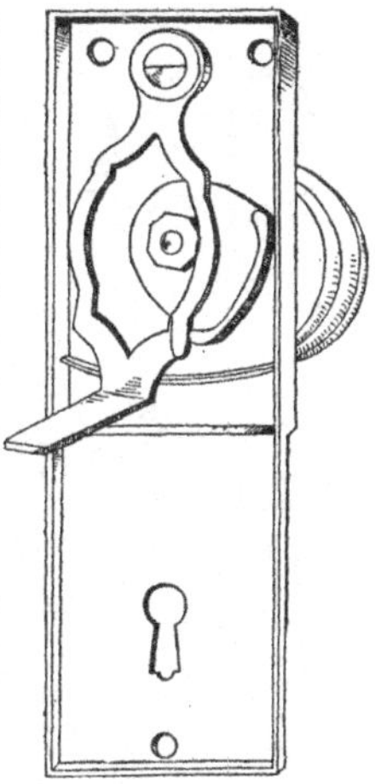

Fig. 382. Gilbert's Lock and Escutcheon. Gilbert Lock Co.

[1] See Figure 313 for an illustration of the "Gilbert" locks.

Chapter XII.

of door without binding. An improvement might be made by so extending the lugs on the spindle plate that when the latch is out both lugs will bear against the operating lever, in order that the latch may move at once, no matter in which direction the knob be turned. This form can, of course, be used only with "Gilbert" locks.

Materials used for knobs.

In regard to appearance, and the materials used, knobs of the following materials are found in the market. In wood, they are made of mahogany, cherry, oak, ash, apple, maple and ebony. Glass knobs are cut, pressed, silvered or of black

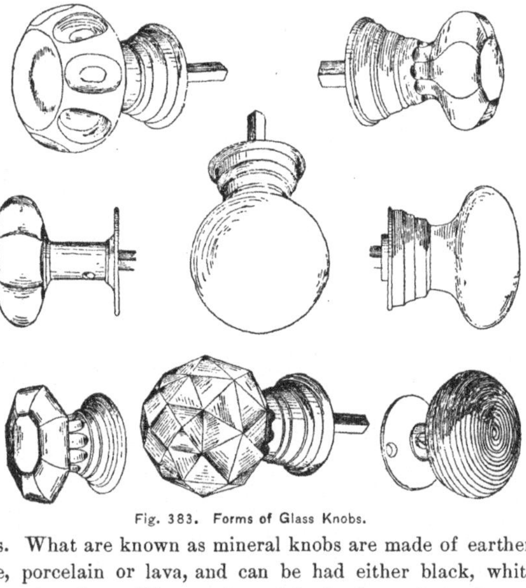

Fig. 383. Forms of Glass Knobs.

glass. What are known as mineral knobs are made of earthenware, porcelain or lava, and can be had either black, white, or grey in color. The metals used for knobs are brass, bronze, silver, nickel and iron. Compositions of celluloid, hemacite, etc., are also used. The shanks in all cases are made of either bronze or iron, the latter only in the cheapest work.

Wooden knobs are generally finished in natural colors, and can be obtained with wooden roses to match. They are very good, strong, and serviceable, and are excellent for interior use. Wood.

Glass knobs are somewhat out of style just at present, but are still made in a great variety of forms, both cut and pressed, and are really very handsome in appearance. The silvered-glass knobs are rather cheap looking, though the cost is somewhat higher. Figure 383 shows a few of the great variety of knobs made in glass. Glass.

Black glass, and what is known as mineral, and white porcelain are all used a great deal for common purposes. They are cheap, clean, and as generally constructed are quite strong. Lava knobs are used but little. Porcelain.

The greatest variety of designs is found in metal knobs. These are made in all shapes and in all colors. Some of the special shapes will be considered subsequently under the head of Artistic Hardware. Some of the styles of iron knobs recently put on the market by the Yale & Towne Mfg. Co., and finished by the Bower-Barff process, are very serviceable and pleasing. Hopkins & Dickinson have a very dark, rich bronze almost as black as gun-metal which they use for some of their hardware. Of late years oxidized silver has come in as a great favorite for knobs and knob-plates, and is now worked up in a great variety of designs and in several different colors. The Yale & Towne Mfg. Co. has a grade of oxidized silver which almost matches the dark bronze of Hopkins & Dickinson. Metal knobs are made either oval, spherical or in a flattened sphere, egg-shaped, and indeed in an almost infinite variety of shapes and designs. In the nicest grades of work the knobs are always made to order. In some of the very choicest work knobs are gold plated. This increases the cost a great deal, to an extent, indeed, which renders it beyond the reach of the ordinary buyer; the advantage, however, is not alone in the looks, for a gold-plate has nearly the same color as some shades of bronze; but gold-plate is absolutely untarnishable, and will not change its color, whereas all the finishes of bronze, Metal. Gold-plate.

Chapter XII

silver, brass or nickel, are more or less liable to change. The various finishes for metal knobs have been previously considered in the introduction.

Celluloid.

The knobs of the Boston Knob Co. are made of composition, presumably celluloid, or at least of that nature. Celluloid plates are bent over a strong metal frame, and held in position by a brass rim which is shrunk on to cover the joints between the two plates. They form a very neat pretty knob, Figure 384. The celluloid is made in a variety of colors, including several shades of blue, garnet, black, malachite, green, drab, slate, yellow, brown and white. In many cases the varied colors will be an at-

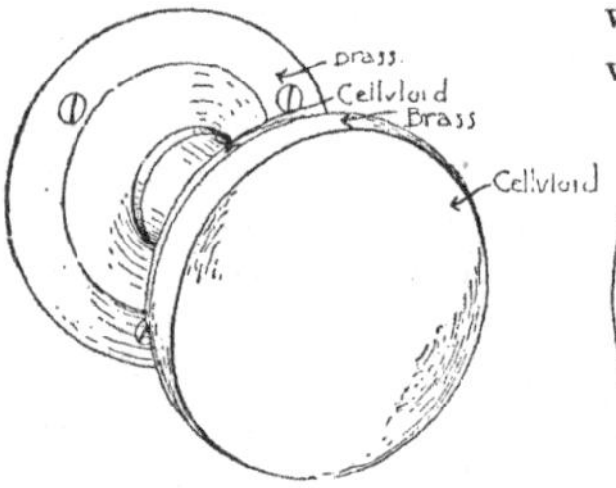

Fig. 384. Celluloid Door-knob. Boston Knob Co.

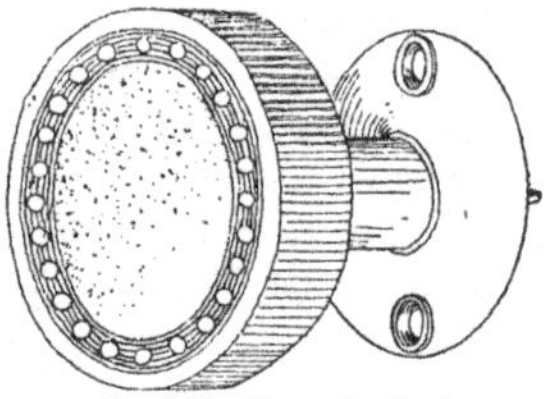

Fig. 385. Hemacite Knob.

traction. One would imagine this form of knob made in white with a simple band of brass around the edge might be used very nicely in connection with rooms that are furnished in the prevailing Old Colonial white-and-gold style.

Hemacite.

Hemacite is a composition which as nearly as can be discovered, consists of blood, glue and sawdust. This is pressed in moulds and finished in several different shades, either jet black or a deep rich brown. Figure 385 shows the commonest form adopted for hemacite knobs. They are usually made with face-plates of brass or bronze inserted in the front of the knob, and the edges of the knob are milled. This composition is most excellent for interior use. It will wear indefinitely and is exceedingly strong and tough; but is not altogether suitable for exterior use as it is said to be affected by the weather.

Besides the ordinary double knobs it is often desirable to have a lever on one side of the door and a knob on the other. Figure 386 shows a typical knob and T-handle. There is, of course, an infinite variety of styles of this sort, some of which will be considered later on. Figure 387 illustrates a so-called ship-handle, consisting of a plain knob at one end of the spindle and a ring-handle at the other. The form shown

Chapter XII.

Knob and Handle.

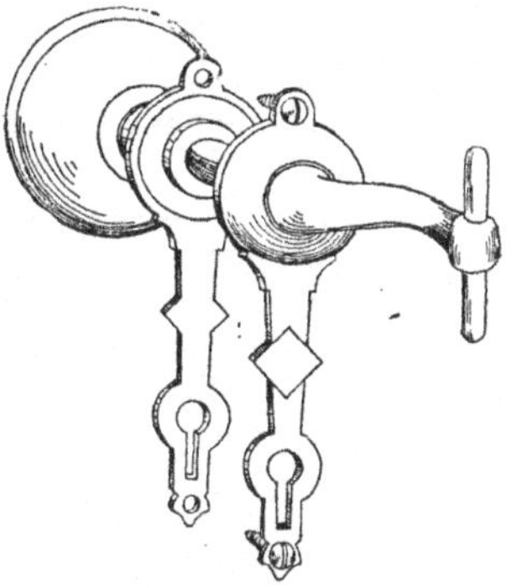

Fig. 386. Knob and T-Handle. Ireland Mfg. Co.

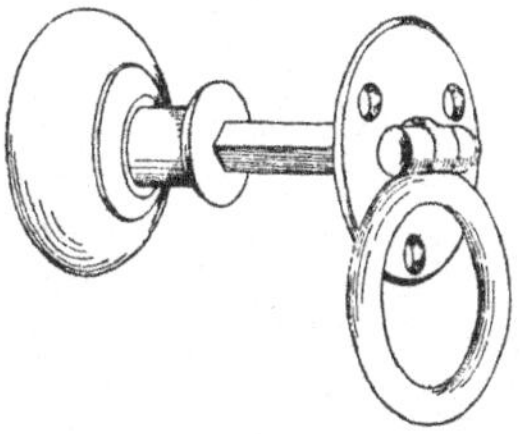

Fig. 387. Ship Handle. J B. Shannon & Sons.

by Figure 388 is termed a crank-handle, being intended for French windows and narrow style doors. The inner knob is

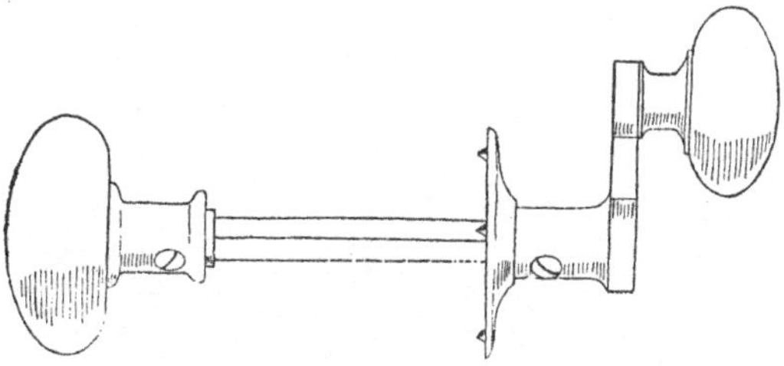

Fig 388. Crank Handle. Nimick & Brittan.

kept away from the jamb so that in opening the door the hand will not be caught. The common forms of pulls or handles employed for sliding-doors have been previously discussed.

Chapter XII.

Bell-pulls.

Bell-pulls are usually similar in appearance to door-knobs, and in order-work are made exactly the same, and to match. The internal construction of the spindle however is a little different.

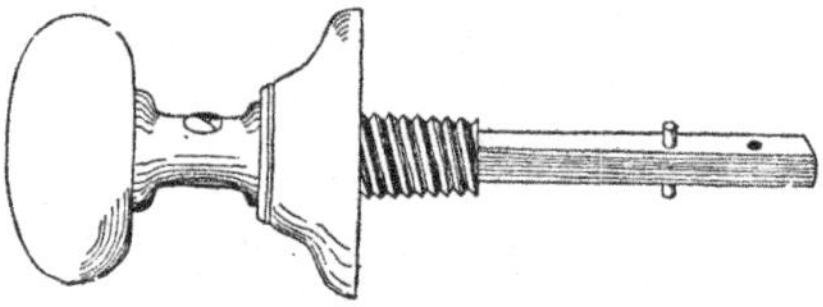

Fig. 389. Common Bell-pull.

Figure 389 shows the commonest form, the rose being provided with a long hub fitting over the spindle, and screwing into the frame of the door. Figure 390 is a form of lever bell-pull suitable for outdoor work. The same form is sometimes used for bells in the interior of the house, although Figure 391 is a better and more common form.

Fig. 390. Lever Bell-pull. Russell & Erwin

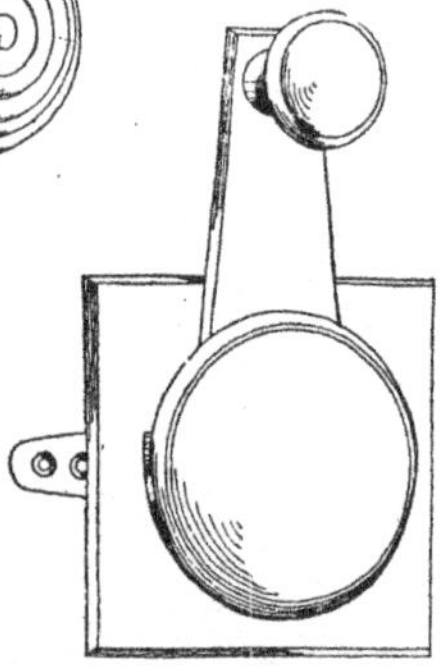

Fig. 391. Parlor Bell-lever.

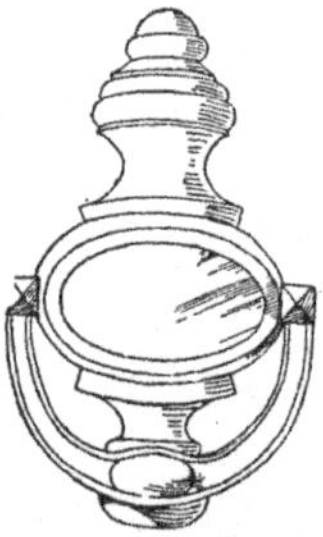

Fig. 392. Old-fashioned Door-knocker.

Knockers.

Door-knockers are made in a great variety of shapes. A few of these will be considered later on in connection with the designs. A single example, Figure 392, will be sufficient to illustrate a typical door-knocker in this connection.

The following table gives the average retail prices of the principal styles of door-knobs. The prices are for a dozen pairs of medium-sized ($2\frac{1}{4}$ inch) knobs, complete, with roses and screws to match. Chapter XII. Prices.

TABLE OF DOOR-KNOBS.

Fig.	Description.	Knob.	Shank and Rose.	Manufacturer.	Price.
367	Knob with concealed Screws	Porcelain....	Bronze Plated....	Russell & Erwin.	$5.75
369	Screwless Knob and Escutcheon combined	Bronze....	Bronze...	P. & F. Corbin.	12.50
372	Morris Patent Door-knob	"	"	Ireland Mfg. Co.	15.00
373	Screwless Knob-shank	"	"	Yale & Towne Mfg. Co.	18.00
378	Phipps Patent Door-knob	Wood	"	Milford Door-Knob Co.	12.00
379	Door-knob, expanding Spindle	"	"	J. B. Johnston.	7.00
381	Niles Door-knob.....	Bronze.....	"	Chicago Hardware Co.	30.00
382	Gilbert Door-knob and Escutcheon......	"	"	Gilbert Lock Co.	40.00
384	Boston Door-knob...	Celluloid...	"	Boston Knob Co.	10.00
385	Hemacite Door-knob	Hemacite...	Brass and Hemacite	Dibble Mfg. Co.	6.00
386	Knob and T-handle..	Bronze.....	Bronze...	Ireland Mfg. Co.	16.00
387	Ship Handles........	Brass.......	Brass....	J. B. Shannon & Sons.	16.80
390	Lever Bell-pulls.....	Bronze.....	Bronze ..	Russell & Erwin.	27.00
391	Parlor Bell-levers...	"	"	"	27.00
392	Antique Knocker—each	"	"	—	3.00
—	Common style Door-knob	Pressed Glass.....	"	—	10.00
—	do.	Cut Glass...	"	—	18.00
—	do.	Porcelain....	Iron......	—	1.00
—	do.	Cherry.....	Bronze...	—	7.50
—	do.	Iron Bronzed..	Iron......	—	4.00
—	do.	Iron Bower-Barffed...	"	—	8.00
—	do.	Plain Bronze...	Bronze...	—	8.00
—	do.	Figured Bronze...	"	—	6.00

There remains but a single door-knob to be considered. Some ingenious person who had been troubled by tramps, or who imagined that everybody else was, devised a burglar door-knob. This consists simply of a knob on the inside of the door, which at the same time is a bell, the mechanism of which years of use will not disarrange. It costs but little more than a common knob and can be applied by any person, the least turn of the outside knob causing the alarm to be rung on the Burglar-alarm Knob.

Chapter XII.

inside so that immediate warning is given of even an attempt to enter. The knob is so constructed that upon being turned from the inside it gives no alarm. It is known as the Burglar Door-knob and Window-alarm, and is manufactured by Wm. C. C. Matthews & Co.

ESCUTCHEONS.

Escutcheons.

The term escutcheon is used to designate the peculiar locking mechanism of a cylinder-lock, as has been explained in the previous chapter. It is also applied to the finish, of metal or other material, about the key-hole of a lock. Escutcheons are made both with and without drops or covering pieces. For inside work the drop had better be omitted, though for front doors both the latch and the lock key-hole should be protected. The common forms of escutcheons are too well-known to require any illustration.

Prices.

The following table gives the average retail prices.

TABLE OF KEY-HOLE ESCUTCHEONS.

Material.	Price per dozen pairs with drop and screws.	Price per dozen pairs without drop, with screws.
Iron bronzed	$ .50	$.25
Porcelain	.75	.35
Wood	.85	.65
Brass	1.50	.60
Bronze — plain	1.50	.60
" figured	1.00	.42

CHAPTER XIII.

Closet–Fittings.

Closet-Fittings.

SOME of the appliances included under the title of Closet-Fittings appertain perhaps more truly to furniture than to Builders' hardware, though they are sometimes used in connection with the finished carpenter work. The designation of closet-fittings is a somewhat arbitrary one, and while not strictly applying to everything considered under this classification, might include many of the articles described in previous

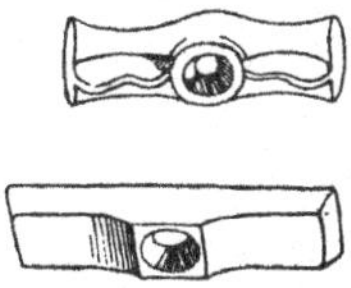

Fig. 393.

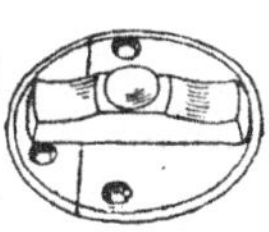

Fig. 394.

Turn-buttons.

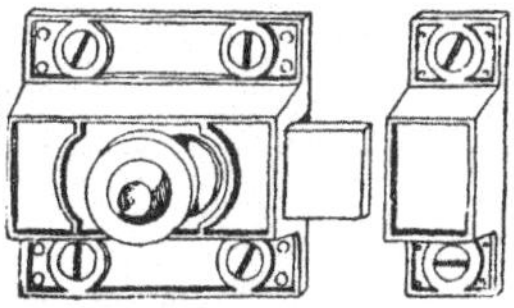

Fig. 395.
Cupboard-bolt.

chapters. The limitations will, however, be sufficiently exact for the present purpose.

Turn-Button.

The simplest appliance for securing the door of a cupboard is what is known as a turn-button. Figure 393 illustrates the cheapest form, consisting of a metal-bar or button which is secured in place by a screw through the centre, the screw being turned in so as to allow the button to rotate freely. An improvement is to have the button pivoted on a plate which is

Chapter XIII.

screwed independently to the door-frame, Figure 394, while a plate is secured to the door for the button to turn upon, or *vice versa*.

A turn-button acts as a bolt, but it is often preferable to use some other form. Any of the flush, sunk, raised, mortise or neck bolts described in a previous chapter will answer for a cupboard, though there are a few styles which are especially designated as cupboard-bolts. Figure 395 is an example. Again, it is often desirable to have a spring-catch on a cupboard, such as that shown by Figure 396, which may be considered as a type of many different styles. Figure 397 shows a lever-cupboard catch, which works by gravity, without springs, the catch being released by raising the handle. Each of these varieties can be used for double or single doors, though with double-doors some form of bolt is necessary in addition. The book-case bolts and catches described in the chapter on bolts might properly be included, also, in the present category.

Cupboard-Bolt.

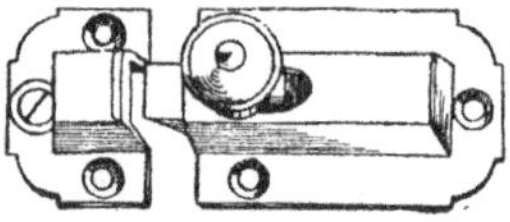

Fig. 396. Cupboard-catch.

Cupboard-Catch.

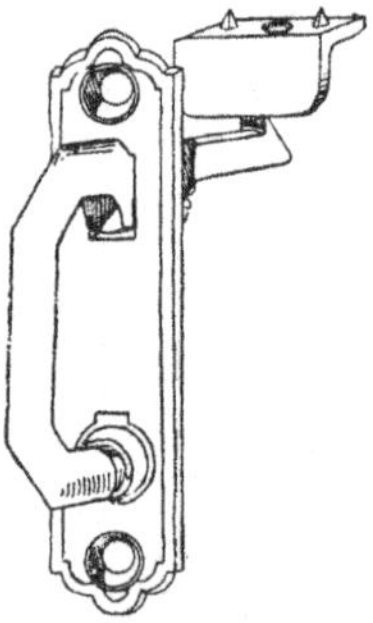

Fig. 397. Cupboard-catch.

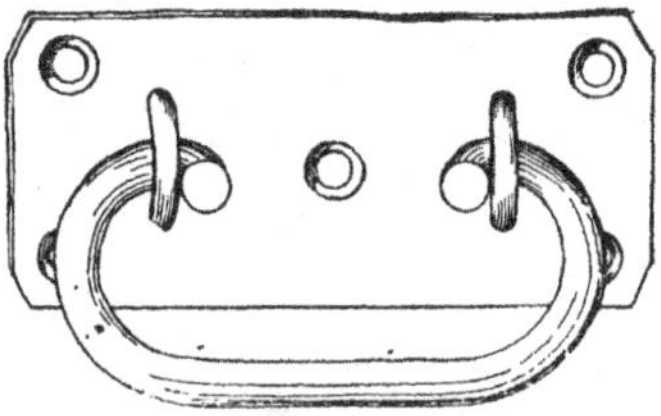

Fig. 398. Chest-handle.

Drawer-Pulls.

Drawer-pulls are made in a great variety of styles, only a few of which need be considered here. Figure 398 is a cheap and very common form of wrought-iron chest or drawer handle, suitable only for rough work. Figure 399 is a very serviceable

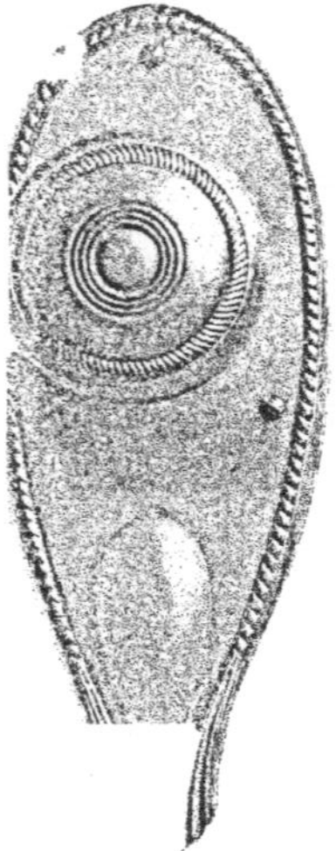

481

477

479

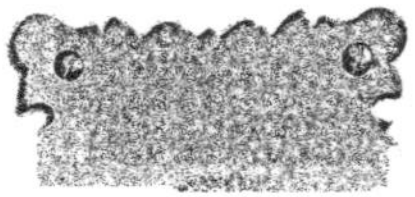

478

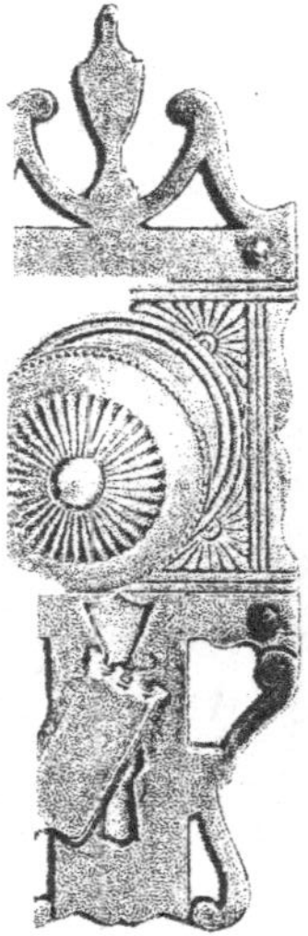

476

475

480

482

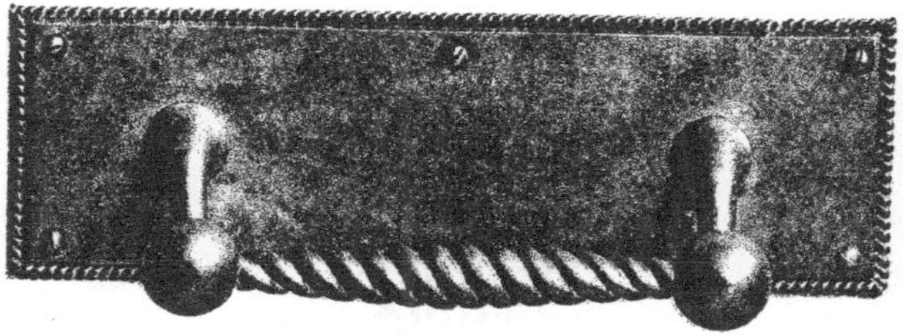

483

··ARTISTIC·HARDWARE···

·MANUFACTURED·BY·A·G·NEWMAN·

Heliotype Printing Co. Boston

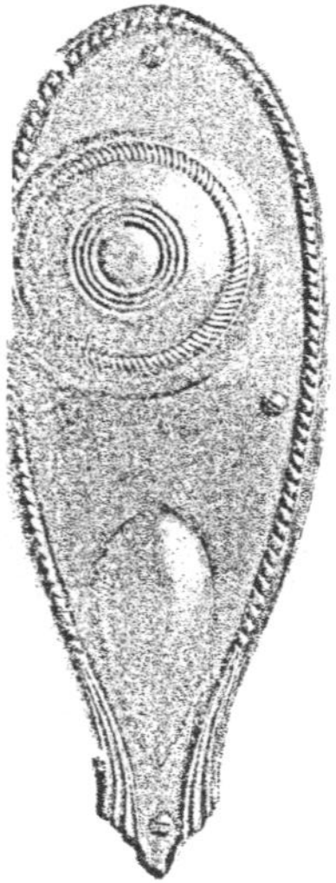

481

477

479

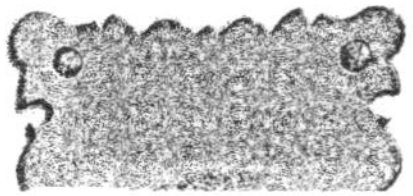

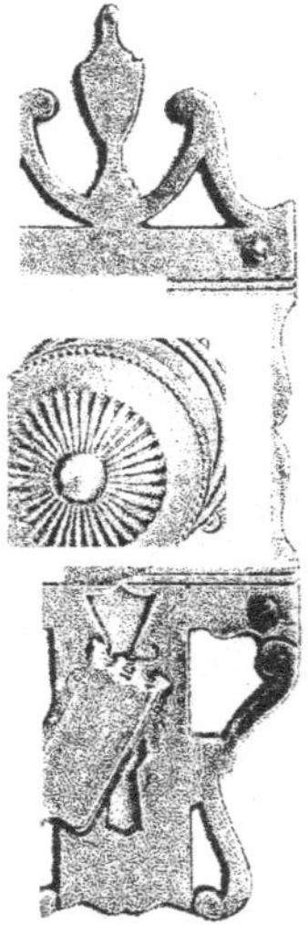

476

475

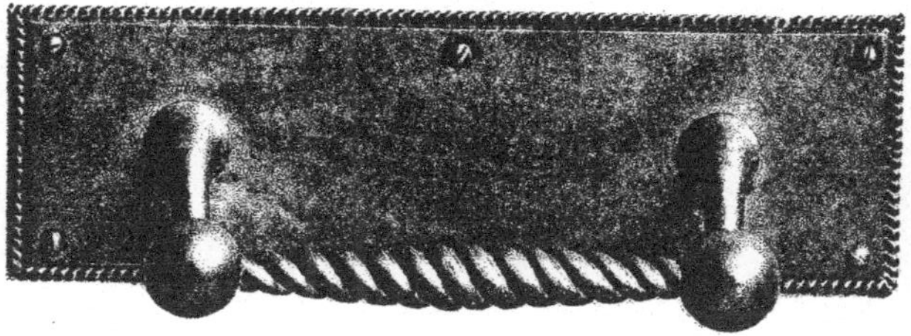

483

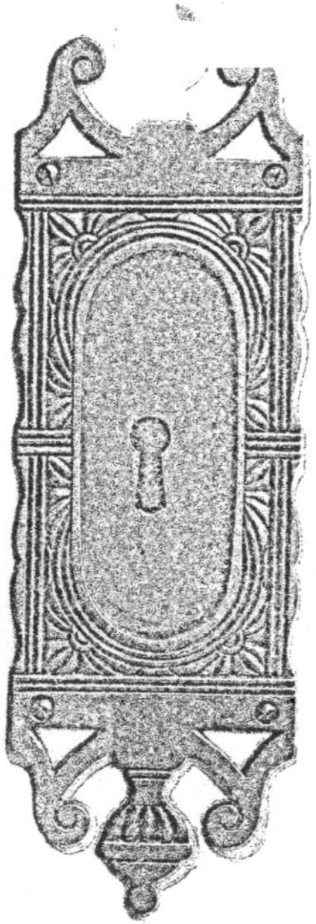

482

··ARTISTIC·HARDWARE··

·MANUFACTURED·BY·A·G·NEWMAN·

Heliotype Printing Co. Boston

drawer-pull, and in plainer form, with sides as well as front rounded in, is what is commonly employed for china-closets, wardrobes, etc. Figures 400 and 401 are drop-handles for Chapter XIII.

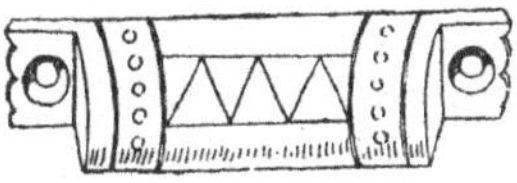

Fig. 399.

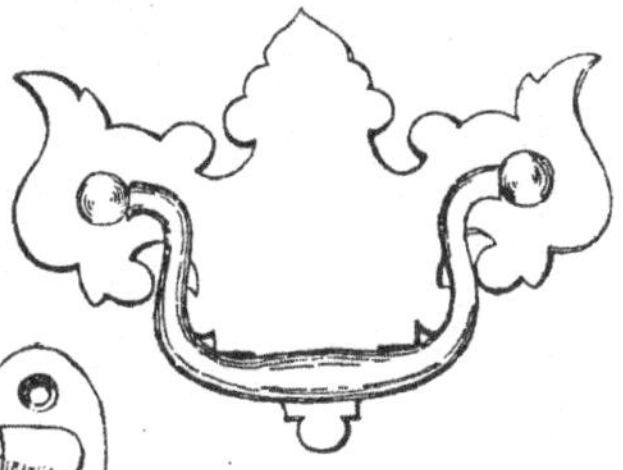

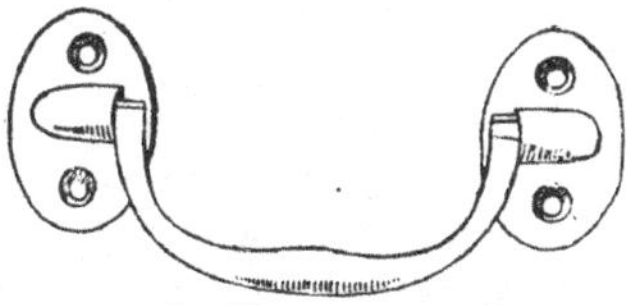

Fig. 401.

Fig. 400. Drawer-pulls.

nice work on the same principle as the first pull illustrated. The latter is a very old pattern, such as is found on most of the antique colonial wardrobes and dressing-cases, and is just now

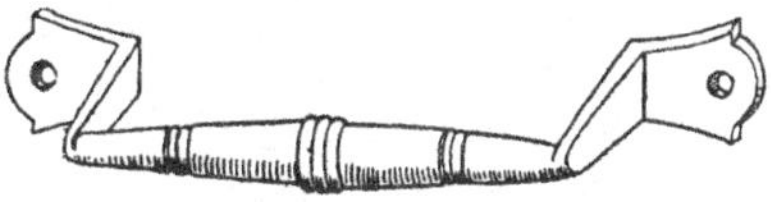

Fig. 402. Bar-pull.

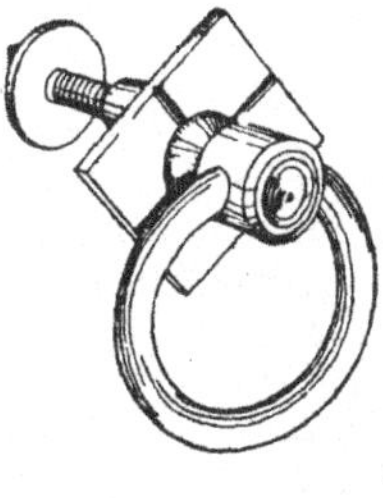

Fig. 403. Ring-pull. Fig. 404. Druggists' Drawer-pull.

quite in fashion. Figure 402 is a straight bar-pull and Figure 403 is a serviceable and easily attached ring drawer-pull

Chapter XIII.

occasionally employed for wardrobes. Figure 404 is a type of what is particularly designated as a druggists' drawer-pull, being on the principle of Figure 399 but with frame and slot on the face to receive a card or label.

Drawer-Knobs.

Drawer-knobs are mostly too simple to require any illustration. They are made with heads of wood, porcelain, mineral, composition or metal, and are usually so shaped as to dispense with a separate shank or spindle, the knob sometimes having a slight metal rose or collar. In the cheaper grades the knob has leaded into it a gimlet-pointed screw-threaded spindle which can be turned directly into the drawer-front; but a more satisfactory form has a spindle extending entirely through the drawer, and secured by a nut and washer on the inside. With a wooden knob the attachment is sometimes made by means of a screw passing through the drawer-front and turning into the knob from behind. Wood or porcelain knobs are most suitable for kitchen and china closet work, though no knob is ever as permanent or satisfactory as a drawer-pull.

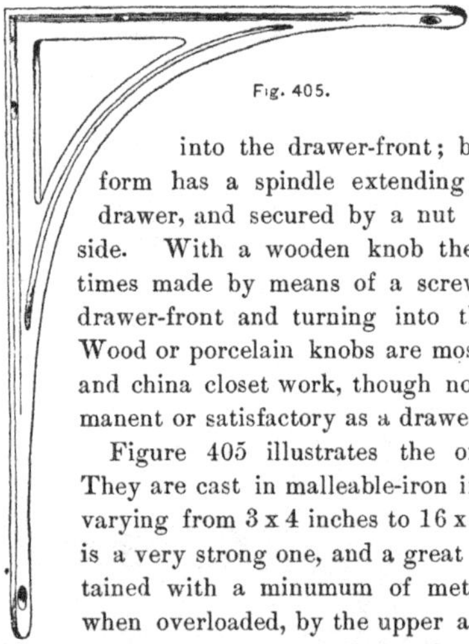
Fig. 405.

Shelf-Bracket.

Figure 405 illustrates the ordinary shelf-brackets. They are cast in malleable-iron in sixteen or more sizes varying from 3 x 4 inches to 16 x 20 inches. The form is a very strong one, and a great deal of stiffness is obtained with a minumum of metal. They usually fail, when overloaded, by the upper arm or flange breaking near the inner screw-holes, but it requires a greater load than one would suppose to break such a bracket. There are many so-called "fancy" forms of shelf-brackets in the market few of which are in the slightest degree artistic, though most of them are stronger than the simple form shown by the figure, on account of having more metal-work between the flanges. Brass brackets are seldom required for ordinary house work and can usually be had only on a special order.

Chapter XIII.

Shelf-Pins.

It is often desirable, in fitting up book-cases or china-closets, to have movable shelves. Shelf-pins of some sort are then used, holes being bored at regular intervals in the sides of the

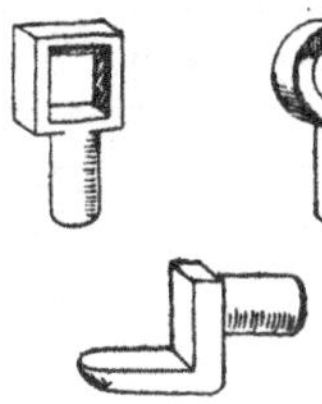

Fig. 406. Book-case Shelf-pegs.

Fig. 407. Coat-hook.

case into which the pins will fit. Ordinary screw-eyes answer very well for most purposes, but are rather conspicuous when proportioned for heavy loads, and are not very easily moved. Figure 406 illustrates three patterns of specially devised shelf-pins. The ones with square and round heads are taken from the catalogue of A. G. Newman. The rebated pattern is manu-

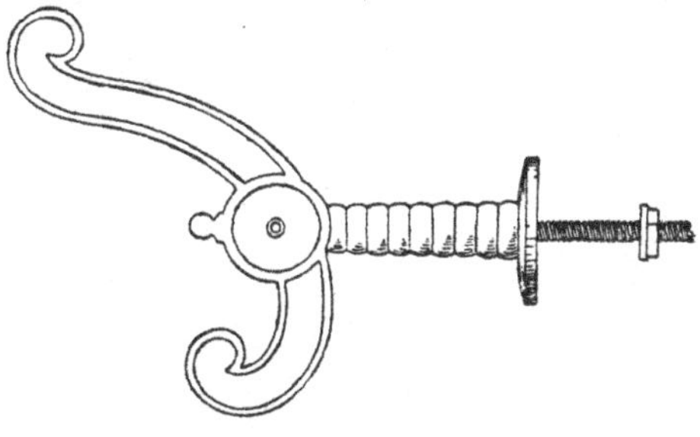

Fig. 408. Coat-hook.

factured by Russell & Irwin, and holds the shelf so that the greater portion of the pin is hidden.

Closet-Hooks.

Closet-hooks are made in so many different styles, and, withal, are so well-known that only a few forms need be considered, which will serve as types for the hundred or more

Chapter XIII.

varieties to be found in the hardware market. The hooks are invariably secured to wooden cleats which are nailed to the wall over the plaster. Figure 407 is a hat-pin with hook beneath, which can be turned directly into the wood by means of the screw-thread on the extension of the shank. Figure 408 is held by a nut turned up from behind, and can, of course, be

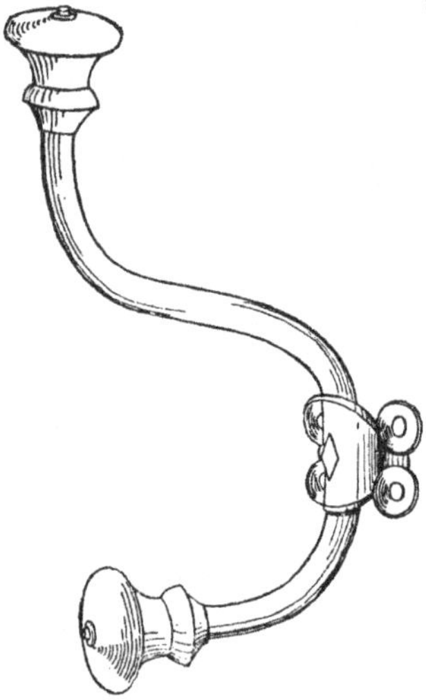

Fig. 409. Coat-hook.

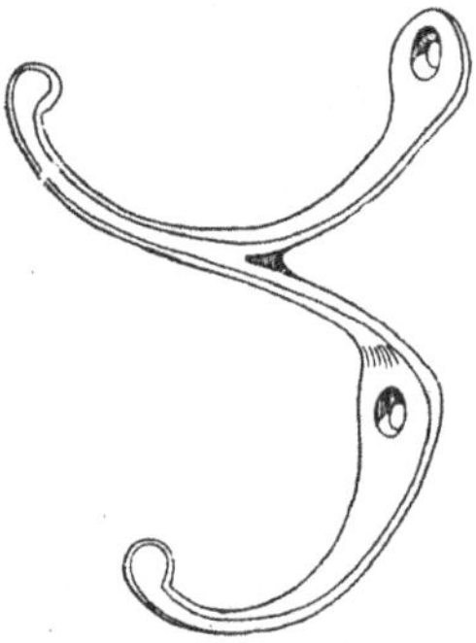

Fig. 410. Coat-hook.

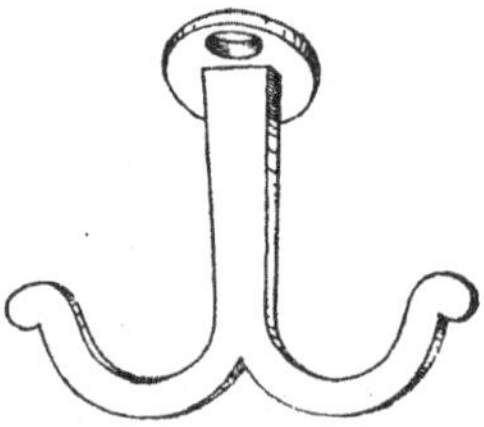

Fig. 411. Wardrobe-hook.

used only on some form of hat-rack. This and the preceding are properly furniture-trimmings. Figure 409 is a familiar, old-fashioned coat-and-hat hook with porcelain knobs, a very serviceable article even though it is not quite in style. Figure 410 is a form of wardrobe hook usually made in brass or bronze, and Figure 411 is a wardrobe hook intended for dresses which are to hang from the ceiling. A similar hook, Figure 412, is fitted with a gimlet-pointed screw-shank, to screw directly into the wood. A very good wooden hook, Figure 413, is made on the same principle as a harness hook, the hardwood pin being inserted from the rear of the iron-base and bevelled, so it cannot work loose or pull out. Figures 414 and 415 are types

Fig. 413.

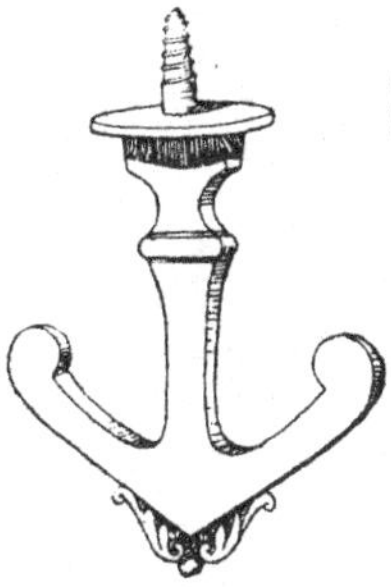

Fig. 412.

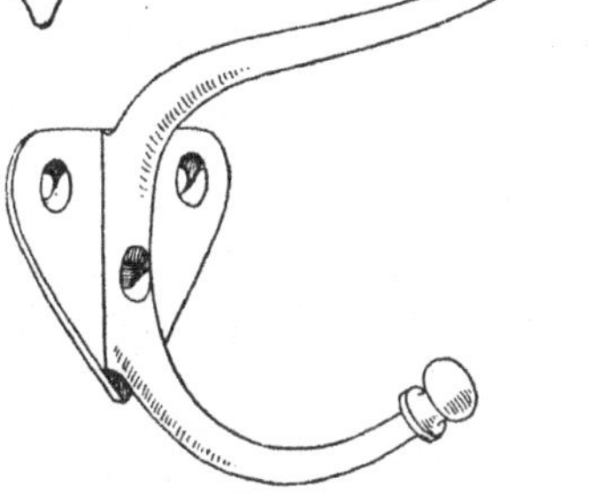

Fig. 414.

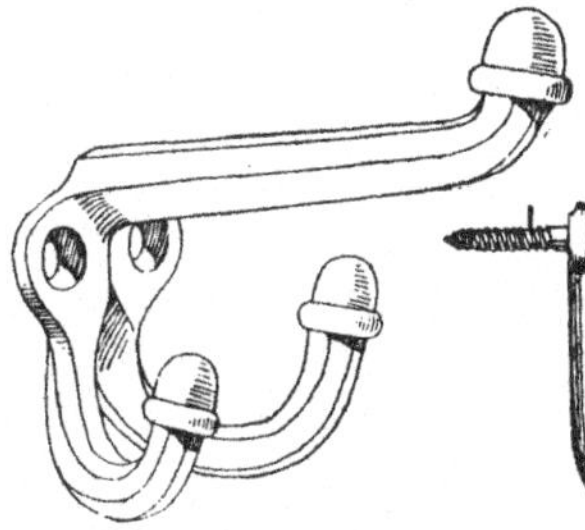

Fig. 415.

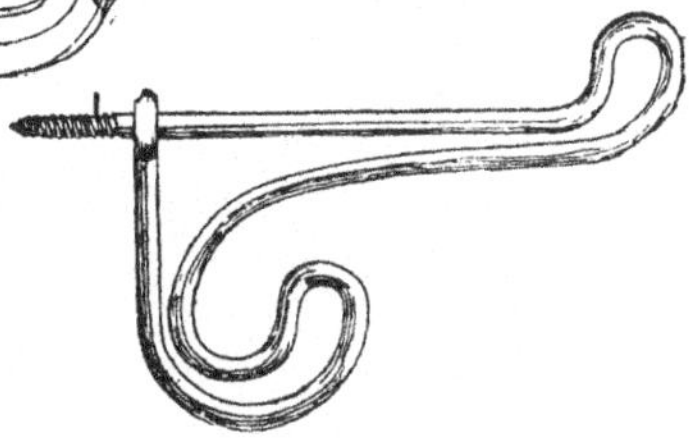

Fig. 416.

of the common hat and coat hook, the latter being specially designated as for school use. Several styles of hooks are also

Chapter XIII. made of bent steel wire, Figure 416, and are very strong, light and serviceable.

All of the foregoing closet-fixtures can be had in various sizes and in different materials. The following table gives some average retail prices which will serve as guides in selecting goods. The prices are for a dozen medium-sized fixtures, complete, with screws.

TABLE OF CLOSET-FIXTURES.

Fig.	Fixture.	Bronze.	Japanned Iron.	Bronzed Iron.	Wood.	Porcelain.
393	Turn-buttons, without plate, 1¾ in.........	$ .33	$ .10	$ .20	$	$
394	" " with plate, 1¾ in...........	.87	.65	.75	—	—
395	Cupboard-bolts.....	2.25	—	.87	—	—
396	" catches..........................	3.00	—	.65	—	—
397	Lever cupboard-catches....................	3.25	—	.75	—	—
398	Chest-handles, wrought....................	—	1.35	—	—	—
399	Drawer-pulls, plain.......................	2.00	.25	.38	—	—
400	Lifting-handles, 3½ in. single swing.......	2.25	.75	1.25	—	—
401	Draw-handles..............................	4.00	—	—	—	—
402	Bar-pulls, 4½ in..........................	2.00	—	.30	—	—
403	Ring-pulls................................	1.50	—	—	—	—
—	Druggist-pulls, plain...........	4.50	—	1.50	—	—
—	Drawer-knobs, screw end...................	2.50	—	—	.15	.35
—	" " bolt and nut................	2.65	—	—	.30	.50
405	Shelf-brackets, 8 x 10, per doz. pairs........	—	3.00	4.00	—	—
406	Shelf-pins................................	1.25	.08	.50	—	—
—	Ordinary coat and hat hooks...............	2.50	.20	.25	—	—
413	Wooden " " " "	—	—	—	.25	—
416	Wire " " " "	—	.20	.25	—	—

CHAPTER XIV.

Miscellaneous Hardware.

HOOKS.

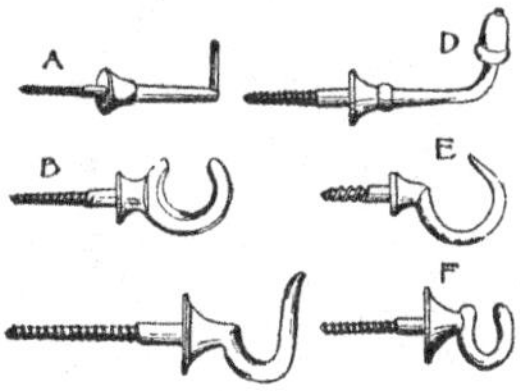

Fig. 417. Screw-hooks.

Fig. 418. Bird-cage Hook.

BESIDES the hooks described in the last chapter, there are other forms which cannot be classed as closet hardware. Figure 417 illustrates several varieties of brass screw-hooks. *A* is made in nine sizes, from $\frac{1}{2}$ inch to 2 inches in length. The same form is made with a sharp-pointed shank instead of a screw, intended to be driven into the wood. *B* is known as a cup-hook, intended to go on strips to receive cups, which are hung by the handle over the hook. This form is made in five sizes, from $1\frac{1}{2}$ to $2\frac{3}{4}$ inches long. *C* is termed a looking-glass hook. *D* is an acorn-hook, made in six sizes, from 2 to $4\frac{1}{2}$ inches long. *E* and *F* are both picture-hooks. The former is made in six sizes, from $\frac{3}{4}$ inch to $1\frac{3}{4}$ inches. Figure 418 represents a hook similar to the preceding, but with a longer shank, being made in seven lengths, from four to ten inches: it is designated as a bird-cage hook. Screw-Hooks.

Picture-moulding hooks are made in quite a variety of Picture-Moulding Hooks.

shapes, a few of which are shown by Figure 419. The most common form is the second one on the upper row, it being made to match the common stock picture-moulding. A very

Fig. 419. Picture-moulding Hooks.

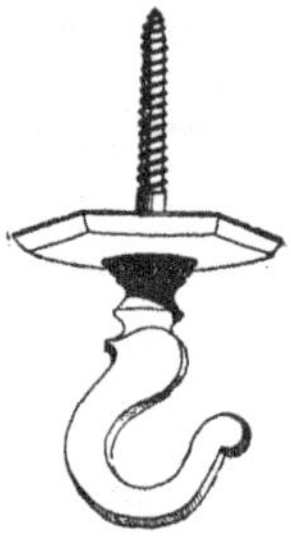

Fig. 420. Chandelier-hook. J. B. Johnston.

serviceable hook, not illustrated here, is made with flat brass, with the ordinary contour, quite broad at the top where it fits over the moulding, but narrowing at the bottom to receive the cord or wire. Moulding-hooks are usually made in three sizes, and are

Fig. 421. Chandelier-hook with Catch. J. B. Johnston.

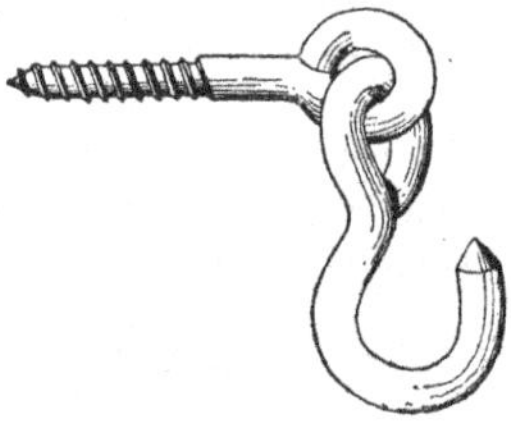

Fig. 422. Hammock-hook.

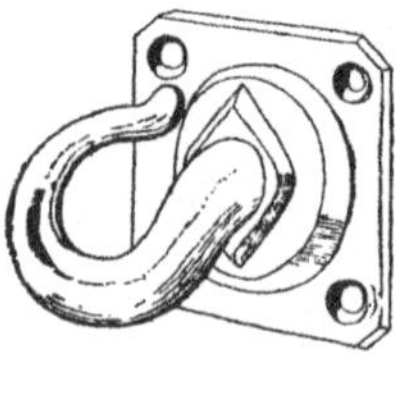

Fig. 423. Clothes-line Hook.

Fig. 424. Awning-hook.

Chandelier-Hook.

always of brass or bronze. Chandelier hooks, Figure 420, are intended to screw through the lath and plaster into the ceiling beams or the furring, the screw part being 2½, 4, 6 or 8 inches long. Figure 421 shows a chandelier-hook provided with a catch, so that nothing can slip out when once hooked.

Chapter XIV.

Hammock and Clothes-line Hooks.

Hammock-hooks, Figure 422, are made of ⅜ inch galvanized or tinned wrought-iron. Clothes-line hooks, Figure 423, are also sometimes used for hammocks, though less suitable on account of the friction of the rope in the hook. A lighter form of clothes-line hook is made to be attached by two screws. These hooks are made in three sizes.

Awning Hooks.

Awning-hooks, Figure 424, are made to drive into the wood, and be caught in eyelets in the awning. They are manufactured in sizes from 1½ to 6 inches.

BRACKETS.

Stair-rail Brackets.

Shelf-brackets have been previously discussed. Some form of inclined bracket is often desirable to support the side-rail of a flight of stairs. One of the simplest consists of a bent plate, Figure 425, screwed to the wall on an angle, so as to bear against the under side of the rail. A better form is screwed to

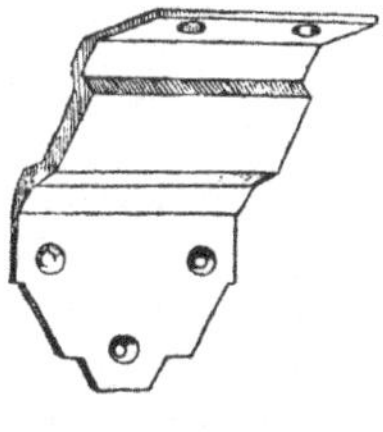

Fig. 425. Stair-rail Bracket. Reading Hardware Co.

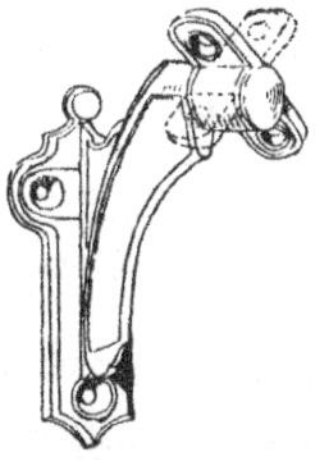

Fig. 426 Stair-rail Bracket. Reading Hardware Co.

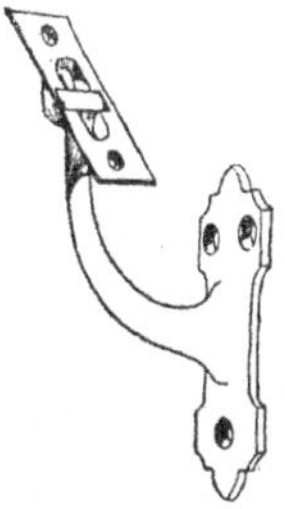

Fig. 427. Stair-rail Bracket. Shepard Hardware Co.

the wall in a vertical position, and has a swiveled bar or plate which adjusts itself to any angle of the stair-rail. Figures 426 and 427 illustrate two styles. Similar brackets are made with fixed rail-plates, and there are a number of varieties in the market differing from those described chiefly in regard to finish.

Chapter XIV.

Bar-rail Brackets.

Bar-rail brackets, Figure 428, are intended to support a round rail such as is usually carried across the front of a bar-room counter. The first form shown is sometimes used to support a

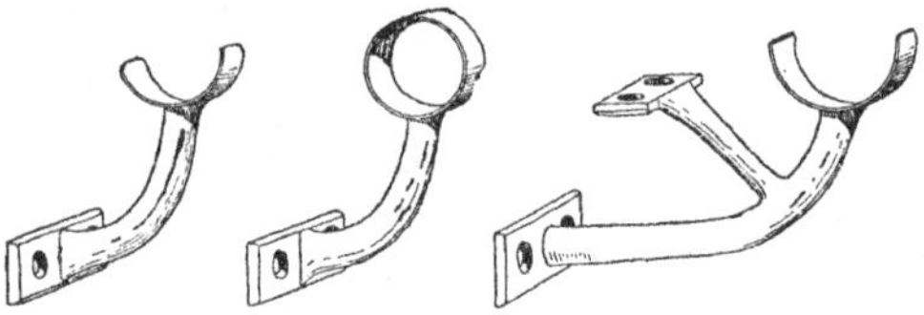

Fig. 428. Bar-rail Bracket. J. B. Shannon & Sons.

round stair-rail, and when made of plain bronze, presents a very good appearance. A bracket like the second form is sometimes used to support an iron foot-rail at the base of a bar or counter.

All these brackets can be had in either bronze or bronzed iron.

LETTERS AND PLATES.

Letters.

Very few styles of letters and numbers are kept in stock by hardware dealers. Plain, Roman characters, Figure 429, are usually the only ones on hand. They are in seven sizes, from ½ inch to 3 inches high, and are secured to the door or the woodwork by blind tacks, soldered to the back of the pieces. They can be had in either bronze, brass or nickel-plate. Letter-plates are often used to cover the letter-slot through office-doors. Figure 430 shows one style, with a recessed slot protected by a hinged flap. This is essentially what is commonly employed. On fly-doors some form of plate is desirable on each face of the door to prevent the paint

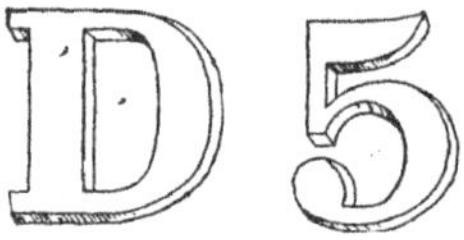

Fig. 429.

Letter-plates.

Fig. 430. Letter-Plate. Hopkins & Dickinson Mfg. Co.

Chapter XIV.

from being soiled, and such plates are often marked "push" or "pull." They may be of porcelain, iron, bronze, brass or nickel-plate, the first material being the cleanest and most easily cared for. They are made in all varieties of design, but are in principle too simple to require any illustration.

Push-plates.

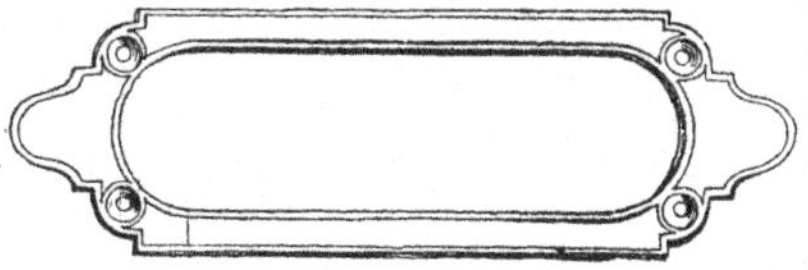

Fig. 431. Label-plate.

Label-plates.

Label-plates are made to order in porcelain quite extensively for druggists' drawers. There are also plates manufactured to go on drawer-fronts and receive card-labels, the upper part of the plate being thinner than the rest, so that the card can be slipped in from above. Figure 431 will illustrate the general form of a label-plate. The neatest style has a plain, rectangular outline in bronze. Label-plates are made in several sizes from about 1 x 2½ inches to 2 x 4 inches.

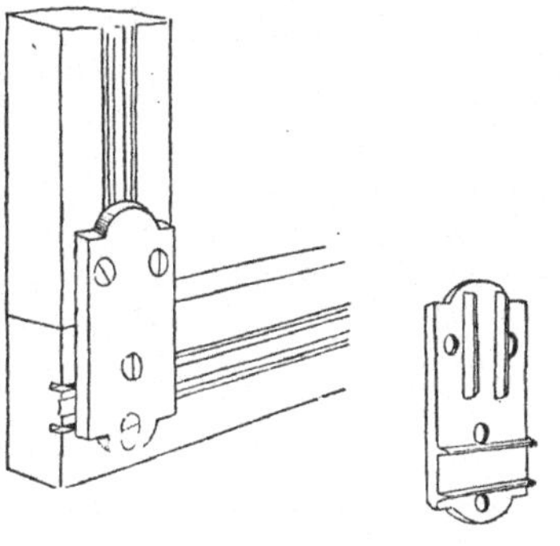

Fig. 432. Screen-door Corner-iron. E. C. Stearns & Co.

Screen-door Irons.

Figure 432 shows a plate a little foreign to the present topic, it being used to stiffen the joints of light screen-doors. It is provided with tongues which enter firmly into the wood in each direction, and prevent any sagging or settling. The plates are sold in sets, each set including six corner irons and a knob or handle, with the necessary screws. The list-price is $5 per dozen sets, in bronzed iron.

Chapter XIV.

FOOT-SCRAPERS.

Foot-scrapers.

Foot-scrapers are used much less than formerly. A simple form, consisting of a thin-plate supported by one or two plain drive-shanks is always advisable, however, for the piazza of a country house. Figure 433 shows a more elaborate scraper, intended to be screwed to the floor or step. A form often seen

Fig. 433. Foot Scraper. J. B. Johnston.

Fig. 434. Foot Scraper. J. B. Johnston.

in some parts of the country, Figure 434, is set in a pan or dish, intended to collect the scrapings. The other varieties found in the market differ only in design or finish, but not in principle. Foot-scrapers are usually of Japanned cast-iron.

BELL HARDWARE.

Bell-fixtures. Gongs.

The subject of bell-fittings is too extensive to be considered very fully in detail, especially as bell-hanging is a trade by itself, and the house-carpenter has usually very little to do beyond hanging the simplest kind of kitchen-bell or fitting a gong to the back-door. The front-door is fitted with a bell-pull, as explained in the chapter on knobs. This is connected with wires which usually are carried down to the cellar-ceiling, and across and up to the kitchen. The corners are turned by the aid of bell-cranks. Figure 435 shows the form of crank generally fitted just inside of the bell-pull, and Figure 436 shows a complete set of bell-hanging fixtures, including the bell, which is secured to the wall by a spike driven through the centre of the spiral coil. The elasticity of the coil and the connected spring is so great, that when the fixtures are properly set, the least pull at the front-door will cause the bell to ring. Figure 437 illustrates a different form of bell-carriage, Chapter XIV.

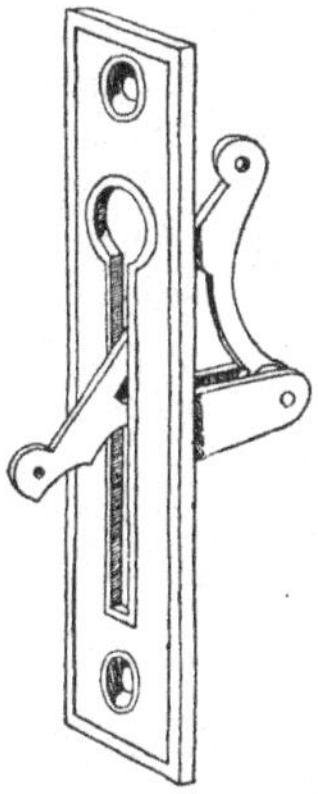

Fig. 435. Bell-crank. Russell & Erwin Mfg. Co.

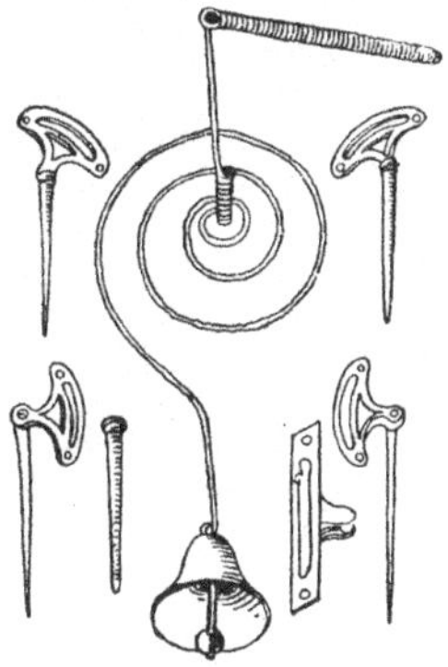

Fig. 436. Bell-hangings.

Fig. 437. Bell-carriage. Russell & Erwin Mfg. Co.

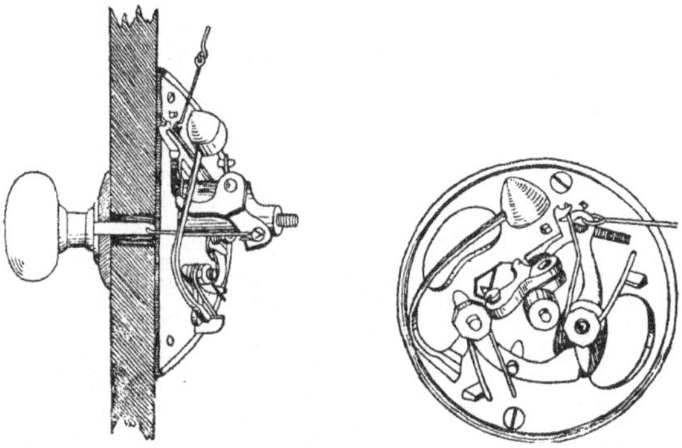

Fig. 438. Double-stroke Pull Gong-bell. Russell & Erwin Mfg. Co.

made by the Russell & Erwin Manufacturing Company.

For the back-door it is customary to use some form of gong

Chapter XIV.

which can be screwed to the inner face of the door. In the cheapest makes the bell-strike is operated by a handle on the outside, which on being drawn down, releases a spring-hammer. Some gongs are made so as to give a double-stroke. Figure 438 illustrates a double-stroke bell which works with a pull instead of a lever. There is, also, in the market a bell provided with a spring escapement which is set by pulling the handle, and gives a continuous ring like that of an electric-bell, lasting about five seconds. This is known as "Bushby's Escapement Bell."

GATE–FIXTURES.

Gate-hinges.

Ordinary strap-hinges are sometimes used for gates, and there are a few forms of heavy wrought-iron butts which also

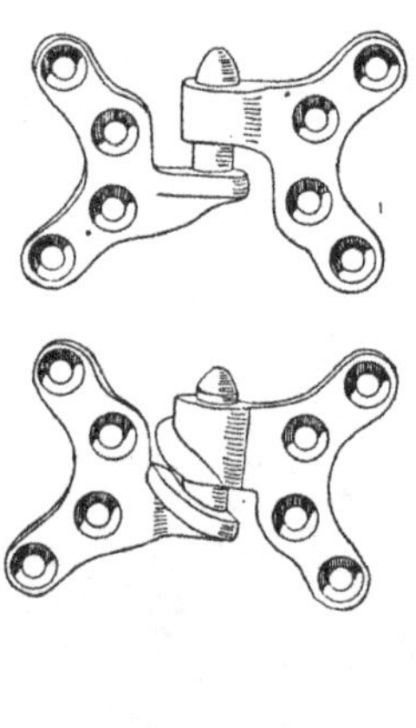

Fig. 439. Self-closing Gate-hinge, St. Louis Pattern. Shepard Hardware Co.

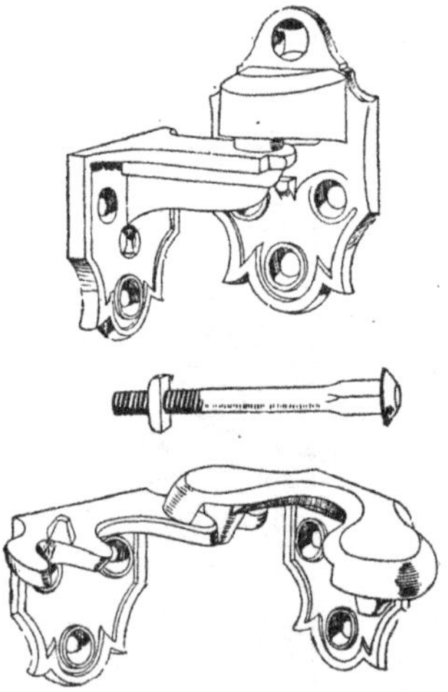

Fig. 440. Seymour's Gate-hinge. P. & F. Corbin.

answer for the purpose; but there is in the hardware market quite a variety of fixtures especially devised for gates, and the special forms are usually preferred. Gate-hinges are always arranged to be self-closing, generally acting by gravity. Shepard's St. Louis pattern, Figure 439, has the bearing-surfaces of the lower hinge made on a sharp incline, so that when the gate is opened, it is lifted bodily, and descends in closing. This principle is embodied in several different patterns. It, of course, permits the gate to open only in one direction. With "Seymour's" hinge, Figure 440, the gate is practically suspended from the upper pivot, and bears laterally against two pivots at the bottom, so spaced, that when the gate is open, the bottom is thrown out more than the top, and its own weight is sufficient to close it. Figures 441 and 442 are variations of the same principle, a simpler applica-

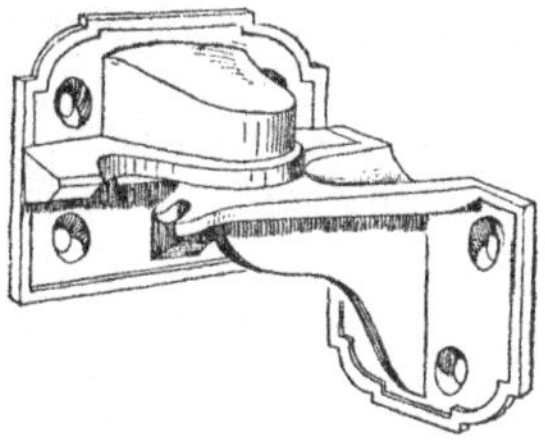

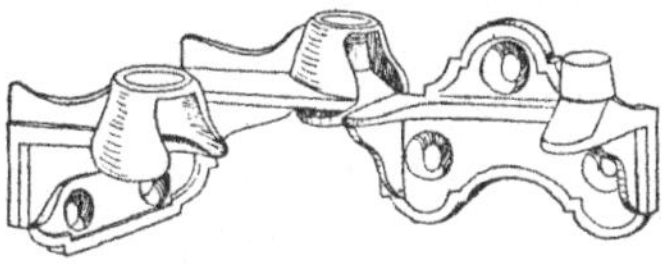

Fig. 441. Gate-hinge. Shepard Hardware Co.

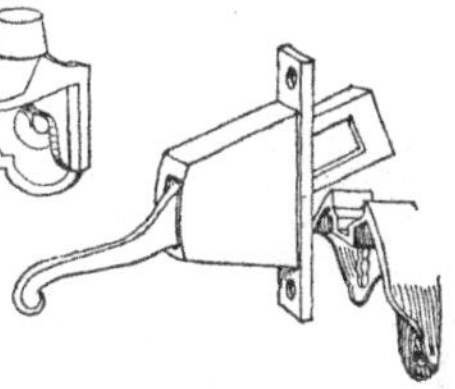

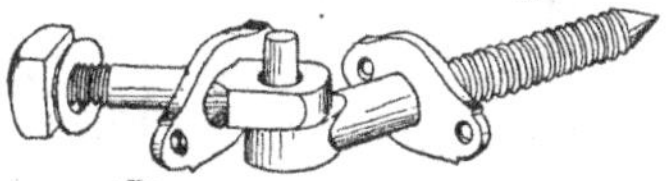

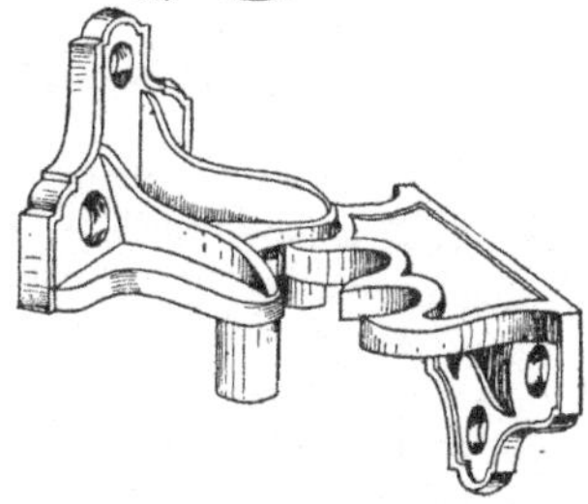

Fig. 442. Gate-hinge. Shepard Hardware Co.

Chapter XIV.

tion of the idea being shown by Figure 443. All of these will open both ways.

GATE-LATCHES.

Gate-Latches.

A very common form of gate-latch is shown by Figure 444. It consists of a bent lever which is mortised through the gate-frame, the bolt catching in a strike on the post. A spring keeps the bolt thrown out, and the beveled strike permits the latch to be self-closing. With a strike which is beveled each

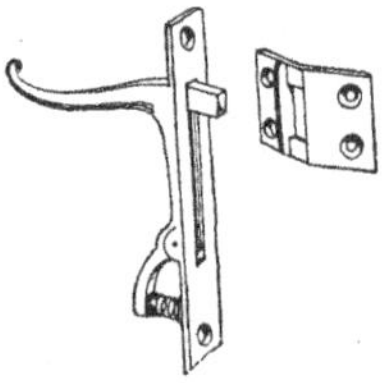

Fig. 444. Mortise Gate-latch. Ireland Mfg. Co.

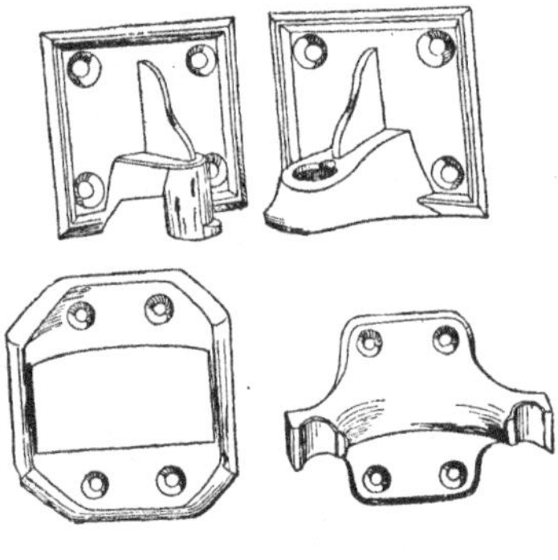

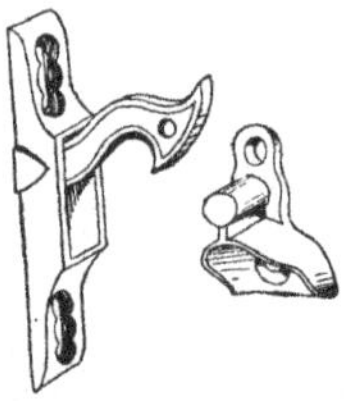

Fig. 443. Gate-hinge No. 20. Shepard Hardware Co.

Fig. 445. Broads's Patent Gate-latch. Ireland Mfg. Co.

way, this latch can be used for a double-swing gate. Figure 445 shows a latch which is very commonly used with gates swinging only one way. The catch acts by gravity alone. Figure 446 represents a gravity, mortise catch. The latch shown by Figure 447 is planted on the face of the gate-frame, and works with a spring, while in Figure 448 it is planted on the edge of the gate-frame, which has to be kept correspondingly away from the post. The Yale & Towne Manu-

486.

488.

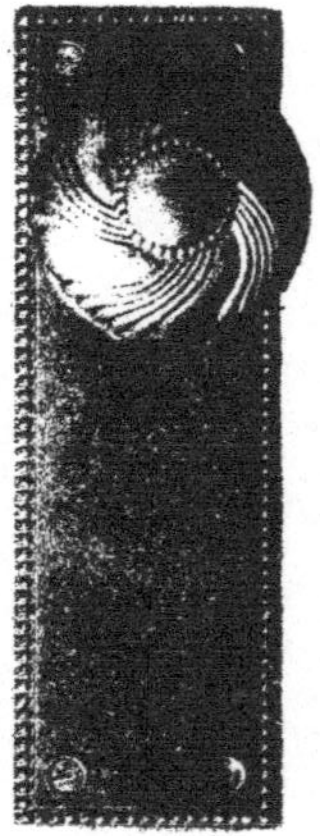

485.

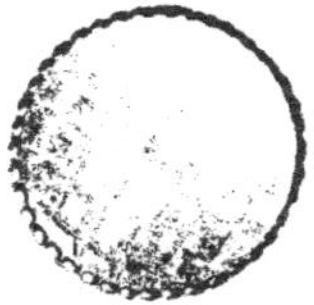
484.

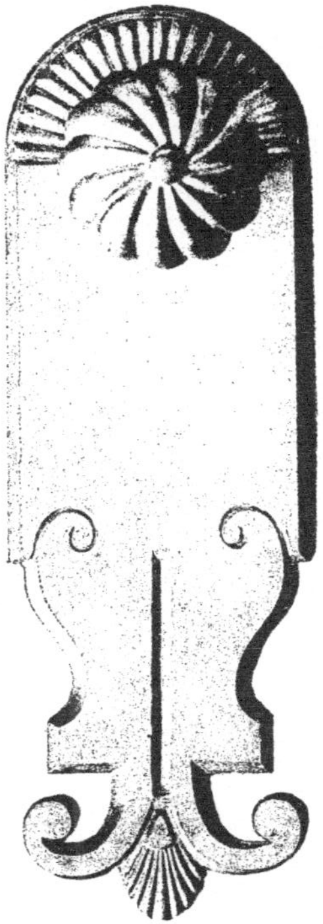
491.

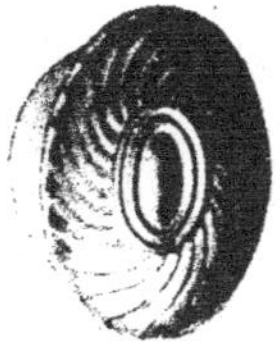
490.

492.

487.

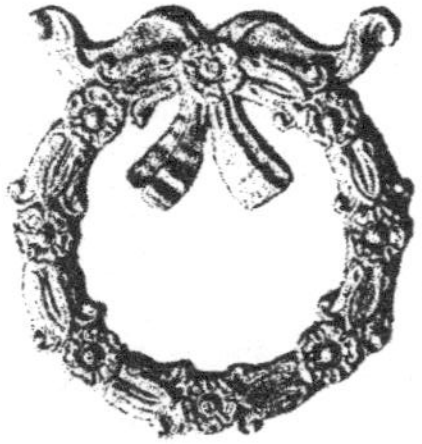
493.

··ARTISTIC·HARDWARE··

facturing Company has a somewhat similar gate-latch, Figure 449. Each is opened by pressing down one of the arms. "Seymour's" cylindrical gate-latch, Figure 450, is mortised

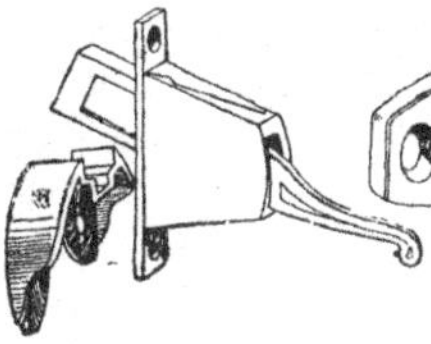

Fig. 446. Gate-latch No. 2. Shepard Hardware Co.

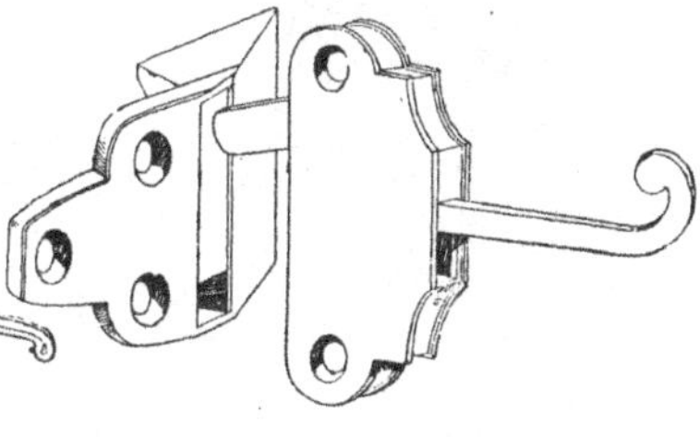

Fig. 447. Seymour's Gate-latch. P. & F. Corbin.

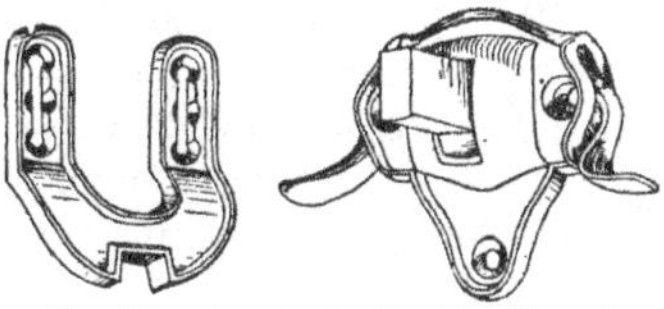

Fig. 448. Gate-Latch, No. 20. Shepard Hardware Co.

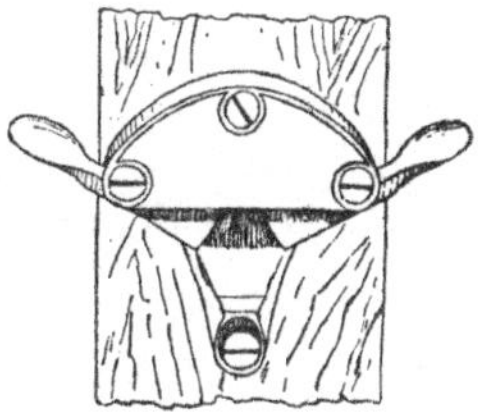

Fig. 449. Leeds's Gate-Latch. Yale & Towne Mfg. Co.

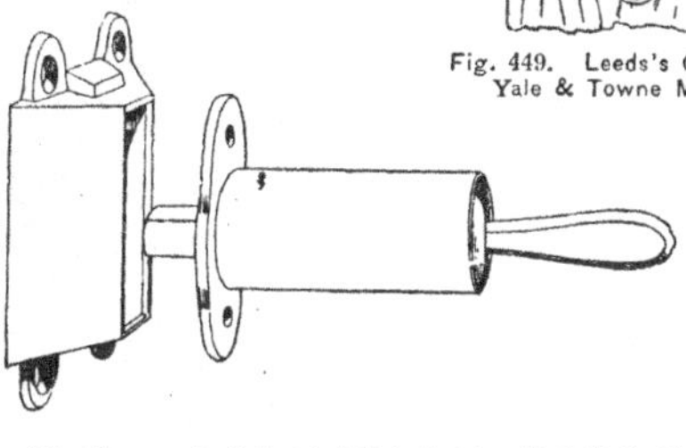

Fig. 450. Seymour's Cylindrical Gate-Latch. P. & F. Corbin.

through the gate-frame, and opens when the handle is depressed.

There are many other styles of gate-hinges, few which differ materially from those we have considered.

Chapter XIV. DUMB-WAITER FITTINGS.

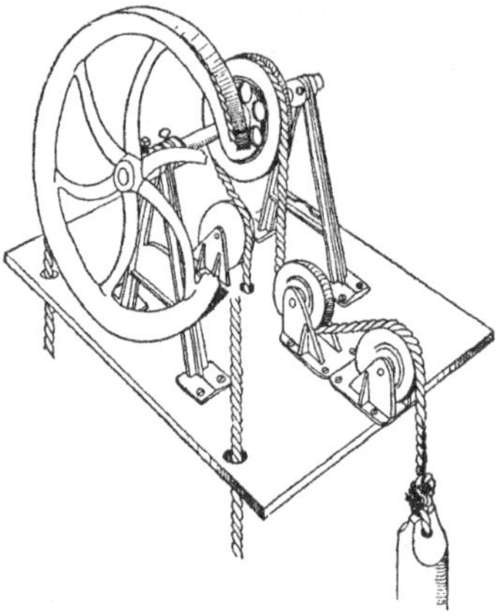

Fig. 451. The Cannon Dumb-waiter.

Common Dumb-waiter.

Where are three styles of dumb-waiters in common use. For the cheapest sort of work, a rope is attached to the top of the car, carried up over a wheel, down one side to the bottom of the well, under a second wheel and up to the bottom of the car, to which the end of the rope is attached. A counter-balance weight is connected with the top of the car by a rope passing over a third wheel. Another style

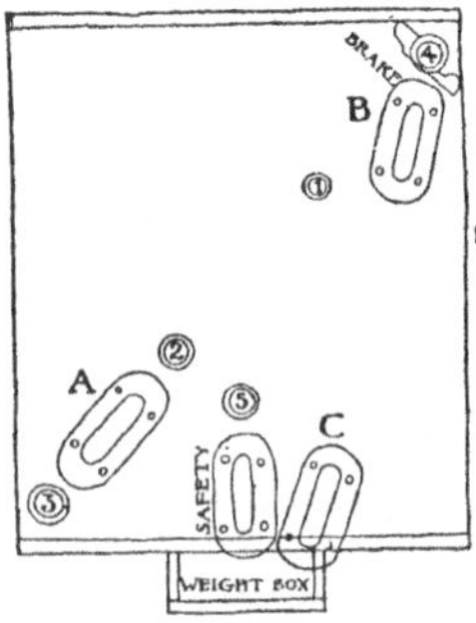

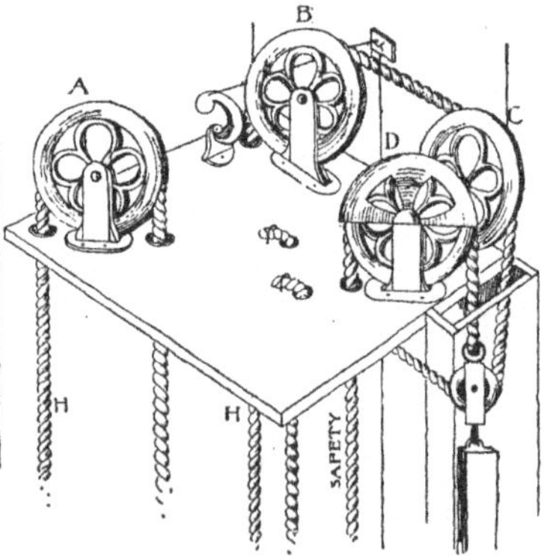

Fig. 452. New York Safety Dumb-waiter. Fig. 453.

Cannon Dumb-waiter.

is shown by Figure 451. A rope is fastened to the top

of the car and passes about the four small wheels to the counterbalance weight on one side. One of these wheels is on the shaft of a large wheel over which a thick rope is carried and continued around a similar wheel at the bottom of the shaft. This holds on the large wheel simply by friction, and in moving, winds up or lowers the hanging-rope.

New York Safety Dumb-waiter.

The third style is illustrated by Figure 454, with a plan of the upper gearing, Figure 452, and a diagonal view of the upper works, Figure 453. A rope is attached to the bottom of the shelf on which the upper wheels rest, starting at 1, Figure 452. Thence it is carried down to and under the wheels *E E* on the top of the car, up through 2, over wheel *A*, and down through 3 to the wheels at the bottom of the shaft. Then it

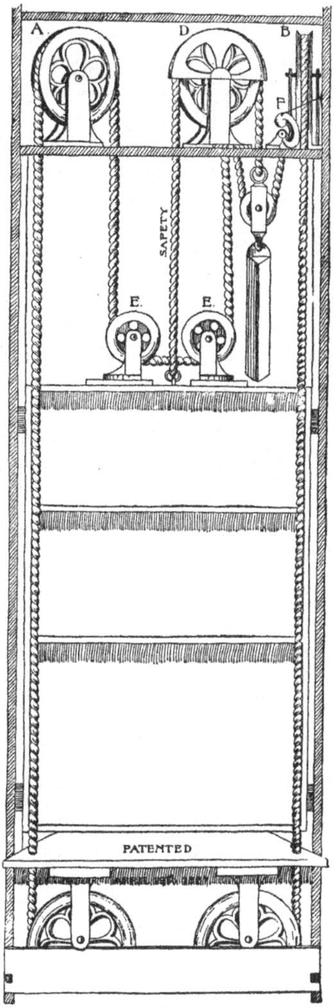

Fi . 454. New York Safet Dumb-waiter. Edw.

Chapter XIV.

4, over wheels *B* and *C*, and there connected with the counterbalance by a pulley, the end of the rope being fastened to the under-side of the top shelf, close by where it started. A safety-rope is attached to the top of the car, carried up through 5, over pulley *D*, and connected with the counterbalance. A cam-brake, Figure 455, on the shelf beside 4 prevents the car from descending when heavily loaded. It will be seen that all the working connections are made with a single rope, so arranged that no matter how much it may stretch, it will always be taut, the slack being taken up by the counterbalance, so that the slightest motion of the rope will start the car. This style is very generally used in good work.

The doors at the openings into the dumb-waiter shaft are usually hung in the same manner as an ordinary window, and are provided with some form of spring-catch like Figure 456,

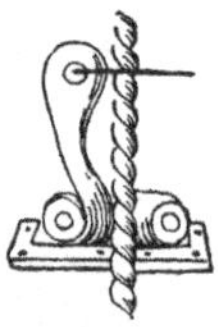

Fig. 455. Brake.

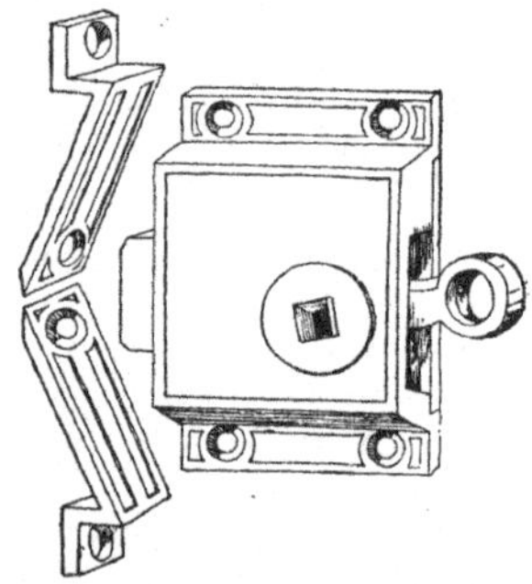

Fig. 456. Dumb-waiter Catch.

which will hold the door either up or down, the catch being released by depressing the arm at the side.

Dumb-waiter Catch.

The following table gives the average retail prices of the miscellaneous articles described in this chapter.

TABLE OF MISCELLANEOUS HARDWARE.

Chapter XIV.

Fig.			
417	Brass Screw-hooks	per doz.	15 to 25 cents.
419	Picture-hooks	per doz.	35 to 42 cents.
420	Chandelier-hooks, 6-inch screw	each	58 cents.
422	Hammock-hooks	per doz.	$ 1.12
423	Clothes-line hooks, 4 screws	per doz.	.42
	Stair-rail brackets	per set	.10
429	Letters and numbers, 2-inch	each	.35
430	Letter-box plate, plain	each	.75
436	Bell-fixtures, complete without wire	per set	1.00
	Bell-gong, bronze handle	each	.75
	Gate-fixtures, hinges or latches	per set	25 to 50 cents.
452	N. Y. Safety Dumb-waiter fittings	per set	15.00

CHAPTER XV.

Artistic Hardware.

THE manufacture of fine hardware for house-trimmings has by no means kept pace with the artistic development which this country has witnessed during the past fifteen years in nearly every other branch of the art industries; so that while the mechanical details of the wares have been brought to the highest degree of excellence, the manufacture seldom rises to the dignity of an art, and seems to be considered in general as requiring no further direction than can be given by the most skilful mechanics. A trained designer, who would treat hardware as a part of a house rather than as an opportunity for mechanical elaboration, who would appreciate the relation it necessarily bears to the architecture, who would consider the fitness of the material for certain forms and the adaptability of certain styles for certain uses, without attempting to combine an Albert Durer cartouche with an Italian Renaissance scroll, crown the product with a Greek honeysuckle and apply the whole to a design for a door-plate, would have chance for employment with but few of the large firms which supply the market with builders' hardware. Consequently we find that while there is a considerable quantity of perfectly plain, unadorned hardware which is good because of its simplicity, the elaborate hardware is nearly always bad, and unfortunately most hardware is elaborate. These conditions are by no means wholly due to the indifference or artistic incapacity of the manufacturers. It is found that cheap patterns sell the best; a fussy, meretricious design is the most profitable to put on the market, and the producers are obliged to cater to the cheap trade. There are a few exceptions to this rule, some manufacturers whose average artistic productions are unimpeachable, but when such have made money it is due more to good business management and mechanical perfection than to artistic capacity. The public buys hardware because it is well-made and reliable, rather than because it is good-looking. At the same time, even the best is disappointing, because being so good, it ought to be a great deal better. With so much to draw from — the wealth of ideas in the European museums, suggestions in silver and gold smith's work, the old Pompeian bronzes, the delicate, antique Etruscan jewelry, besides the quantities of architectural ideas which might be adapted; and having in addition the best of mechanical workmanship, together with the element of color, which the work of former times seldom possessed, the hardware designer ought never to be at a loss what to do. No one of the art industries is capable of so wide an æsthetic expansion or presents so varied a field for the play of individual fancy, and few have been so persistently misapplied and misunderstood. It is not fair to say there is no good hardware to be had unless it be specially designed, but the general average of the goods which are kept in stock by the leading manufacturers, is, to say the least, commonplace; and this, too, notwithstanding some very notable exceptions.

Materials used.

At one time, iron was almost the only material employed for artistic effects in hardware. This was before the day of patterns, when the individual fancy was applied directly to the metal. But now, when the artist has ceased to be the workman and an article is salable only when it can be readily reproduced, cast brass and bronze are almost the only metals used. Quite recently cast-iron has been employed to a considerable extent, and when finished with the Bower-Barff process, sells at about the same prices as bronze. A defect in nearly all cast-

Chapter XV.

Cast-work.

work, however well-designed, is that the patterns, instead of being modelled by hand, so as to show every touch of the artist and permit of an exact reproduction of the feeling which a design should have when worked out in a metal like bronze or iron, is first carved in wood. One who does not appreciate the difference between a carving and a casting, has only to compare a Greek *stèle* with some of the exquisite bronzes which are taken out of Pompeii; and no workman, however skilful, can give a wood-carving the easy flow and metal-like appearance of a bronze casting. Again, our castings are, in one sense, too nice. That is to say, after they are taken from the mould they are pickled in acids and then hand-chased, a process that could spoil even the best of modelling. It is to be hoped that some day fine hardware will be cast only by the *cire-perdu* or some similar process, which reproduces every touch of the artist and leaves the casting perfect as it comes from the mould.

Glass.

Glass is used more now than it was a few years ago, and for knobs is peculiarly suitable. It is not usually exhibited by the manufacturers as artistic hardware, though surely when such ugly faceted abominations are tolerated in cut-glass tableware, a cut-glass knob ought to please every one.

Change in Forms.

A rather interesting change has taken place in the forms in which artistic designs in hardware are expressed. During the Middle Ages, the greatest amount of work was expended upon the hinges and their accessories, an example of which is afforded by the hinges of the western doors of the Cathedral of Paris. Knobs and latches were quite subordinate; indeed, knobs, as we apply them now, were unknown at that time. During the Renaissance period the ingenuity of the artisans was devoted to locks and keys, some of which are most surprising examples of hardware. At present neither hinges nor locks nor keys are elaborated to any extent, the greatest amount of thought being given to the door-knobs and the plates by which they are secured to the woodwork. In collecting the samples which are illustrated in this chapter, the various manufacturers were requested to indicate which of their goods they considered as most typical of their artistic possibilities; and out of some fifty pieces so designated, all but three were door-knobs. This might be considered as indicative of a belief that knobs alone are susceptible of artistic treatment, though, in reality, there are plenty of well-designed pulls, butts, etc.; but it shows that the popular taste, when craving for art, looks at present no farther than the door-knob. And yet the first impulse of nine out of ten designers would be towards ornamental hinges rather than elaborate knobs, though with our present forms of butt-hinges there is little opportunity, and no real necessity for the long-strap hinges which form such a delightful feature of the old-fashioned doors and casement-windows. Butts are cheaper, more easily applied, and for some styles of work are more suitable, still one rather regrets the opportunities which the old style of hinges afforded. Chapter XV.

Expense.

Artistic hardware usually implies an unnecessary expense to the minds of most buyers, and unfortunately the implication is a correct one, even with very simple designs. The perfectly plain bronze, which is so much in favor in some parts of the country, is more expensive than any of the mean, contemptible figured-bronze goods which form the stock of nearly every hardware store. Plain bronze must be perfect, as every imperfection will show, whereas a criss-cross pattern will hide a multitude of sand-holes and imperfect casting. Until the general average of figured-bronze goods is higher, the architect will, therefore, have considerable difficulty in persuading his client that artistic hardware is worth all it costs. Still there is plenty of good, plain hardware which is within the means of most clients, and the aggregate cost above what the common figured work would be, is quite inconsiderable for an ordinary house. A difference of two dollars per pair on knobs would not amount to more than fifty dollars on an entire house, and the enjoyment of seeing well-chosen hardware about one's home is surely worth more than the interest on fifty dollars, or twice that sum. We, as a nation, are fast appreciating the fact that beauty is worth more than mere utility; and when the public taste

Chapter XV. demands a better average quality of house-hardware, there will be no lack of proper artists to furnish the right kind of supply.

Comparisons are always dangerons, and are apt to be invidions. There are upwards of a hundred firms in this country who make a business of supplying hardware in one form and another. It would be impracticable, were it desirable, to fairly represent the work of any considerable portion of this number, nor is it necessary in order to show the state of the market. Accordingly, a limited selection has been made from the goods of a few of the manufacturers whose reputation for fine work is not questioned, and the pieces illustrated can, at least generally, indicate what are the market possibilities. It was intended that the selection should be left to the manufacturers, so that the samples would in a measure indicate their own views as to what constitutes good hardware, but this idea was only partly carried out, it being necessary to make some personal selection so that the illustrations should not be confined entirely to knobs.

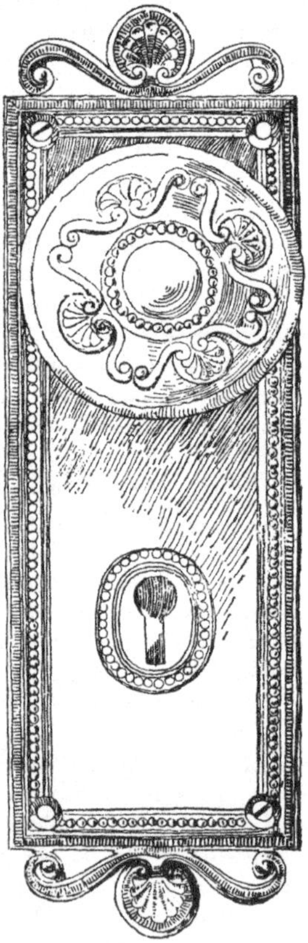

Fig. 461. Coor-knob. Yale & Towne Mfg. Co.

1. The Yale & Towne Manufacturing Company. Figures 457, 458, 459 and 460 [See Illustrations] are all executed in

Bower-Barffed iron. The first design, while not particularly handsome, is thoroughly appropriate to the material and would be suitable for the inside door of a public building, though rather heavy for a dwelling and not heavy enough for an outside door. The knob is perhaps a little too suggestive of wrought-iron. The next example is in a style which might be used to advantage for hardware, much more than it is, though the fish scales in the ground about the key-hole are not altogether appropriate to hardware. The next design is intended for a front door. The foliated pattern, Figure 460, is one of the best which has been made in this material. The design is well arranged and worked out, and excepting the smoothness of the surface, which is too suggestive of the carved pattern, it is very satisfactory. Bower - Barffed iron hardware makes a very effective finish for either oak, or white paint, and has the advantage of not changing its tone with age. It is unfortunate that the illustrations cannot give an idea of the color of these samples, as with some it counts for nearly as much as the form.

Fig. 465. French Window-knob.
Yale & Towne Mfg. Co.

Figure 461 is a very daintily designed knob and escutcheon, finished with gold-plate, the plain surfaces being slightly dulled, and the raised portions polished so as to give brighter reflections. Another knob is finished in one of the most durable colors that is given to bronze hardware, known as old brass, having the soft yellow of an old binnacle which has been scoured and rubbed down year after year to a smooth dull but not polished surface, and has big nail-heads studded very effectively over the escutcheon. Figure 461 would be suitable only for the most elaborate room, preferably one finished in ivory-white and gold, while the old-brass design would answer for much humbler apartments, looking best against dark oak, mahogany or white.

Figures 463 to 467 inclusive, are all in bronze. Figure 465 is a French window-knob finished in deep copper color which is

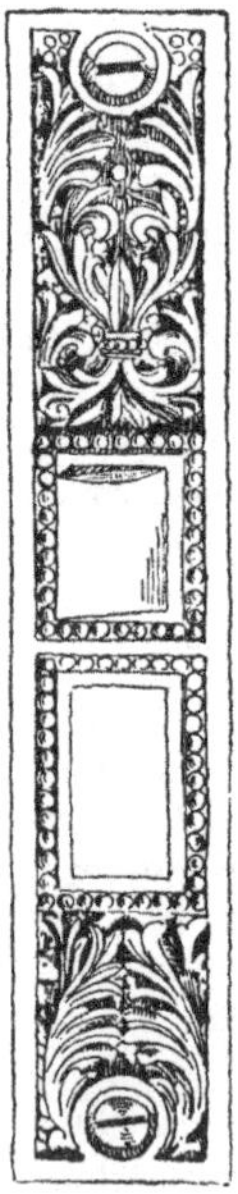

Fig 470.

Fig. 472. Oxidized-silver Hardware. Yale & Towne Mfg. Co. Fig. 473.

especially well designed both for looks and for wear, the numerous rows of beads hiding any inequalities of tone and giving a rich effect at slight cost. Figure 467 is for inside work. The three others are for front doors. Chapter XV.

Figure 468 is an elaborate drawer-pull in silver-plate, slightly oxidized. Figure 469 is in bronze, copper-plated and oxidized, the high lights being buffed down to show the copper. The design is made to match the iron escutcheon Figure 460. The shutter-knob, Figure 470, is a very neat design in a semi-Greek style which we ought to see used more frequently. The sample is gold-plated. The shutter-bar, Figure 471, the lock and key, Figure 472, and the hinge, 473, are all carried out in the same semi-Byzantine style, in oxidized-silver. The key is unique, and the only one of its kind found in modern work. Few persons would care to go to the expense implied by such keys and lock-plates, but where the money is not an important consideration, it would surely add to the finished effect of a room to have such keys as this in the lock instead of the flat utilitarian structure one is usually content with. Figure 474 is a number-plate in oxidized-copper, the leaf work showing in dull copper tones against a very dark old-bronze ground. The number is gold-plated.

The Yale & Towne Manufacturing Company has made its name chiefly on its locks, but of late years it has included the manufacture of fine hardware. The work done by it is excellent in finish and mechanical details and the foregoing samples fairly illustrate the character of the designs.

2. A. G. Newman. Figure 475 is a fine example of a door-handle, having every appearance of being made from a pattern which was modelled instead of being chiselled or carved in wood. The leaf at top and bottom of the plate grows naturally from the single roll at the sides, and the handle is made to look just what it is — a door-pull — without being tortured into any unique or startling form. It is a thoroughly well-designed piece of hardware and as artistic as anything that is found in the market. Figure 476 is less pleasing in the result, though the motive is a good one. The top of the escutcheon is borrowed from an architectural form which needs relief and some

Chapter XV.

fine mouldings, both of which could be added to advantage in this design. It is executed in silver plate, slightly oxidized and buffed to a clear surface on the high lights. Figure 477 illustrates a knob which serves admirably for a parlor or a music-room carried out with colonial fittings, — a Chippendale knob in fact, if the expression may be allowed, and very handsome by reason of the irregular reflections and high lights on the curved flutings. Figure 478 is a quieter design, but very satisfactory except for a little heaviness in the flutings. Figure 479 is a very careful and judicious design, the knob being one of the very few in which a broad, milled-edge forms a marked feature. It is a natural device and is worked in very prettily, the dainty ornament at the top of the knob being equally appropriate for its place. Figures 480 and 481 are other forms, both finished in dark antique bronze. Figure 482 is an escutcheon-plate for a sliding-door, and matches the plate of Figure 476. Figure 483 is a simple but very effective door-pull, in light bronze.

Newman has the reputation of doing a great deal of fine order work for the New York architects from special designs. Some of his best work cannot be reproduced here as the designs are not public. It should be remembered in considering all of this hardware that when a successful design is put on the market it is apt to be copied indiscriminately by rival manufacturers, which is naturally discouraging to those who are conscientiously seeking to do the best work. The rule, almost, is, that the better manufacturers make their designs, and the cheap men steal them.

3. Enoch Robinson. Figure 484 is a cut-glass knob set in a fluted, cup-shaped shank, with a simple Queen Anne escutcheon-plate. Knobs of this material are more commonly attached so as to show glass on all sides. Figures 485 and 486 show two varieties of plain, rectangular escutcheon-plates which are used a great deal by some architects and are always satisfactory. The knob of Figure 486 is well designed, in that the shape is suited to the purpose, the few mouldings simply used

to give lines rather than detail, and the rows of beads are just sufficient to add a sparkle to the whole. Generally, the hardware manufacturer will put too much work on a knob, rather than too little. It is so easy to add leaf work or convolutions or tortured detail which passes for richness that simplicity seldom finds expression in that which dealers are most apt to put forward as artistic hardware.

The knobs shown by Figures 487 and 488 are great favorites about Boston, especially the former, which in various slight modifications is assumed to be peculiarly adapted to white paint and colonial finish. These are often used to advantage with perfectly plain rectangular escutcheons. The plates shown here are a trifle heavy for domestic work, especially Figure 488, which is more suited to a public building.

Fig. 489.

Figures 489 and 489[A] illustrate two elaborate designs which were made on a special order, goods of this description being seldom kept in stock. The work is excellent in detail, though there is less purity of style than one would expect in so ambitious an attempt. Figure 489 is arranged very daintily, but with a little flatter treatment, less relief to the foliage, perhaps, and a mask which would be less literal in its modelling, this design could be rendered much more charming. Figure 491 is for a front-door, and naturally calls for a bolder treatment than Figure 489.

Figure 490 is a variation on the twisted-knob pattern, with an oval form and bevelled cuts instead of flutings, giving a prismatic effect to the surface.

Figures 491 to 493, inclusive, illustrate some pieces of hard-

Chapter XV. ware which can hardly be classed with house-fittings. Every one is familiar with the delightful old mahogany furniture of the early part of this century, ornamented with brass rosettes and wreaths, contrasted with plain surfaces and large, swelling mouldings. This style of the First Empire is beginning to be appreciated again, and the pieces shown by the figures are intended to be used in connection with such work. The patterns are all copied directly from old French furniture, and for the purpose are artistic and very effective. Only rarely can the architect use such pieces as these, but it is easy to imagine a room carried out in the Empire style, with ornaments of this sort used not only on the furniture, but also on the doors and along the architraves, and with charming effect.

Robinson carries a comparatively small stock, his business being very largely on orders, with a great deal of fine hand-work. His goods are thought very highly of by the Boston architects.

4. The Hopkins & Dickinson Manufacturing Company manufactures an escutcheon-plate on a little different scheme from any which have been previously considered, the design being simply a flat piece of metal cut or stamped in an open pattern, the only ornamentation aside from the form, being in the shape of large nail-heads, which are studded about the plate. This sort of design can be made very effective in a great variety of forms. Figure 494 is an example of a good effect obtained by the judicious use of a minimum amount of ornamentation: the convoluted pattern on the knob is enough to add life and interest without marring the simple shape, and just a few touches of the same pattern at the corners of the plate and single dots on the key-hole cover are enough to consistently carry out the design and intensify the simplicity. Whoever designed this pattern knew just where to stop, a rare qualification nowadays. This piece is executed in bronze of a clear, old brass color.

Figure 495 is a design in oxidized-silver, with some excellent detail, notably the mask or grotesque at the top and the arabesque on the face of the knob. The shape of the knob

497

499

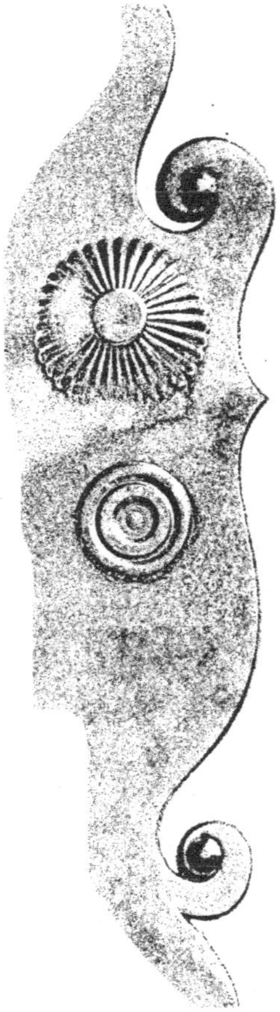

500

494

501

498

496

503

502

495

··ARTISTIC·HARDWARE··

·MANUFACTURED·BY·HOPKINS·&·DICKINSON·

Heliotype Printing Co. Boston

might be better if it were more spherical. So flat a form looks well in elevation, but does not appear to advantage in perspective. Even the other extreme of an elongated sphere, Figure 496, is preferable in some respects, though this form seems to call for a rectangular escutcheon, rather than the cut plate which is shown with it.

Fig. 491a.

Figure 497 is a variation of a familiar design which has already been illustrated in other makes, and is always pleasing. Figure 498 is a neat arrangement in which rope moulding is used with good effect; and another sample shows a very pleasing knob, not unlike that of Figure 494, set on a plain bronze plate, relieved only by slight, open cuttings at top and bottom.

Figures 499 to 503, inclusive, illustrate a few miscellaneous forms which speak for themselves. The hinge and the escutcheon are in oxidized-silver, the others in bronze.

The Hopkins & Dickinson Manufacturing Company bears an excellent reputation in the hardware trade, turning out a very satisfactory class of goods with clean, sharp castings and effective chased work, special care also being given to the colors of the metal, and with exceptional results.

The four manufacturers whose hardware has been illustrated are considered among the best in the country, and, given the same design, one would do quite as good work as the other. Mechanical excellence has been carried to a pitch where there

Chapter XV. is no difficulty in having the work done well. The great lack always is in ideas, and these are supplied only by careful, intelligent study on the part of the designers, and an appreciation of artistic work on the part of those who purchase and use the hardware; and it is quite possible that any deficiency in the æsthetic quality of our modern hardware may be due as much to lack of appreciation as to any lack of proper artists. But it must not be supposed that these are the only manufacturers who are capable of turning out good hardware. The illustrations may be considered as general rather than special, and as showing the possibilities of not only the four firms named, but of the majority of the hardware manufacturers.

CHAPTER XVI.

Hardware Specifications.

THERE are three methods of specifying the hardware to be used about a building. The first, followed very generally in Boston, is by allowances, the general specification containing clauses to the effect that the contractor is to allow so much per door and so much per window for hardware. It is well, in order to prevent mistakes or disputes, to add that the allowances are to cover the net cost to the builder, and that all the hardware is to be set by the contractor in addition to the prices named. This method conveniently disposes of the subject of hardware at the time of writing the specification, the whole matter being relegated to the time when the hardware is actually needed, the architect then making such selection as he sees fit, within the allowances.

Boston method.

The amounts of the allowances depend, of course, entirely upon circumstances. Very fair hardware for inside-doors can be had as low as $1.75 per door. This allows for bronzed iron butts, 60 cents; wooden knobs, escutcheons and roses, 65 cents, and a one-lever lock, 50 cents. This means, however, a cheaper grade of goods than would often be advisable. An allowance of $5 per door will permit of solid bronze hardware throughout, including bronze butts, $2.50; a good, three-lever, machine-made lock at $1; bronze knobs, $1, and bronze escutcheon and rose, 50 cents. For office-work an allowance of $6 per door will provide a "Yale" lock with corrugated key, a

Prices for allowances.

Chapter XVI.

spring latch, and bronze butts and knobs. For front-doors the allowances must be considerably increased. The lock ought to be of a grade costing not less than $4; the bolts, $2.25; the knobs, $2, and the butts, $6, or a total allowance of $14.25. For closet-doors, bronze butts and knobs and a small lock would average $3.50 per door, only one knob being necessary. $4 each is sufficient for sliding-door hardware, exclusive of the hangers, with latch only, or $5 with latch and lock. For fly-doors add $9 per door to the hardware allowance if the butts are of bronze, or $4.50 if of bronzed-iron. This is for house-work. For office fly-doors $12 should be added for bronze butts, and if push-and-pull plates are required an additional sum of $2 will be required.

For French windows $4 will buy two pair of bronze butts, and $2.25 answers for the bolts and the lock, making the total allowance $6.25 per pair. Door-transoms should have an allowance of $4 each, for nice work, which would purchase pivots or 3½ inch butts, $1.75; chains and catch, 75 cents, and lift, $1.50. Transoms are often hung with painted butts, which are less conspicuous than bronze, and the total allowance can be scaled down as low as $1 for very simple work.

Sashes are trimmed for 50 to 75 cents each, though the allowance price is usually per dozen fixtures, say $4.50 for fasts and $2.50 for pulls.

A specification which would, therefore, provide for bronze hardware throughout an ordinary dwelling, of a good, average quality, might read somewhat as follows.

Typical Specification.

HARDWARE. — Allow in the contract the following sums to be expended as the architect directs upon hardware:

Front-doors,	$15.00 each pair.
Back-doors,	6.00 each.
Inside-doors,	5.00 each leaf.
French-windows,	5.00 each leaf.
Sliding-doors,	4.00 each leaf.
Closet-doors,	3.50 each.
Transoms,	2.50 each.
Sash-fasts,	4.50 per dozen.
Sash-pulls,	2.50 per dozen.

These prices are the net cost to the builder, and the owner is to have the benefit of any reduction therefrom. Beyond these allowances list and set all the hardware in proper manner; also provide and set such other hardware (in pantries, closets, etc.) as are called for or implied by the plans or specifications.

Chapter XVI.

Sliding-doors to be hung by (Richards, Prindle, etc.) hangers running on (wooden, iron) tracks, properly adjusted, with proper centre stop.

The hardware for blinds, cellar bulk-heads, etc., is best specified in connection with the individual items, and should be described exactly, as the prices of the different makes vary greatly.

The allowance method is a very simple one for the architect, and when carried out according to the true intent and meaning of the specification, answers very well for ordinary house-work, especially as each contractor usually purchases his hardware of one firm, and the architect can then simply select from a single assortment of goods. The difficulty of such a system, however, is that the goods seldom cost as much as the allowances, and the difference goes to the contractor instead of to the owner, where it belongs. No architect has yet fully mastered all the intricacies of a hardware manufacturer's price-list, and as the price of each article is based upon so flexible a sliding-scale that even the builders are sometimes charged unevenly, there might be a dozen grades of knobs or locks, all at the same nominal price, though varying greatly in quality and in the real amount for which the dealer sells them.

Objection to Boston method.

Also, the system of allowances as ordinally followed gives too much liberty to the contractor, who usually submits to the architect one or two styles of knobs and locks which he will furnish for the allowances, so that the hardware often expresses the builder's rather than the architect's ideas of door-trimmings. There may be a hundred things in the market far better and at less cost than those offered, and while such a specification as the one quoted gives the architect the right to choose his hardware anywhere, in practice he is generally quite willing to shove an irksome job onto the shoulders of the

Chapter XVI.

contractor. In fact, the system of allowances is good only at its best.

New York method.

In New York it is customary to be much more exact in the hardware specifications. No prices are given, the various articles being designated by sizes and by reference to some particular class of goods which is to serve as a standard. Thus: —

For inside-doors provide and set solid bronze 4½ inch loose-pin butts, two to each fold. Outside-doors to have three 6 inch loose-pin bronze butts.

Likewise for locks: —

Closet-doors to have 3½ inch locks; main inside-doors, 4 inch upstairs, 4½ inch downstairs; front-doors, 5 inch. Closet-door locks to have one lever, other doors all three levers; front-door in addition a night-latch and anti-friction strike. The works of all locks to be of (brass, steel), with steel keys and (brass, phosphor-bronze) springs. Door-knobs in service portions to be plain, white porcelain, with black enamelled-iron roses and escutcheons; elsewhere to be plain, solid (or spun) bronze, with bronze roses and escutcheons, etc.

It is usual to describe knobs and miscellaneous fittings by referring to a catalogue or to samples in the architect's office, adding "or of a similar quality satisfactory to the architect."

It will easily be understood that in order to write a proper specification for hardware on such a basis as this, the architect must not only know and keep in mind all the numerous fittings necessary for a house, but must also be thoroughly posted on the possibilities of the market. As a matter-of-fact it is generally easier to call in a hardware man and ask him to write the specification, and the way in which such a method of specifying will fail will be in not being sufficiently general to cover everything, so that the contractor will have an excellent opportunity to bring in a big bill of extras at his own prices. Besides, a specification by sizes referred to a catalogue, "or something equally good," is by no means absolute. The difference between really good and absolutely bad hardware is often so effectually concealed by a superficial finish that, although the architect may refer to the best in the market and mean to have it, an unscrupulous contractor might run in inferior goods which only the closest, technical inspection would detect, though the difference would be speedily manifested in the wear of a few months. Even restricting the choice to the goods of a single firm is not always sufficient to prevent imposition, as there are, unfortunately, some manufacturers who make goods to suit the consciences of all sorts of customers, and have no fixed standards of either quality or price. In time the architect will discover these points by sad experience, and will grow very cautious; but the younger men, the householder and those who are either too indolent or too busy to properly attend to such details, will be more imposed upon by the New York than by the Boston method of specifying the hardware. Chapter XVI.

Government method.

The third method is the only one which is really absolute, or by which the architect is sure of what he obtains. It has been adopted very largely by the Government in obtaining bids for public work, and, though quite clumsy at first, is the surest and easiest way in the end, relieving the architect from a great deal of bother and entirely obviating any discussion as to either prices or qualities. There is no reason why the same system should not be followed in connection with every building of sufficient size to warrant it. The specification calls for locks, butts, etc., of certain sizes, descriptions and weights, and requires that a full line of samples shall be submitted, with prices of each. These samples are then examined carefully, a selection made therefrom or others substituted if none are exactly suitable, and the contract is based directly upon the samples. It remains, then, only for the superintendent to insist upon having the identical fixtures referred to, allowing no leeway for anything "equally good," or "of similar quality."

For small dwellings the method of allowances is very convenient, giving the architect full liberty to change his ideas to suit the work or to introduce any article which seems advisable, without being obliged to determine every detail in advance. But for all heavy buildings the Government method is better and fairer for both architect and contractor. The former can

Chapter XVI.

decide just as well first as last what he wants, and the hardware being all selected several months in advance, the contractor has ample time in which to make the best terms for the wares, and any delay in purchasing is then due solely to himself, while the question of price is definitely settled before any contract is signed. It is then the contractor's business alone how much he pays or how much discount he obtains, or what commission he receives for placing his order with the retailer, and so long as he matches the samples agreed upon, no one has any just cause for complaint.

Conclusion.

In concluding this series of papers upon the subject of Builders' Hardware, the writer desires to acknowledge many sins of omission and commission which have been almost unavoidable, considering the complexity of the subject, the difficulty of ascertaining the condition and scope of the market, and the uncertainty of definite deductions. It has been written with special regard to the possibilities of the Boston market, but, at the same time, the goods of nearly all the manufacturers in the country have been studied and compared, so that it is believed the series will have more than a local significance. The most the writer can hope is that he may help to relieve a little of the drudgery inseparable from the architect's profession, and that the book may have a reference value to both architect, builder and house-furnisher.

INDEX.

NO. III. AMES MEMORIAL BUILDINGS

AT NORTH EASTON, MASS. H. H. RICHARDSON, Architect. 22 plates, 13 x 16, also two lithographs. In portfolio. $6.00.

General View of the Town Hall and Memorial Library, from the Southwest — General View, from the Northeast — Front View of Town Hall — Arcade of Town Hall — Detail of Arcade — Interior of Loggia — West End of Town Hall — View from Southeast — East End — Detail of East End of Town Hall — General View of Library — Entrance Archway — Details of Library Front (Two Plates) — East End — Chimney-piece in Reading-Room — Interior of Book-Room — The Gate-Lodge, from the Grounds of F. L. Ames, Esq. — The Gate-Lodge, from the Southwest — The Gate-Lodge, from the Street — The Railroad Station from the Track — Plans of Town Hall and Memorial Library — Plans of Gate Lodge and Railroad Station.

NO. IV. THE MEMORIAL HALL

AT HARVARD UNIVERSITY. WARE AND VAN BRUNT, Architects. 13 plates, 13 x 16, and one Photo-Lithograph. In portfolio. $5.00.

General View of Cambridge, from the West — Harvard Memorial Hall, from the Southeast — The Main Entrance — The Memorial Vestibule — Entrance of Dining Hall — Southwest Porch — Cloister and Memorial Tablet — Views of Memorial Hall, from the West. with Statue of John Harvard — Dining Hall, looking West — Dining Hall, looking East — East End — Sanders' Theatre — Plan.

NO. V. TRINITY CHURCH, BOSTON, MASS.

22 Gelatine Views and 1 Heliochrome, 13 x 16. $10.00.

Portrait of H. H. Richardson — West Front of Trinity — General View, from Southeast — Apse and Tower, from Southeast — General View, from Northwest — Main Entrance — Detail of West Front — Tower from South — Details of Tower — Cloister and Chapel — Cloister and Tracery Window — Cloister Garth — Chapel Stairway — Entrance to North Transept — Chancel — North Side of Nave — Font — Bust of Dean Stanley — Nave, from Gallery — Interior of Tower — Cartoon for Window — Plan of the Church.

Indispensable to Architects:

MODERN PERSPECTIVE.

A Treatise upon the Principles and Practice of Plane and Cylindrical Perspective, by WILLIAM R. WARE, Professor of Architecture in the School of Mines, Columbia College. 1 vol. 12mo. 321 pages, with 27 plates in a portfolio. $5.00.

This is by far the most exhaustive of modern works on the subjects relating to perspective, plane and panoramic, and will be of great value to all architects and artists, and others interested in the problems of art. The scientific and pictorial aspects of these investigations are carefully and thoroughly considered, both independently and in their connection with drawing; and the propositions of the author are illustrated by plates of architectural objects and perspective plans. An invaluable book for artists, architects, draughtsmen, and civil engineers.

"The book is written in clear English, free from unnecessary technicalities, and in a much more felicitous style than such text-books usually are. The plates required a prodigious quantity of careful work and are correspondingly valuable." —*New-York World.*

ILLUSTRATIONS OF SPIRES AND TOWERS

Of the Mediæval Churches of England. Preceded by some observations on the Architecture of the Middle Ages and its Spire-Growth. By CHARLES WICKES. One vol. Folio. Nearly one hundred plates, with text, $15.00.

Years ago, when Mr. Charles Wickes was engaged in the preparation of his famous work on "Spires and Towers" he had the good sense to print before the completion and issue of the more elaborate colored drawings, an uncolored edition of his work for the special use of architects. Thanks to this action, he placed before the profession one of the most useful and beautiful works which have found a place in architectural libraries.

This work Ticknor & Co. have reprinted. Owing to the great skill and care exercised by the Heliotype Printing Company, these prints have lost nothing of their original force and delicacy in reproduction, and the reprint is quite as valuable and useful as the original. The present edition contains, also, in full, the notes and criticisms of the original, making forty pages of text and notes, not less valuable than the plates.

A Valuable Text-book:

BUILDING SUPERINTENDENCE.

A MANUAL: For young Architects, Students, and others interested in Building Operations as carried on at the present time. By T. M. CLARK, Fellow of the American Institute of Architects. 1 vol. 8vo. 336 pages. Illustrated with 194 Plans, Diagrams, etc. Price, $3.00.

Introduction — The Construction of a Stone Church — Wooden Dwelling-Houses — A Model Specification — Contracts — The Construction of a Town Hall — Index.

"This is not a treatise on the architectural art, or the science of construction, but a simple exposition of the ordinary practice of building in this country, with suggestions for supervising such work efficiently. Architects of experience probably know already nearly everything that the book contains, but their younger brethren as well as those persons not of the profession who are occasionally called upon to direct building operations, will perhaps be glad of its help."

There is hardly any practical problem in construction, from the building of a stone town-hall or church to that of a wooden cottage, that is not carefully considered and discussed here; and a very full index helps to make this treasury of facts accessible. Every person interested in building should possess this work, which is approved as authoritative by the best American architects.

This volume has been used for years as a text-book in the chief Architectural Schools of the United States.

AMERICAN MANSIONS AND COTTAGES.

By CARL PFEIFFER, F. A. I. A. 100 folio plates. In handsome portfolio. $10.00.

During his last years the late Carl Pfeiffer, F. A. I. A., of New York, prepared for publication a work on American Domestic Architecture.

The drawings represent designs for dwelling-houses of various classes, with all their details carefully worked out, and perspective views of each house from several points.

One hundred of these drawings carefully selected as the best, have been chosen for publication, forming the most unique and useful work ever issued upon the subject for either the profession or the public.

Spotted something that isn't quite right?

Let us know and we'll fix it for future reprints.
And as a thank you for your diligence, we'll
send you a free Sothis book of your choice.

publisher@sothis.press

This page intentionally left blank

Made in the USA
Monee, IL
08 November 2026

67812162R00215